JOHN DEERE

T0257327

SHOP MANUAL JD-203

Series ■ 3010 ■ 3020 (before SN. 123000) ■ 4010
 ■ 4020 (before SN. 201000) ■ 5010
 ■ 5020

Series ■ 3020 (SN. 123000 & up) ■ 4000
 ■ 4020 (SN. 201000 & up) ■ 4320
 ■ 4520 ■ 4620

Series ■ 6030

Information and Instructions

This shop manual contains several sections each covering a specific group of wheel type tractors. The Tab Index on the preceding page can be used to locate the section pertaining to each group of tractors. Each section contains the necessary specifications and the brief but terse procedural data needed by a mechanic when repairing a tractor on which he has had no previous actual experience.

Within each section, the material is arranged in a systematic order beginning with an index which is followed immediately by a Table of Condensed Service Specifications. These specifications include dimensions, fits, clearances and timing instructions. Next in order of arrangement is the procedures paragraphs.

In the procedures paragraphs, the order of presentation starts with the front axle system and steering and proceeding toward the rear axle. The last paragraphs are devoted to the power take-off and power lift systems. Interspersed where needed are additional tabular specifications pertaining to wear limits, torquing, etc.

HOW TO USE THE INDEX

Suppose you want to know the procedure for R&R (remove and reinstall) of the engine camshaft. Your first step is to look in the index under the main heading of ENGINE until you find the entry "Camshaft." Now read to the right where under the column covering the tractor you are repairing, you will find a number which indicates the beginning paragraph pertaining to the camshaft. To locate this wanted paragraph in the manual, turn the pages until the running index appearing on the top outside corner of each page contains the number you are seeking. In this paragraph you will find the information concerning the removal of the camshaft.

More information available at Clymer.com
Phone: 805-498-6703

Haynes Publishing Group
Sparkford Nr Yeovil
Somerset BA22 7JJ England

Haynes North America, Inc
859 Lawrence Drive
Newbury Park
California 91320 USA

ISBN-10: 0-87288-360-4
ISBN-13: 978-0-87288-360-4

© Haynes North America, Inc. 1989
With permission from J.H. Haynes & Co. Ltd.

Clymer is a registered trademark of Haynes North America, Inc.

Printed in Malaysia
Cover art by Sean Keenan

JOHN DEERE

Series ■ 3010 ■ 3020 (before SN. 123000) ■ 4010 ■4020 (before SN. 201000) ■ 5010 ■ 5020

Previously contained in I & T Shop Service Manual No. JD-49

SHOP MANUAL
JOHN DEERE

SERIES

3010	4010	5010
3020 (Before	4020 (Before	5020
Serial No. 123000)	Serial No. 201000)	

(General Purpose, Hi-Crop & Standard)

Tractor serial number located on rear of transmission case
Engine serial number located on front right side of engine block

INDEX (By Starting Paragraph)

2

CONDENSED SERVICE DATA

GENERAL

	3010	3020	4010	4020	5010	5020
Engine Make	Own	Own	Own	Own	Own	Own
No. of Cylinders	4	4	6	6	6	6
Bore, Inches						
Diesel	$4^1/_8$	$4^1/_4$	$4^1/_8$	$4^1/_4$	$4^3/_4$	$4^3/_4$
Non-Diesel	4	$4^1/_4$	4	$4^1/_4$	...	...
Stroke, Inches						
Diesel	$4^3/_4$	$4^3/_4$	$4^3/_4$	$4^3/_4$	5	5
Non-Diesel	4	4	4	4	...	...
Displacement, Cu.In.						
Diesel	254	270	380	405	531	531
Non-Diesel	201	227	302	341	...	...
Compression Ratio						
Diesel	16.5:1	16.4:1	16.5:1	16.4:1	16.1:1	16.5:1
Gasoline	7.5:1	7.5:1	7.5:1	7.5:1	...	...
LP-Gas	9.0:1	9.0:1	9.0:1	9.0:1	...	...
Pistons Removed From?	Above	Above	Above	Above	Above	Above
Cylinder Sleeves	Wet	Wet	Wet	Wet	Wet	Wet
Forward Speeds	8	8	8	8	8	8

TUNE-UP

	3010	3020	4010	4020	5010	5020
Firing Order	1-3-4-2	1-3-4-2	———————	1-5-3-6-2-4	———————	
Compression Pressure @ Cranking Speed (psi)						
Diesel	450-475	450-475	450-475	450-475	450-475	450-475
Gasoline	160	160	160	160	...	...
LP-Gas	170	170	170	170	...	...
Valve Tappet Gap (Intake)						
Diesel	0.018H	0.018H	0.018H	0.018H	0.015H	0.018
Non-Diesel	0.015H	0.015H	0.015H	0.015H	...	...
Valve Tappet Gap (Exhaust)						
Diesel	0.018H	0.018H	0.018H	0.018H	0.022H	0.028
Non-Diesel	0.028H	0.028H	0.028H	0.028H	...	...
Ignition Distributor						
Make	D-R	D-R	D-R	D-R	...	...
Model	1112577	See Text	1112581	See Text	...	...
Breaker Contact Gap	0.022	0.022	0.022	See Text	...	...
Ignition Timing						
Retard (Gasoline)	5° ATC	5° ATC	5° ATC	5° ATC	...	...
Advanced (Gasoline)	20° BTC	20° BTC	20° BTC	20° BTC	...	...
Retard (LP-Gas)	TDC	TDC	TDC	TDC	...	...
Advanced (LP-Gas)	25° BTC	25° BTC	25° BTC	25° BTC	...	...
Timing Location	Flywheel	Flywheel	Flywheel	Flywheel	...	...
Spark Plug Size	18 mm	18 mm	18 mm	18 mm	...	...
Spark Plug Electrode Gap						
Gasoline	0.025	0.025	0.025	0.025	...	...
LP-Gas	0.015	0.015	0.015	0.015	...	...

TUNE-UP (Cont'd)	3010	3020	4010	4020	5010	5020
Carburetor Make	Marvel-Schebler or Zenith				......	......
Injection Pump						
Make	Roosa	Roosa	Roosa	Roosa	Roosa	Roosa
Model	DBG	DBG	DBG	DBG	DBG	DBG
Timing	10° BTC	TDC	14° BTC	TDC	TDC	TDC
Injectors						
Make	Bosch, Bendix or Roosa					
Engine Rated Speeds......................	Refer to paragraph 64B, 66 or 68					
Horsepower At PTO (Manufacturer's Observed Rating @ ASAE 540-1000 RPM PTO Speed)						
Diesel, Synchro-Range	55.2	64.3	76.7	86.8	108.6	122.7
Power Shift	......	59.2	......	83.5	......	......
Gasoline, Synchro-Range	50.5	64.7	73.9	85.8	......	......
Power Shift	......	59.3	......	79.9	......	......
LP-Gas, Synchro-Range	50.8	62.8	72.8	84.2	......	......
Power Shift	......	56.3	......	81.1	......	......

SIZES—CAPACITIES—CLEARANCES

(Clearances in Thousandths)	3010	3020	4010	4020	5010	5020
Cooling System (Gallons)	4¾	4¾	6	6	9¼	8¼
Crankcase Oil (Quarts)	8	8	8	8	12	20
Transmission & Hydraulic						
System (Gallons)	11	See Note	11	See Note	16	16
Crankshaft Journals						
Diameter	3.372	3.372	3.372	3.372	3.7485	3.7485
Diametral Clearance	3.1-6.1	3.1-6.1	3.1-6.1	3.1-6.1	3.1-6.1	3.1-6.1
Crankpin Journals						
Diameter	2.999	2.999	2.999	2.999	3.4985	3.4985
Diametral Clearance	1.5-4.5	1.5-4.5	1.5-4.5	1.5-4.5	1.5-4.5	1.5-4.5
Piston Pin Diameter						
Diesel	1.500	1.500	1.500	1.500	1.750	1.750
Non-Diesel	1.250	1.250	1.250	1.250	......	......
Piston Skirt Clearance						
Diesel	5-7	2.8-4.2	5-7	2.8-4.2	4.8-7.2	2.8-5.2
Non-Diesel	4.8-6.8	4.8-6.8	4.8-6.8	4.8-6.8	......	......
Crankshaft End Play	2.5-8.5	2.5-8.5	2.5-8.5	2.5-8.5	2.5-8.5	2.5-8.5

NOTE: Capacities on Series 3020 and 4020 are 11 gallons if equipped with Syncro-Range Transmission; or 14 gallons with Power Shift.

TIGHTENING TORQUES—FT.-LBS.

	3010	3020	4010	4020	5010	5020
Cylinder Head	115	115	115	115	180	180
Connecting Rod Cap Screws						
Diesel	120-130	120-130	120-130	120-130	155-165	155-165
Non-Diesel	85-95	85-95	85-95	85-95	......	......
Main Bearings	150	150	150	150	170	170-180

FRONT SYSTEM

Tractors are available with three tricycle type front ends and two axle types. Tricycle systems consist of a fork mounted single wheel, dual wheel tricycle, or "Roll-O-Matic" dual wheel tricycle, which attach directly to the center steering spindle. Axle types consist of a fixed tread, heavy duty axle or an adjustable, row-crop axle, which attach to a removable pivot bracket.

SPINDLE EXTENSION (PEDESTAL)
Tricycle Models

1. **REMOVE & REINSTALL.** The spindle extension (pedestal) attaches directly to steering spindle by four cap screws. To remove the unit, jack up the tractor, remove the four cap screws and roll the assembly forward away from tractor. When reinstalling, tighten the retaining cap screws to a torque of 300 ft.-lbs.

2. **SINGLE WHEEL TRICYCLE.** The fork mounted single wheel is supported on tapered roller bearings as shown in Fig. JD1. Bearings should be adjusted to provide a slight rotational drag by means of adjusting nut (39).

3. **DUAL WHEEL TRICYCLE.** An exploded view of the dual wheel tricycle pedestal is shown in Fig. JD2. Horizontal axles are not renewable. Service consists of renewing the extension assembly (40).

4. **"ROLL-O-MATIC" UNIT.** The "Roll-O-Matic" unit can be overhauled without removing assembly from tractor. To overhaul, support front of tractor and remove wheel and hub units. Remove knuckle caps (45—Fig. JD3), then unbolt and remove thrust washers (25). Pull knuckle and gear units from housing and remove felt washer (21).

Series 3000 is available only with the regular "Roll-O-Matic" unit, while Series 4000 is available with either Regular or Heavy Duty units. Service procedures are the same for both types but parts are not interchangeable. On late models, the Regular Duty "Roll-O-Matic" unit is equipped with a lock which may be installed for rigidity when desired.

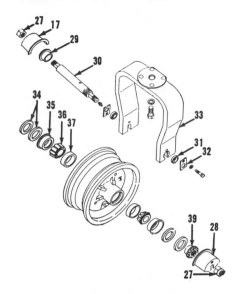

Fig. JD1—Exploded view of front wheel fork and axle assembly used on single wheel tricycle models. See legend under Fig. JD3.

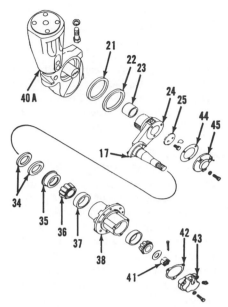

Fig. JD3—"Roll-O-Matic" spindle extension showing component parts.

17. Dust excluder
21. Felt washer
22. Retainer
23. Bushing
24. Knuckle
25. Thrust washer
27. Nut
28. Dust shield
29. Spacer
30. Axle
31. Washer
32. Lock plate
33. Yoke
34. Felt washer
35. Retainer
36. Bearing cone
37. Bearing cup
38. Hub
39. Adjusting nut.
40. Pedestal extension
40A. Pedestal extension
41. Nut
42. Gasket
43. Hub cap
44. Gasket
45. Cap

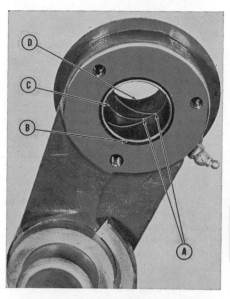

Fig. JD4—Method of bushing installation in "Roll-O-Matic" knuckle unit. Bushings should be installed with open end of oil groove to center as shown at (A). See text for dimensional location of bushings.

Fig. JD5—Make sure timing marks (M) are aligned when installing knuckles in "Roll-O-Matic" unit.

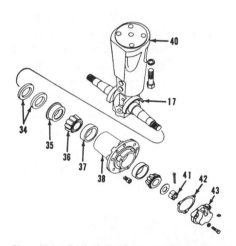

Fig. JD2—Exploded view of dual wheel tricycle spindle extension (pedestal) and wheel hub assembly. See legend under Fig. JD3.

Check the removed parts against the values which follow:

Regular "Roll-O-Matic"

Knuckle Bushing ID 1.873-1.875
Knuckle Shaft OD 1.870-1.872
Thrust Washer
    Thickness 0.156

Heavy Duty "Roll-O-Matic"

Knuckle Bushing ID 2.127-2.129
Knuckle Shaft OD 2.124-2.126
Thrust Washer
    Thickness 0.187

Bushings (23—Fig. JD3) are presized and contain a spiral oil groove which extends to one edge of bushing only. When installing new bushings, use a piloted arbor and press bushings into knuckle arm so that OPEN end of spiral grooves are to the center of arm as shown at (A—Fig. JD4).

Bushing at spindle end of Regular "Roll-O-Matic" unit should be pressed into arm so that outer edge (B) is flush with machined surface. There should be a gap (C) of 1/32 to 1/16-inch between bushings, and outer end (D) of opposite bushing should be ¼ to 17/64-inch below machined edge of gear.

Bushing at spindle end of Heavy Duty "Roll-O-Matic" unit should be pressed into arm so that outer edge (B) is 1/32-inch below machined surface. There should be a gap of 1/32 to 1/16-inch between bushings as shown at (C), and outer edge of opposite bushing should be 3/16-inch below machined edge of gear.

Soak felt washers (21—Fig. JD3) in engine oil prior to installation. Install one of the knuckles so that wheel spindle extends behind vertical steering spindle. Pack the "Roll-O-Matic" unit with wheel bearing grease and install the other knuckle so that

timing marks on gears are in register as shown at (M—Fig. JD5). Tighten the thrust washer attaching screws to a torque of 56 ft.-lbs.

AXLE AND SUPPORT

Axle Models Except Series 5000

5. **HOUSING & PIVOT BRACKET.** Front axle housing (center section) (4—Fig. JD6) attaches to tractor frame by means of pivot bracket (1). To remove the axle assembly, support front of tractor, disconnect tie rods from center steering arm, remove the two clamps (7) and roll axle assembly forward away from tractor. Pivot pins (5) are positioned in pivot bracket by means of dowel pins (8). Refer to specifications in paragraph 5B.

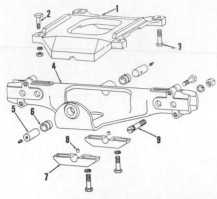

Fig. JD6—Exploded view of axle housing (center section) and pivot bracket used on all adjustable axle models except series Hi-Crop.

1. Pivot bracket
2. Bolt
3. Cap screw
4. Axle housing
5. Pivot pin
6. Bushing
7. Pivot clamp
8. Dowel pin
9. Lock bolt

5A. SPINDLES AND BUSHINGS. Steering arm (5—Fig. JD7) is splined to spindle (13) and retained by a bolt. Spindle bushings (10) are pre-sized. Thrust washers (11) are secured to spindle and axle extension by means of dowels (12). When renewing, be sure ground surfaces of thrust washers are together and adjust end play of assembled unit to 0.036 by means of shims (7).

5B. Specifications and tightening torques are as follows:

Pivot Pin OD 1.494-1.495
Pivot Bushing ID 1.504-1.506
Spindle Vertical OD 1.494-1.495
Spindle Bushing ID 1.504-1.506
Spindle Thrust Washers
  Thickness 0.167-0.172
Spindle End Play ....0.036 Max.

Adjusting Shims
  Thickness 0.036
Toe-In Adjustment ....⅛-⅜ In.
Tightening Torques, ft.-lbs.
  Pivot Bracket To Frame Bolts..150
  Pivot Brkt. To Support 300
  Center Steering Arm Bolts 300
  Housing Clamp Screws 300
  Tie Rod Clamp Bolts 30

Series 5000

6. AXLE AND PIVOT BRACKET. When removing front axle or pivot bracket, raise front of tractor with a hoist attached to tractor frame. DO NOT attempt to raise tractor by placing a jack under oil pan. The manufacturer provides a Front Hoist Bracket (Tool No. JDG7) which attaches to front of front support by cap screws.

Front axle and pivot bracket may be removed as a unit after disconnecting steering cylinders at front ends and removing pivot bracket retaining cap screws. NOTE: Steering cylinder pivot pins contain ½-13 threads for installing puller screws. Support front of pivot bracket and axle assembly to prevent axle from tipping, when assembly is removed. When reinstalling, tighten pivot bracket retaining cap screws to a torque of 300 ft.-lbs. and make sure the step-washers are properly positioned in steering cylinder rod ends.

Front axle can be detached from pivot bracket after detaching tie rods and removing pivot pin retaining cap screws, hollow dowels and pivot pins.

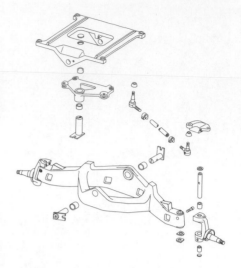

Fig. JD7A—Exploded view of Fixed Tread Axle and associated parts used on some Series 5000 tractors. Refer to Fig. JD7B for adjustable tread axle.

Steering bellcrank is retained to pivot bracket by a pivot pin and can be removed after front axle is out. Pivot pins and bushings are serviced individually; normal diametral clearance of axle pivot pins is 0.004-0.020; bellcrank pivot pin 0.0015-0.0065. New axle or bellcrank pivot pin is 1.998-2.000 in diameter.

Exploded views of the two front axle types are shown in Figs. JD7A and JD7B.

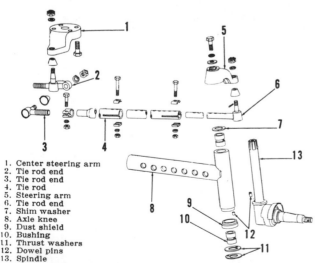

1. Center steering arm
2. Tie rod end
3. Tie rod end
4. Tie rod
5. Steering arm
6. Tie rod end
7. Shim washer
8. Axle knee
9. Dust shield
10. Bushing
11. Thrust washers
12. Dowel pins
13. Spindle

Fig. JD7—Steering spindle, steering linkage and axle knee used on standard models and 3010 utility with short wheel base. Other adjustable axle models are similar except that steering arm (5) is installed pointing forward and both tie rods attach directly to center steering arm (1).

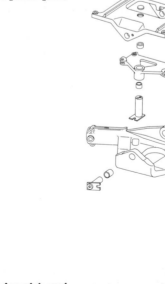

Fig. JD7B — Exploded view of Adjustable Tread Axle and associated parts used on Series 5000 tractors so equipped.

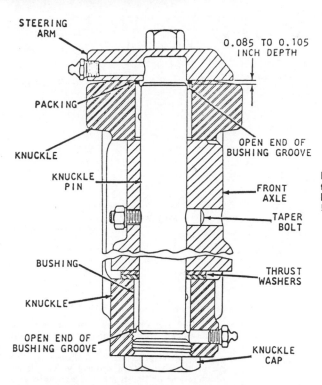

Fig. JD7C — Cross sectional view of steering knuckle used on Series 5000. Refer to text for overhaul procedures.

knuckle thrust washers are each 0.103-0.107 in thickness and must be installed with square edges engaging axle and knuckle.

On adjustable tread axles, steering arm and spindle splines are tapered. Remove retaining cap screw and washers, then reinstall cap screw. Rap cap screw sharply with a heavy hammer to free the splines.

TIE RODS & TOE-IN

All Axle Models

6B. The automotive type tie rods should be adjusted equally to provide ⅛-⅜ inch toe-in. Tighten tie rod clamp bolts to a torque of 35 ft.-lbs. with clamps pointing downward.

RADIUS ROD

Hi-Crop Models

7. **ADJUSTMENT.** The special Hi-Crop models are equipped with radius rods which clamp to spindle housing on axle extension. Rear pivot is unbushed; if clearance is excessive, renew the parts. When installing, make sure radius rods are of equal length, and that no more than 1⅜ threads are exposed.

6A. SPINDLE (KNUCKLE) PINS & BUSHINGS. Refer to Fig. JD7C for cross section of unit used on fixed tread models. Knuckle pin can be pushed or driven from bore after removing steering arm, lower knuckle cap and the retaining taper bolt. The 1.234-1.235 knuckle pin should have 0.002-0.005 clearance in the renewable knuckle pin bushings. Bushings must be installed in knuckle with open end of lubrication groove toward grease fittings. Press upper bushings 0.085-0.105 below flush with top face of knuckle to provide room for the square-section packing ring. New

POWER STEERING SYSTEM

All models are equipped with a full power steering system. No mechanical linkage exists between steering wheel and front unit; however, steering can be manually accomplished by hydraulic pressure when tractor hydraulic unit is inactive. Power is supplied by the same hydraulic pump which powers the lift and brake systems. A pressure control (priority) valve is located in outlet line from main hydraulic pump. Valve gives steering system first priority on hydraulic flow.

OPERATION
All Models

8. The power steering system consists of the tractor hydraulic supply system described in paragraph 114, plus the steering control unit and steering cylinders or motor described in this section (See Fig. JD8).

The control unit contains a double acting piston (3) of approximately equal displacement to the two operating pistons (16). In addition, it contains two pressure valves, two return valves, a relief valve and an unloading valve which actuate the power assist.

Series 5000 uses two, single acting displacement type steering cylinders; all other models use a steering motor, consisting of a main housing, steering spindle (17) and two rack-type double piston assemblies (16).

When the control unit is in the neutral position, there is no fluid flow but oil at pump pressure is available at the pressure line (15). When the steering wheel is turned for a right-hand turn, the first movement of the steering shaft (1) and operating col-

lar (5) causes the lower operating valve lever (7) to unseat the inlet valve (13) and return valve (12). Fluid at pump pressure then enters the area below steering valve piston (3) forcing the trapped fluid above piston through right hand steering motor line (2) into right-rear and left-front steering cylinders. At the same time, the unseating of return valve (12) allows the fluid in the opposite ends of steering motor cylinders to return to the sump.

When the steering wheel is turned for a left-hand turn, the first movement of steering shaft (1) and operating collar (5) causes the upper operating valve lever (6) to unseat inlet valve (8) and return valve (9). Fluid at pump pressure then passes through inlet valve (8) and left hand

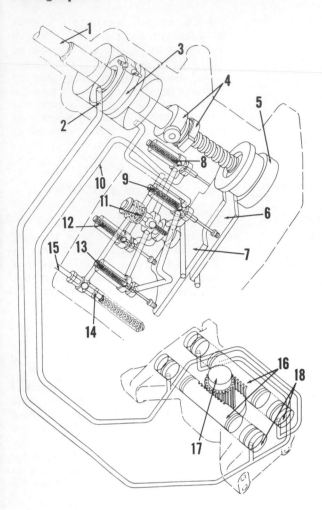

1. Steering shaft
2. Steering line
3. Control piston
4. Operating collar
5. Actuating collar
6. Operating lever
7. Operating lever
8. Inlet valve
9. Return valve
10. Steering line
11. Relief valve
12. Return valve
13. Inlet valve
14. Unloading valve
15. Inlet pressure line
16. Operating pistons
17. Steering spindle
18. Piston plug

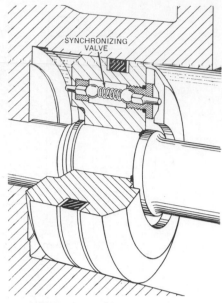

Fig. JD8B—Steering valve piston must be synchronized with steering motor for full turning action. This is automatically accomplished when pressure is available by the pictured valve. When control valve piston reaches end of its stroke, the extended rod unseats the ball check valve allowing fluid to flow through piston until motor and valve are synchronized.

steering motor line (10) into left-rear and right-front steering cylinders. Fluid at opposite ends of cylinders (16) is forced back through line (2) to area above steering valve piston (3). The fluid below the piston returns to the sump through the unseated return valve (9).

On early models, whenever the main hydraulic system supply pressure drops below approximately 600 psi, the steering system unloading valve (14) is returned to its seat by spring pressure. The steering system is thus blocked from remainder of

hydraulic system and any steering action is accomplished manually, the trapped oil recirculating around the check valve ball located beneath relief valve (11).

On late models, unloading valve (14) is replaced by a ball-type check valve which blocks a reverse flow to the main system. No definite pressure exists at which the system reverts from power to manual control; the action taking place whenever manually exerted pressure exceeds system pressure.

BLEEDING
All Models

9. To bleed the steering system, first remove the cowling and attach a small hose from bleed machine screw (B—Fig. JD9) on left side of control valve cylinder housing. Hose must be long enough to reach transmission filler opening. Start engine and run tractor at slow idle. Turn steering wheel to full right, then full left to allow steering valve to synchronize with front wheels.

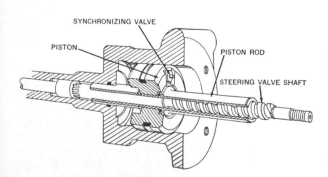

Fig. JD8A — Cross sectional view of steering valve operating piston, cylinder and steering shaft. Piston is moved up or down in cylinder by helical thread on steering shaft. Synchronizing valve corrects for internal leaks. See Fig. JD8B.

Fig. JD9—To bleed the steering system, remove cowl and attach bleed line to bleed screw (B). See text for details.

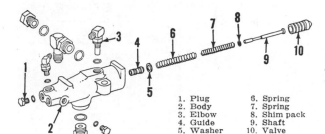

Fig. JD9A — Exploded view of steering priority valve used on early models.

1. Plug	6. Spring
2. Body	7. Spring
3. Elbow	8. Shim pack
4. Guide	9. Shaft
5. Washer	10. Valve

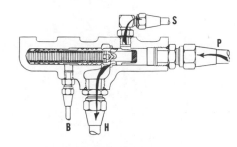

Fig. JD9C—Cross sectional view of latest type steering priority valve. Fluid from main hydraulic pump enters at (P). Steering line (S) is always open to pump pressure; hydraulic systems line (H) is closed off when system pressure drops below 1800-1900 psi.

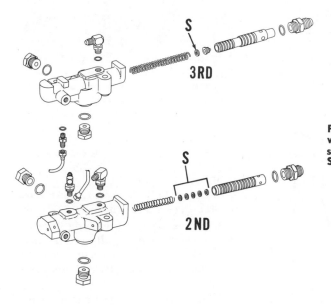

Fig. JD9B — Exploded views of two, later type steering priority valves. Shims (S) adjust operating pressure.

Leave steering wheel and front wheels in full left turn position, loosen bleed valve lock nut and back out bleed valve approximately ½ turn. With the engine at slow idle, and without turning front wheels, turn steering wheel very slowly to full right position. Close bleed valve and allow front wheels to turn to full right position.

Repeat the procedure, if necessary, until air-free fluid is being returned to reservoir.

PRIORITY VALVE

All Models

9A. **OPERATION.** Most tractors are equipped with a Pressure Control (Priority) Valve which is mounted under engine cowl on right side. The pressure control valve cuts off hydraulic flow to all units other than steering and brakes whenever system pressure drops below 1800-1900 psi.

Three types of valves have been used. Refer to Figs. 9A and 9B for exploded views and to Fig. 9C for a cross sectional view of latest valve.

9B. **R&R AND OVERHAUL.** To remove the priority valve, first drain the cooling system and remove cowl and hood. Detach temperature gage sending unit from cylinder head, disconnect electrical wiring and tachometer cable from instrument panel; then remove steering wheel and instrument panel. Disconnect pressure and bleed lines and unbolt and remove the priority valve assembly. Refer to Fig. JD9A for exploded view of early valve and Fig. JD9B for views of later units.

On early valve, steering line elbow (3—Fig. JD9A) must be removed before valve can be withdrawn from housing bore. Outer spring (6) should have a free length of $3\frac{5}{32}$ inches and should test 2¼-2½ lbs. when compressed to a height of 2 inches. Inner spring (7) should have a free length of approximately 2½ inches and should test 44-54 lbs. when compressed to $2\frac{1}{16}$ inches. On all other valves the single spring should have an approximate free length of $4\frac{5}{8}$ inches and test 45-55 lbs. when compressed to a height of 3½ inches. Pressure adjusting shims are shown at (8—Fig. JD9A) or (S—Fig. JD9B).

STEERING CONTROL UNIT

All Models

10. **REMOVE AND REINSTALL.** To remove the steering control unit, first remove steering wheel using a suitable puller. Remove cowling, hand throttle and instrument panel; then unbolt and remove throttle linkage from steering column. Note: Instruments may be disconnected from panel by removing the attaching screws, without disconnecting instruments from tractor. Disconnect steering fluid lines, being sure to cap all exposed couplings, then unbolt and remove the complete steering control unit.

When reinstalling, bleed the steering system as outlined in paragraph 9 and adjust throttle linkage as in paragraph 66 or 68. Tighten steering wheel nut to a torque of 50 ft.-lbs.

11. **OVERHAUL.** To disassemble the removed unit, first remove lower cover (42—Fig. JD10) and, on early models, withdraw unloading valve spring (37) and shaft (36). Remove cotter pin and nut (24) from lower end of steering shaft, then unbolt and withdraw control valve housing (33) and operating collar (23) from steering shaft. On late models, withdraw and save the loose check valve stop (43—Fig. JD10A), spring (44) and shaft (45) as valve housing is separated from cylinder housing (6—Fig. JD10).

Remove nut (20), spring (19), rollers (16) and pins (17), then withdraw collar (18) from lower end of piston rod. Temporarily install steering wheel and turn wheel counterclockwise to force cylinder cover (15) from cylinder housing, then withdraw cover, piston (13) and steering shaft

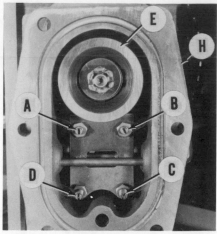

Fig. JD11—To adjust the steering valve, it is necessary to remove the unit and remove the lower cover as shown. See text for details.

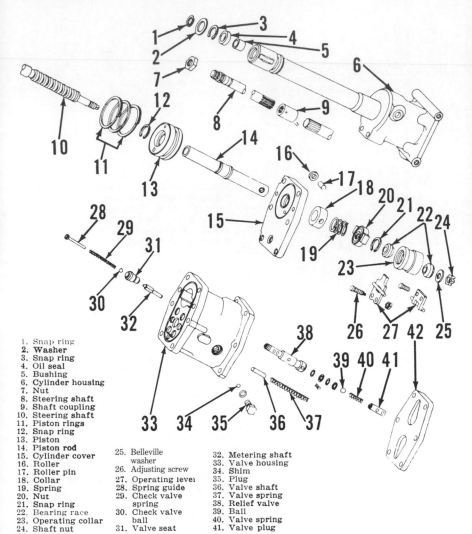

1. Snap ring
2. **Washer**
3. Snap ring
4. Oil seal
5. Bushing
6. Cylinder housing
7. Nut
8. Steering shaft
9. Shaft coupling
10. Steering shaft
11. Piston rings
12. Snap ring
13. Piston
14. **Piston rod**
15. Cylinder cover
16. Roller
17. Roller pin
18. Collar
19. Spring
20. Nut
21. Snap ring
22. Bearing race
23. Operating collar
24. Shaft nut
25. Belleville washer
26. Adjusting screw
27. Operating lever
28. Spring guide
29. Check valve spring
30. Check valve ball
31. Valve seat
32. Metering shaft
33. Valve housing
34. Shim
35. Plug
36. Valve shaft
37. Valve spring
38. Relief valve
39. Ball
40. Valve spring
41. Valve plug

Fig. JD10—Exploded view of early steering valve assembly showing component parts. No mechanical linkage exists between steering valve and the hydraulically actuated steering spindle.

from cylinder housing. Steering shaft (8) is retained in steering column by snap ring (3). Remove shaft if service on oil seal (4), bushing (5), shaft or housing is indicated. Withdraw springs from operating valves in valve housing, remove valve balls and inspect balls and seats for line contact. Renew any parts that are damaged or worn.

Specifications and tightening torques of the components of steering control unit are as follows:

Steering Valve Piston OD . . 3.370-3.372
Steering Valve Cyl-
  inder ID 3.374-3.377
Piston Rod OD 1.122-1.123
Cylinder Cover ID 1.126-1.128
Control Valve OD 0.748-0.749
Control Valve Bore ID 0.750-0.752

Steering Wheel Shaft
  Bushing ID 0.880-0.884
Steering Wheel Shaft OD . . 0.874-0.876
Valve Spring Test Specifications —
Lbs. @ In.
  Relief Valve
    Spring . . . . 101-123 @ 1 1/16

Unloading Valve
  Spring . . . . 36-44 @ 3 7/8
Tightening Torques—Ft.-Lb.
  Valve Hsg to Cyl.
    Cover Bolts . . . . 85 ft.-lbs.
  Steering Wheel Nut . . . . . 50 ft.-lbs.

Clean all parts by washing in clean solvent and immerse all parts including "O" rings and back-up washers in clean hydraulic fluid before assembly.

When reassembling, tighten spring loaded nut (20) on operating shaft until a gap of approximately 5/16-inch exists between nut (20) and collar (18). This tension provides the friction which gives a feeling of stability to the steering effort. Shims (34) beneath operating lever plugs (35) are provided to limit end play of shafts on levers (27) to 0.005. Shims are available in thicknesses of 0.005, 0.010 and 0.030. Position spring washers (25) with dished sides together and tighten nut (24) to a torque of 10 ft.-lbs., then turn nut until nearest

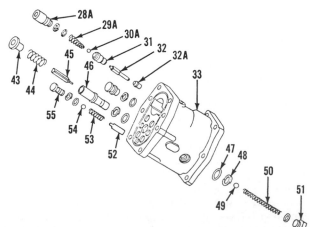

Fig. JD10A — Exploded view of late steering valve assembly. Refer also to Fig. JD10.

28A. Plug
29A. Spring
30A. Ball
31. Control valve
32. Metering shaft
32A. Shaft end
33. Housing
43. Piston stop
44. Seat spring
45. Valve shaft
46. Valve seat
47. "O" ring
48. Backup ring
49. Ball
50. Spring
51. Plug
52. Guide
53. Spring
54. Ball
55. Valve seat

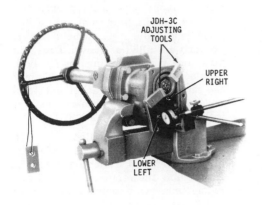

JDH-3C
ADJUSTING
TOOLS

UPPER
RIGHT

LOWER
LEFT

STEERING MOTOR
All Models Except Series 5000

12. REMOVE AND REINSTALL. To remove the steering motor, remove the grille screens and left side frame. On tricycle models, support front of tractor and remove wheels and pedestal assembly as outlined in paragraph 1. On axle models remove center steering arm. Disconnect the two fluid supply lines and vent line leading to oil cooler. Remove cap screws securing steering motor to right side frame and front frame, tilt steering motor and withdraw from left side of tractor.

To install, reverse the removal procedure and bleed system as outlined in paragraph 9. Tightening torques are as follows:

Motor To Side Frame......215 ft.-lbs.
Motor To Front Frame....150 ft.-lbs.
Steering Spindle Flange
    Bolts ....300 ft.-lbs.

Fig. JD11A — Steering valve housing with adjusting tools installed. See text for details of adjustment.

12A. OVERHAUL. To disassemble the removed steering motor, turn unit upside down on bench and remove the cap screws securing spindle retainer (7—Fig. JD12) to motor housing. Tap spindle retainer from it's doweled position and withdraw spindle (8), retainer (7) and bearing (5) as a unit. Spindle bearing is retained to spindle by snap ring (4) and can be removed if service is indicated on any of the components. Install spindle bearing (5) with shielded side up.

castellations are aligned and insert cotter pin. Operating collar (23) must be free to turn by hand on bearing.

To adjust valve operating levers (27), install special adjusting tools, JDH-3C, on housing as shown in Fig. JD11A. Use side of tools marked "3000-4000-5000." The tools will hold lower edge of operating collar (E–Fig. JD11) 0.030 inch from machined faced of housing (H), which is neutral position. Temporarily install steering wheel, then turn wheel clockwise until operating collar contacts special tools. Hang a weight on wheel as shown in Fig. JD11A to hold operating collar securely in this neutral position for steering valve adjustment.

Loosen the locknuts on adjusting screws (A, B, C and D–Fig. JD11) and make sure both operating levers have at least 0.003 inch free movement. Position a dial indicator on housing as shown in Fig. JD11A so operating lever clearance can be checked at each adjusting screw (A, B, C and D–Fig. 11). Adjust each screw to specified clearance using the following adjusting sequence: Adjust upper left return valve screw (A) to provide 0.003 inch lever movement, then adjust upper right pressure valve screw (B) to obtain 0.001 inch movement. Adjust lower right return valve screw (C) to provide 0.003 inch lever movement, then adjust lower left pressure valve screw (D) until lever movement is 0.001 inch. This adjustment allows steering pressure valves to open slightly before return valves to ensure smooth steering response.

Note that valve adjusting screw locations are as viewed from front side of housing with operating collar positioned at top as shown in Fig. JD11A. Also, adjustment can be more accurately made if operating lever shaft is pulled outward with light pressure to eliminate shaft free play as lever movement is being measured.

Be sure to install unloading valve shaft (36—Fig. JD10) and spring (37), then install packing and cover (42). Tighten the retaining cap screws securely.

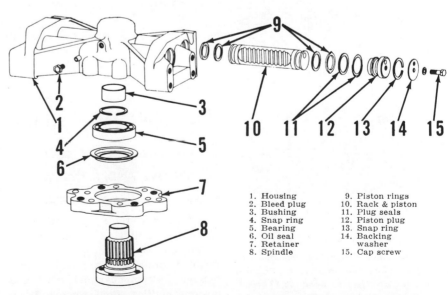

1. Housing	9. Piston rings
2. Bleed plug	10. Rack & piston
3. Bushing	11. Plug seals
4. Snap ring	12. Piston plug
5. Bearing	13. Snap ring
6. Oil seal	14. Backing
7. Retainer	washer
8. Spindle	15. Cap screw

Fig. JD12—Exploded view of steering motor showing component parts. The hydraulically actuated steering motor can be manually turned when power is not available.

To disassemble the piston assemblies, remove cap screws (15) and backing washers (14), then remove snap ring (13). Cap screws may be reinstalled to assist in pulling end plugs (12). Remove pistons after end plugs are removed using a brass drift or other suitable tool.

Overhaul data for steering motor components are as follows:

Spindle OD
    (Bushing End) 2.5005-2.501
Bushing ID 2.5025-2.5065
Retainer ID (At
    Seal Bore) ....5.677 -5.697
Motor Piston OD ......1.995 -1.997
Motor Cylinder ID .....1.999 -2.001

When reassembling, install two diagonally opposite end plugs and secure with snap rings, backing washers and cap screws. Install piston assemblies, teeth toward center of motor housing, until they bottom against previously installed end plugs. Reassemble spindle retainer, spindle and bearing. Then carefully insert spindle in housing so that V-mark on spindle flange is aligned with the appropriate limit-mark on retainer as shown in Fig. JD13. When properly assembled and timed, spindle should rotate 120 degrees between the scribed limit marks, and V-mark on spindle should point directly to the rear in mid-position. Complete the assembly by reversing disassembly procedure, then install unit as outlined in paragraph 12.

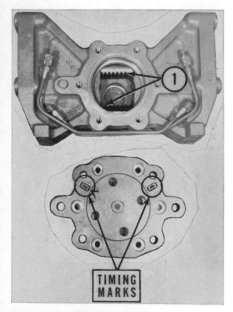

Fig. JD13—When assembling the steering motor, make sure that pistons are at opposite ends of cylinders as shown at (1) and that the appropriate timing marks are aligned on spindle and retainer.

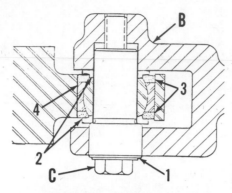

Fig. JD13A — Cross sectional view of steering cylinder attaching pin and associated parts used on Series 5000. Refer to Fig. JD13B for legend except for bellcrank (B).

STEERING CYLINDER

Series 5000

13. **REMOVE AND REINSTALL.** To remove either steering cylinder, first disconnect the pressure hose at cylinder. Remove the cap screws (C—Fig. JD13A) which retain pins (1) to frame bracket and steering bellcrank; then remove pins using a suitable puller. NOTE: ID of pin has ½-13 screw threads for attaching puller.

When reinstalling, make sure felt washers (3) and step washers (2) are properly installed as shown. Draw the pins into position using longer bolts in place of cap screws (C). Tighten cap screws in frame bracket and steering bellcrank and secure by bending locking plate (L—Fig. JD13B).

13A. **OVERHAUL.** Refer to Fig. JD13B for an exploded view of steering cylinder and attaching parts. To disassemble the removed steering cylinder, unscrew extension (10) using a suitable spanner. Remove snap ring (6) and washer (7) from piston rod (14) and withdraw extension (10) from piston. Bushings (9), packing (11) and seal (13) can be renewed at this time. Renew piston rod (14) if scored or otherwise damaged.

Assemble by reversing the disassembly procedure and install as outlined in paragraph 13.

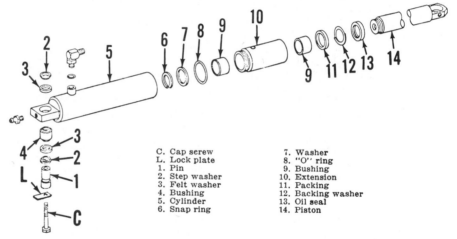

C. Cap screw	7. Washer
L. Lock plate	8. "O" ring
1. Pin	9. Bushing
2. Step washer	10. Extension
3. Felt washer	11. Packing
4. Bushing	12. Backing washer
5. Cylinder	13. Oil seal
6. Snap ring	14. Piston

Fig. JD13B—Exploded view of steering cylinder and associated parts used on Series 5000.

ENGINE AND COMPONENTS

REMOVE AND REINSTALL

All Models Except Series 5000

14. To remove the engine and clutch assembly as a unit, first drain cooling system and, if engine is to be disassembled, drain oil pan. Unbolt and remove instrument panel cowl and upright muffler; then remove hood, grille screens and engine side panels.

Disconnect tachometer cable, oil pressure switch, and on diesel engines, the throttle rod and shut-off cable. Remove hydraulic line clips and radiator brace rods and disconnect wire from fuel gage sending unit. Disconnect hydraulic fluid lines at front junction and at firewall, disconnect hydraulic line spacer bracket and remove lines, spacer and bracket from tractor.

Shut off fuel and remove fuel line. On diesel models, remove leak-off line leading to fuel tank. On LP-Gas tractors disconnect fuel shut-off control rods at tank and pull them back through control panel. On all models, disconnect wiring harness, temperature gage sending unit, and air cleaner and coolant hoses. Detach hydraulic pump drive from crankshaft pulley and remove the pump support cap screws from engine block. Support the tractor beneath transmission housing and frame side rails at rear end. Disconnect side rails from engine block and roll tractor and engine assembly rearward away from front end unit.

CAUTION: On some models, unit may be heavy in front, therefore unstable. If tractor is equipped with front end weights remove weights before separating the units. Use care to prevent damage or personal injury when splitting the tractor.

Support engine in a hoist, remove cap screws attaching engine to clutch housing and separate engine from transmission.

When installing engine, the manufacturer recommends that engine be first attached to front frame, then the two units joined at clutch housing. Tightening torques are as follows:

Hydraulic Pump Drive.... 32 ft.-lbs.
Hydraulic Pump Support.. 85 ft.-lbs.
Side Frame To Cyl. Block..240 ft.-lbs.
Cyl. Block to Clutch Hsg. ..275 ft.-lbs.

Series 5000

14A. To remove engine and clutch as a unit, first drain cooling system and, if engine is to be disassembled, drain oil pan. Remove side shields, side grille screens, cowl and hood. Remove ether tube from cold starting unit and manifold. Disconnect main hydraulic pump inlet pipe at oil filter base on left side of transmission, and allow pump and oil cooler to drain. Disconnect lower hydraulic return pipe from upper return pipe at hose coupling. Remove clamps securing steering tube and hydraulic return pipe to left hand side frame, disconnect steering tube at both ends and remove tube. Disconnect steering cylinder from left side frame bracket. Disconnect battery ground straps and remove starter. Remove step plates and the cap screws securing left side frame to clutch

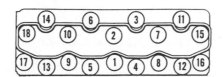

Fig. JD14—When installing the cylinder head on series 3000 tractors, tighten the cylinder head cap screws to a torque of 115 ft.-lbs. using the sequence shown.

housing. Remove air intake pipe and lower radiator hose.

Disconnect main hydraulic pump drive coupling from crankshaft pulley and remove cap screws securing pump support to cylinder block. Shut off fuel. Remove fuel leak-off pipe, upper radiator hose and coolant by-pass tube. Disconnect wiring harness from fuel tank sending unit, regulator and generator; disconnect harness clamps and lay harness rearward over firewall. NOTE: Regulator and bracket can be detached from cylinder block and left attached to wiring, if desired.

Disconnect main hydraulic pressure tube and right steering tube at couplings behind right side frame below engine oil cooler. Disconnect fuel pump inlet tube and cap the connections to prevent dirt entry. Disconnect steering cylinder from right side frame bracket and remove cap screws securing side frame to clutch housing.

Support tractor beneath transmission housing and engine from a hoist. Block up between front axle and pivot bracket on each side and support front of front frame on a rolling floor jack. Remove cap screws securing side frames to engine block and roll front unit and side frames forward as a unit.

Remove cap screws securing engine to clutch housing and separate engine from transmission.

Install by reversing removal procedure. Tighten cap screws which secure engine to clutch housing, and side frames to block and clutch housing to a torque of 300 ft.-lbs. Other tightening torques are as follows:

Hydraulic pump drive ......32 ft.-lbs.
Hydraulic pump support ....85 ft.-lbs.

Fig. JD15—Use the indicated sequence to tighten the cylinder head cap screws on 6 cylinder model tractors.

CYLINDER HEAD

All Models

15. To remove the cylinder head, drain the cooling system and remove muffler, hood, side panels and grille screens. Remove fan and water pump as outlined in paragraph 72. On diesel models, remove injector assemblies as outlined in paragraph 57 or 57A.

On all models, disconnect generator wires, remove voltage regulator and bracket and let generator down against side of engine block. Remove manifolds, disconnect coolant hoses and remove coolant by-pass line. Remove ventilator tube, rocker arm cover, rocker arms assembly and push rods, then unbolt and remove cylinder head.

When installing head, coat both sides of gasket with thin coat of Permatex or similar sealant. Make sure flat washers are installed under all cap screws and tighten to torque indicated in table. Use tightening sequence shown in Fig. JD14 for 3000 series or Fig. JD15 for 4000 or 5000 series. Tighten rocker arm clamp bolts to a torque of 55 ft.-lbs.

Tightening Torques

All 3000 Series ....115 ft.-lbs.
4000 Series:
 "F" grade cap screws
  (6 line marks)...........115 ft.-lbs.
 "G" grade cap screws
  (12.9 marks) ....130 ft.-lbs.
5000 Series:
 "F" grade cap screws
  (6 line marks)...........180 ft.-lbs.
 "G" grade cap screws
  (12.9 marks) ....240 ft.-lbs.

Before installing rocker arm cover, warm engine to operating temperature, retighten cylinder head cap screws and adjust tappet gap as outlined in paragraph 17A.

16. **MANIFOLDS (GASOLINE ENGINES).** Gasoline models are equipped with heat exchangers which regulate the temperature of the incoming fuel-air mixture. Four cylinder engines in 3000 series are equipped with two heat exchanger valves. Six cylinder engines have one valve as shown in

FRONT ▶

Fig. JD16. Adjustment is made by removing valve (or valves), turning them 180° and reinstalling. When outside temperatures are below 32 degrees F. align the "HOT" stamping on manifold and valve. When temperatures are above 32 degrees F., install valve so that "COLD" markings are aligned.

VALVES AND SEATS
All Models

17. On Series 3000 and 4000 tractors, intake valves seat directly in cylinder head. Exhaust valves on LP-Gas models are equipped with seat inserts; on all other models, exhaust valves seat directly in cast iron cylinder head.

On late Series 5000 tractors, seat inserts are used for both intake and exhaust valves and new, hardened valves are used. Do not use the earlier valves with seat inserts.

On 5020 tractors, valves should be recessed a minimum of 0.054 for intake and 0.094 for exhaust; refer to Fig. JD16A. On Model 5010 when late camshaft is installed, intake valves must be recessed a minimum of 0.044 and exhaust valves a minimum of 0.074 as shown. On 4020 tractors after serial no. 200999, intake valve heads must be recessed a minimum of 0.036 and exhaust valve heads must be recessed a minimum of 0.054.

Fig. JD16—Intake manifolds on gasoline models are equipped with heat exchanger valves which can be installed for hot or cold operation. Pictured is the single valve used on series 4000.

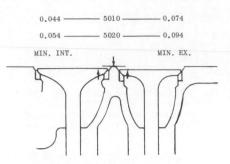

```
0.044 ——— 5010 ——— 0.074
0.054 ——— 5020 ——— 0.094
MIN. INT.              MIN. EX.
```

Fig. JD16A — On Series 5000 tractors, valves must be recessed the indicated amount when late type camshaft is used.

On all models, valve face angle is 44½ degrees and seat angle is 45 degrees for both intake and exhaust. Recommended seat width is 0.083-0.093. Seats can be narrowed using 15 and 70 degree stones.

Intake and exhaust valve stem diameter is 0.3715-0.3725 for 3000 and 4000 series, with a recommended clearance of 0.002-0.004 in stem guides. On series 5000, intake and exhaust valve stem diameter is 0.4335-0.4345 with a recommended stem to guide clearance of 0.003-0.0055.

Valve tappet gap should be adjusted using the procedure outlined in paragraph 17A.

17A. TAPPET GAP ADJUSTMENT. On all models, the two-position method of valve tappet gap adjustment is recommended. Refer to Figs. JD16B and JD16C for four cylinder models or Figs. JD16D and JD76E for six cylinder models, and proceed as follows:

Turn engine crankshaft by hand until "TDC" mark on flywheel is aligned with reference mark on clutch housing, then check the valves to determine whether front or rear cylinder is on compression stroke. (Exhaust valve on adjacent cylinder will be partly open). Use the appropriate diagram and adjust half the valves; then turn crankshaft one complete turn until "TDC" timing mark is again aligned. Adjust remainder of valves using the other diagram.

Recommended valve tappet gaps are as follows:

All Series 3000-4000

Intake Valves
Non-Diesel 0.015
Diesel 0.018

Exhaust Valves
Non-Diesel 0.028
Diesel 0.018

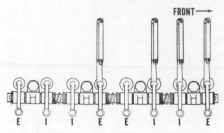

Fig. JD16B—On Series 3000 tractors, with No. 1 piston at TDC on compression stroke, adjust the indicated valves to clearance shown in paragraph 17A. Refer also to Fig. JD16C.

Model 5010
Intake valves 0.015
Exhaust valves 0.022

Model 5020
Intake valves 0.018
Exhaust valves 0.028

VALVE ROTATORS
All Models

18. Positive type exhaust valve rotators are used on all models. Normal service consists of renewing the complete unit. Rotators can be considered satisfactory if the exhaust valves turn a slight amount each time the valves open.

VALVE GUIDES AND SPRINGS
All Models

19. Intake and exhaust valve guides are interchangeable for any one model, however guides used in diesel engines differ from those used in non-diesels. To renew the guide, press old guide out top of head using a piloted arbor. Press new guide in cylinder head from top, so that smaller OD of guide will be toward valve springs. Guide should be installed so that distance from port end of guide to gasket surface of cylinder head is 2⅛-inches for non-diesel models, 1¾-inches for series 3000 and 4000 diesels; or 1⅞-inches for series 5000.

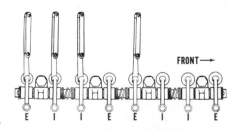

Fig. JD16C—On Series 3000 tractors, with No. 4 piston at TDC on compression stroke, adjust the indicated valves to clearance shown in paragraph 17A. Refer also to Fig. JD16B.

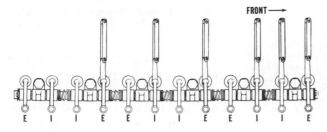

Fig. JD16D—On Six Cylinder Models, adjust the indicated valves when No. 1 piston is at TDC on compression stroke. Refer to paragraph 17A for recommended clearance and to Fig. JD16E for remainder of valves.

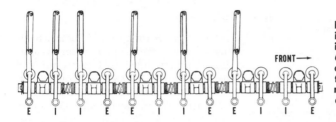

Fig. JD16E—On Six Cylinder Models, adjust the indicated valves when No. 6 piston is at TDC on compression stroke. Refer to paragraph 17A for recommended clearance and to Fig. JD16D for remainder of valves.

Guides are pre-sized and will not require reaming if carefully installed. Normal stem to guide clearance is 0.002-0.004 for series 3000 and 4000; or 0.003-0.0055 for series 5000. Renew valve and/or guide if clearance exceeds 0.006.

Intake and exhaust valve springs are interchangeable for any one model. Renew any spring which is distorted, rusted or discolored, or does not meet the test specifications which follow:

Series 3000-4000
Pounds Test @ Length
Diesel:
    Closed length.....52-64 @ 1 13/16
    Open length....129-157 @ 1 23/64
Non-Diesel
    Closed length.....36-44 @ 1 13/16
    Open length......81-99 @ 1 3/8
Series 5000
Pounds Test @ Length
    Closed length.....61-75 @ 2 7/16
    Open length....190-232 @ 1 29/32

ROCKER ARMS
All Models

20. The rocker arm shaft attaches to bosses which are cast into cylinder head and is held in place by clamps. Shaft rotation is prevented by a roll pin in cylinder head which enters a hole in shaft for positive positioning of lubrication passages.

Rocker arms are right hand and left hand assemblies. On series 3000 and 4000, rocker arms have a recommended clearance of 0.0005-0.0035 on the ¾-inch shaft. On series 5000, suggested clearance is 0.003-0.006 on the 0.999-1.000 shaft. Bushings are not available; if clearance is excessive, renew rocker arms and/or shaft.

When reassembling, make sure roll pin aligns with locating hole in shaft, tighten clamp cap screws to a torque of 55 ft.-lbs. and adjust tappet gap as outlined in paragraph 17A.

CAM FOLLOWERS
(VALVE TAPPETS)
All Models

21. The cylindrical type cam followers can be removed from above with a magnet or hooked wire after removing cylinder head as outlined in paragraph 15. The cam followers operate in unbushed bores in engine block, and are available in standard size only.

TIMING GEAR COVER AND CRANKSHAFT FRONT OIL SEAL
All Models

22. To remove the timing gear cover, first drain cooling system and separate tractor between engine and

front end by following the general procedure in paragraph 14 or 14A.

Remove crankshaft pulley using a suitable puller, loosen oil pan screws, remove cover retaining cap screws and withdraw the cover.

The front oil seal on early models was combined with oil slinger and pressed on crankshaft in front of crankshaft gear. On late models, a seal sleeve is pressed on crankshaft and a lip-type oil seal pressed into timing gear cover bore. The lip type seal can be installed on early tractors if timing gear cover is also renewed.

To renew the early (face type) seal, pry old seal from crankshaft and install new seal with a driver which contacts full area of seal next to shaft. The lip-type seal kit also includes the seal sleeve. To remove the old sleeve, slightly distort the sleeve on seal surface using a blunt chisel then carefully pry seal from shaft. Coat inner diameter of new sleeve with Permatex and install with screw-type puller such as JDE-3. Install seal in cover from inside, with sealing lip to rear.

TIMING GEARS
All Models

23. Timing gear train is identical on all series 3000 and 4000 engines and consists of crankshaft gear (2—Fig. JD17), camshaft gear (1), idler gear (3) and, on diesel engines, the ventilator pump drive gear (4). On non-diesel engines, the ventilator pump is combined with engine governor and unit is driven by gear (4). Refer to Fig. JD17A for timing gear train used on series 5000. Timing gear train is similar to other models except for addition of injection pump drive gear (3) and use of double (two-piece) camshaft gear (2). Tim-

Fig. JD17 — Timing gear train used on series 3000 and 4000. Camshaft is correctly timed when timing marks (TM) are aligned as shown. Gear (4) drives the ventilator pump on diesel models, and the combined ventilator pump and governor on non-diesels.

Fig. JD17A—Timing gear train used on Series 5000. Engine is correctly timed when both timing marks (TM) are aligned as shown.

1. Crankshaft gear
2. Camshaft gear
3. Injection pump gear
4. Idler gear
5. Ventilator pump gear
6. Front seal

supply holes are aligned in block and bushing. Specification data are as follows:

Camshaft Journal OD...2.3745-2.3755
Camshaft Bearing ID....2.3775-2.3795
Valve Lift 0.2750-0.2850
End Play 0.0025-0.0085
Thrust Plate Thickness..0.1185-0.1215

ing gears are available in standard size only, if backlash is excessive renew the parts concerned.

24. ENGINE TIMING. The engine is properly timed on Series 3000 and 4000 when "V" mark on camshaft gear tooth space is aligned with marked gear tooth on crankshaft gear as shown in Fig. JD17. On Series 5000 both timing marks must be aligned as shown at (TM—Fig. JD17A).

25. CAMSHAFT GEAR. To remove the camshaft gear first remove camshaft as outlined in paragraph 29, remove the retaining snap ring, then remove gear by using a press. Inspect camshaft thrust plate for wear or scoring while gear is off. Thrust plate thickness is 0.1185-0.1215 and specified end play of installed shaft is 0.0025-0.0085.

On Model 5000, injection pump drive gear is attached to rear face of camshaft gear with cap screws and may be renewed separately.

26. CRANKSHAFT GEAR. To remove the crankshaft gear first remove timing gear cover as outlined in paragraph 22, then pull gear and crankshaft front oil seal at the same time using a suitable puller. Heat new gear to approximately 350 degrees F. on a stove or hot plate to facilitate installation and align timing marks when gear is installed, as shown in Fig. JD17 or Fig. JD17A.

27. IDLER GEAR. The idler gear (3—Fig. JD17 or 4—Fig. JD17A) serves to drive the ventilator pump on diesel models, or the governor and ventilator pump on non-diesel. The gear thus has no bearing on engine timing and will only need to be serviced if damaged or noisy. To remove

the gear, remove timing gear cover as outlined in paragraph 22. Remove the cap screw and lift off gear and idler stud.

Gear is fitted with a renewable bushing which is pre-sized. When installing, make sure that the fiber thrust washer behind gear is in place and prevented from rotation by the dowel.

28. VENTILATOR PUMP DRIVE gear (4—Fig. JD17 or 5—Fig. JD17A) is also the governor drive gear on non-diesel engines. To remove the gear, unbolt and remove the ventilator pump or governor. When renewing gear on any model, press gear on shaft until an end play of 0.006-0.012 exists in the assembly.

CAMSHAFT AND BEARINGS
All Models

29. Camshaft is identical for gasoline, LP-Gas and diesel models. To remove the camshaft first remove timing gear cover as outlined in paragraph 22 and remove crankshaft front oil seal. Remove rocker arms assembly and push rods, then lift cam followers from bores using a hooked wire. Remove oil pan, oil pump and fuel lift pump. Working through the openings provided in camshaft gear, remove the three cap screws retaining camshaft thrust plate, then withdraw camshaft and gear from engine.

The 2.3745-2.3755 camshaft journals should have a clearance of 0.002-0.005 in bushings. The pre-sized copper lead camshaft bushings are interchangeable. To install bushings after camshaft is out, detach cylinder block from clutch housing, remove clutch, flywheel and camshaft bore plug, then pull bushings into block using a piloted puller. Make sure the oil

ROD AND PISTON UNITS
All Models

30. Connecting rod and piston units are removed from above after removing cylinder head, oil pan and rod bearing caps. On 3000 series tractors, the engine balancer must be removed as outlined in paragraph 36 before No. 2 and No. 3 pistons can be removed.

When reinstalling, the correlation numbers on rod and cap must be in register and face toward camshaft side of engine. Rods are stamped "FRONT" for proper installation. On series 3000 & 4000 tighten cap screws to a torque of 85-95 ft.-lbs. for non-diesel engines, or 120-130 ft.-lbs. for diesels. On series 5000, tighten connecting rod cap screws to a torque of 155-165 ft.-lbs.

PISTONS, RINGS AND SLEEVES
All Models

31. Five ring pistons were used in all new 3010, 4010 and early 5010 engines. Later engines use four ring pistons. Four ring pistons are available for all engines as a service item. Original pistons are stamped to correspond with cylinder from which removed. Gasoline or diesel pistons are marked "F" or "FRONT" for proper installation. LP-Gas pistons are not

Fig. JD18—To remove the camshaft, first remove the crankshaft front oil seal, then remove the three retaining bolts (S).

marked and may be installed either way; however, if piston is re-used it should be installed in the same position from which removed. Pistons and rings are available in standard size only. Piston rings are marked with a dot, which must be installed toward top of piston.

Wet type cylinder sleeves are sealed at the top of head gasket. Bottom of sleeve on 5020 is sealed by a square section packing ring contained in a step in block and sleeve and slightly compressed by sleeve installation. On all other models, two O-rings are used which fit into grooves in engine block. A third groove, located between the two sealing rings, is vented to outside of block as shown at (V—Fig. JD19).

When properly installed, cylinder sleeve should extend 0.001-0.004 above gasket surface of block as shown in Fig. JD19A. Stand-out should be measured with sealing rings removed. Copper shims are available in thicknesses of 0.004, 0.015 and 0.030 for proper positioning of sleeve.

When installing sleeves, make sure that seal ring grooves and vent grooves are clean and vent holes open. Lubricate sealing rings with liquid soap. On engines equipped with O-rings, install red ring in upper groove and black ring in lower groove. Work sleeves gently into place by hand as far as possible, then seat the sleeve using a wooden block and hammer.

Check the pistons, rings and sleeves against the values which follow:

Cylinder Sleeves

(Series 3010-4010)

Inside Diameter,
   Non-Diesel 3.9995-4.0005
Inside Diameter,
   Diesel 4.1245-4.1255

(Series 3020-4020)
Inside Diameter, All ..4.2493-4.2507

(Series 5000)
Inside Diameter 4.7493-4.7507

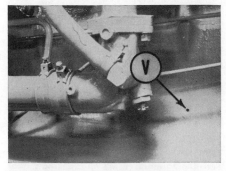

Fig. JD19—The vent holes (V) in side of engine block are ported into an annular groove between the two sleeve sealing rings.

(All Engines)
Maximum Wear............0.005
Maximum Taper 0.005

Piston Skirt Clearance
(Measured at bottom of piston, right angle to piston pin).
5010 Diesel 0.004-0.0065
5020 Diesel 0.002-0.005
3000-4000 Diesel 0.003-0.005
All Non-Diesel Models ..0.0045-0.007

Piston Rings

End Clearance ....0.013 min.
Compression Ring Side
  Clearance 0.0035-0.005
Maximum Wear Limit.......0.008
Oil Control Ring Side
  Clearance 0.0025-0.004
Maximum Wear Limit.......0.007

Pin Bore
(Series 3000-4000)
Non-Diesel 1.2503-1.2509
Diesel 1.5003-1.5009
Diametral Clearance ..0.0000-0.0012
Series 5000
Bore in Piston 1.7506-1.7512
Diametral Clearance ..0.0003-0.0015

PISTON PINS

All Models

32. The full floating type piston pins are retained in piston pin bosses by snap rings. Pins are available in standard size as well as oversizes of 0.003 and 0.005.

The recommended fit of piston pins is a hand push fit in piston and a slip fit in connecting rod bushings. Standard piston pin clearances and diameters are as follows:

(Series 3000-4000)
Diam., Non-Diesel ...1.2497-1.2503
Diameter, Diesel 1.4997-1.5003
Clearance (Rod) 0.0007-0.0023
Clearance (Piston) ...0.0000-0.0012

(Series 5000)
Diameter 1.7497-1.7503
Clearance (Rod) 0.0007-0.0023
Clearance (Piston) ...0.0003-0.0015

CONNECTING RODS AND BEARINGS

All Models

33. Connecting rod bearings are of the precision type, renewable from below after removing oil pan, engine balancer (3000 series) and bearing

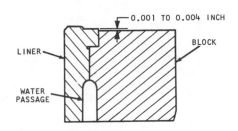

Fig. JD19A—Cylinder sleeves should stand out 0.001-0.004 inch when properly installed. Refer to text.

caps. Bearings are available in standard size and undersizes of 0.002, 0.003, 0.020, 0.022, 0.030 and 0.032.

Connecting rods used in diesel engines are center drilled for lubrication of piston pin. Non-diesel connecting rods are not drilled. When servicing engine, make sure oil passages are open and clean. Mating surfaces of rod and cap have milled tongues and grooves which positively locate cap and prevent it from being reversed during installation. Original rods are stamped with cylinder number on side toward the camshaft. Connecting rods are marked "FRONT" for proper installation.

Check the connecting rods, bearings and crankpin journals against the values which follow:

(Series 3000-4000)
Crankpin diameter 2.998-2.999
Regrind if out-of-round.......0.004
Regrind if tapered.............0.004
Crankpin diametral
  clearance 0.0015-0.0045
Rod bolt tightening torque
  Non-Diesel ....85-95 ft.-lbs.
  Diesel ....120-130 ft.-lbs.

(Series 5000)
Crankpin diameter 3.498-3.499
Regrind if out-of-round 0.004
Regrind if tapered.............0.004
Crankpin diametral
  clearance 0.0015-0.0045
Rod bolt tightening
  torque ....155-165 ft.-lbs.

CRANKSHAFT AND MAIN BEARINGS

Series 3000

34. The crankshaft for non-diesel engines is supported in three renewable, precision type main bearings. Diesel engines have five main bearings. Crankshaft end play of 0.0025-0.0085 is controlled by the flanged

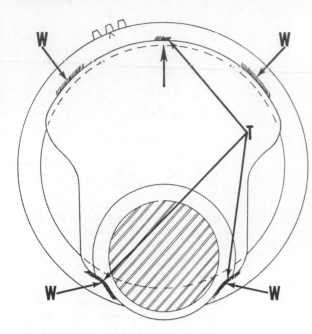

Fig. JD20 — Engine balancer drive gear on series 3000 engines is positively located to crankshaft by a dowel pin at point of arrow. When installing a new gear, tack weld in three places at points indicated by (T), then weld where indicated by (W). See text for details.

center main bearing. Upper and lower bearing shells are not interchangeable, the upper half containing an oil hole.

All engines are equipped with a Lanchester type engine balancer which is driven at twice crankshaft speed by a gear welded to crankshaft just forward of the center main bearing. See Fig. JD20.

All main bearing caps can be removed from below after removing oil pan and engine balancer. On diesel engines, it will be necessary to remove oil pump for access to second main bearing. When renewing the main bearings, make sure that locating lug on bearing shell is aligned with milled slot in cap and block bore, and tighten the retaining cap screws

Fig. JD20A—Block, main bearing caps, connecting rods and rod caps are marked with correlation numbers (CN) as shown. Crankshaft end play is controlled by the flanged main bearing insert (TF).

to a torque of 150 ft.-lbs. Main bearings are available in undersizes of 0.002, 0.003, 0.020, 0.022, 0.030 and 0.032 as well as standard.

To remove the crankshaft, first remove engine as outlined in paragraph 14 and proceed as follows: Remove flywheel and crankshaft rear oil seal retainer. Remove crankshaft pulley, timing gear cover, oil pan and engine balancer. Remove oil pump and rod and main bearing caps, then lift out crankshaft. The hardened, crankshaft rear oil seal slinger is a press fit on flywheel flange and is renewable. When installing, drive slinger on shaft with a suitable driver until rear edge is flush with rear surface of flywheel flange.

To renew the engine balancer drive gear, remove the welds with a lathe or a small grinding wheel and tap gear from shaft. Gear is positively located by a dowel pin at point of arrow, Fig. JD20. Dress down weld surfaces to conform with rest of flange and install gear on shaft, chamfered edge forward. Align groove in gear over dowel pin in shaft, then tack in place at three points as indicated at (T—Fig. JD20).

NOTE: Use ⅛-inch, iron powder, low hydrogen rod and approximately 100 amperes of current. Be extremely careful not to apply more heat than is required. Protect gear teeth and crankshaft journals from weld slag during welding process.

After gear is secured by tacking, weld to shaft in four places as shown at (W) using the old weld marks as

a guide. Beads should be approximately 1-inch long. Thoroughly clean all weld slag from shaft before reinstalling.

Check crankshaft and bearings against the values which follow:
Crankpin diameter .....2.998 -2.999
Main journal diameter..3.3715-3.3725
Regrind if out of round.......0.004
Main bearing diametral
  clearance 0.0031-0.0061
Cap screw torque........150 ft.-lbs.

Series 4000—5000

35. The crankshaft for series 4000 non-diesel engines is supported in four renewable, precision type main bearings. Diesel engines have seven main bearings. Crankshaft end play of 0.0025-0.0085 is controlled by the flanged third main bearing for non-diesel engines; or the flanged fifth main bearing for diesels. Upper and lower bearing shells are not interchangeable, the upper half containing an oil hole.

All main bearing caps can be removed from below after removing oil pan and oil pump. When renewing

Fig. JD21—When installing the Lanchester type engine balancer on series 3000 make sure timing marks (TM) on crankshaft gear and driven gear are aligned as shown in upper view. The two balance weights must be timed as shown in lower view.

bearings, make sure that locating lug on bearing shell is aligned with milled slot in cap and block bore, and tighten the retaining cap screws to a torque of 150 ft.-lbs. Main bearings are available in undersizes of 0.002, 0.003, 0.020, 0.022, 0.030 and 0.032 as well as standard.

To remove the crankshaft, first remove engine as outlined in paragraph 14 or 14A and proceed as follows: Remove flywheel and crankshaft rear oil seal retainer. Remove crankshaft pulley, timing gear cover, oil pan and oil pump. Remove rod and main bearing caps and lift out crankshaft. The hardened, crankshaft rear oil slinger is a press fit on flywheel flange and is renewable. When installing, drive slinger on shaft with a suitable driver until rear edge is flush with rear surface of flywheel flange. Check crankshaft and bearings against the values which follow:

Series 4000

Crankpin diameter .....2.998 -2.999
Main journal diameter ..3.3715-3.3725
Regrind if out of round........0.004
Main bearing diametral
   clearance 0.0031-0.0061
Cap screw torque........150 ft.-lbs.

Series 5000

Crankpin diameter 3.498-3.499
Main journal diameter 3.748-3.749
Regrind if out of round 0.004
Main bearing diametral
   clearance 0.0031-0.0061
Cap screw torque ....170 ft.-lbs.

ENGINE BALANCER

Series 3000

36. REMOVE & REINSTALL. To remove the engine balancer, first remove oil pan as outlined in paragraph 40. Remove the four attaching cap screws and withdraw balancer assembly, being careful not to damage the oil pump intake line.

When installing, turn crankshaft until No. 1 and No. 4 pistons are at bottom center, then install balancer so that timing marks are aligned as shown in Fig. JD21. Tighten the retaining cap screws securely and lock in place by bending the lock plates.

37. OVERHAUL. To overhaul the removed engine balancer, first remove the oil shield plates (5—Fig. JD22), remove snap rings (1), then remove shafts with an arbor press.

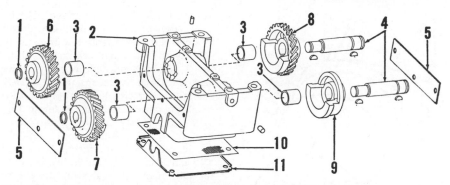

Fig. JD22—Exploded view of the Lanchester type engine balancer used on series 3000. Balancers are used on all models in series, but diesel and non-diesel units are not interchangeable.

1. Snap ring	5. Shield	8. Driven weight
2. Frame	6. Timing gear	9. Balance weight
3. Bushing	7. Timing gear	10. Screen
4. Balancer shaft		11. Cover

NOTE: Support at driven gear (8) or balance weight (9) while applying pressure to prevent damage to balancer housing.

Bushings (3) are pre-sized and must be carefully installed with a piloted arbor such as JDE 16. Bushings contain a spiral oil groove which extends to one edge only. Open end of oil grooves must be installed toward inside of housing as shown at G—Fig. JD23. Press bushings into housing until outer edge is 3/64-inch below machined surface of bore as shown at A.

To assemble, reverse the disassembly procedure. Make sure timing marks are aligned on balancer gears (6 and 7—Fig. JD22) and press the shaft through balance weight and gear at the same time. Install unit as outlined in paragraph 36.

CRANKSHAFT REAR OIL SEAL

All Models

38. The crankshaft rear oil seal is contained in a retainer plate which is doweled and bolted to rear face of cylinder block. To renew the seal first detach (split) engine from clutch housing as outlined in paragraph 82, 82A or 93A, and remove clutch and flywheel.

If seal only is to be renewed, it can be pried from retainer without removing retainer. Seal seats on hardened seal of oil slinger which is a press fit on crankshaft flywheel flange. To renew the slinger, remove oil seal retainer plate and pry slinger from crankshaft. Install slinger with a suitable driver so that rear edge is flush with rear face of crankshaft flange. Slinger can be heated in oil to facilitate installation.

FLYWHEEL

All Models

39. Flywheel is doweled to crankshaft flange and retained by cap screws which have nylon locking inserts. Flywheel can be removed by using forcing screws in the tapped holes provided.

To install flywheel ring gear, heat evenly to approximately 550 degrees F. and position gear so that chamfered end of gear teeth face toward front of engine. Make sure ring gear is firmly seated against shoulder of flywheel.

The locating dowels and cap screws are arranged so that flywheel can be installed in only one position. Make sure machined surfaces of flywheel and crankshaft flange are clean and

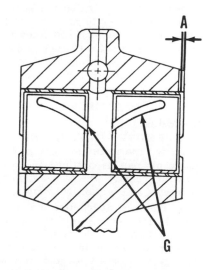

Fig. JD23—When installing bushings in balancer frame, make sure open end of oil grooves (G) are together and that bushing is 3/64 inch below bore surface as shown at (A).

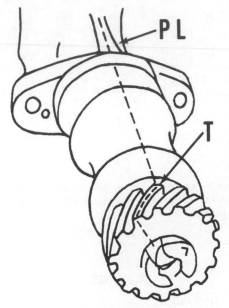

Fig. JD24—Ignition distributor or injection pump on series 3000 and 4000 is driven from camshaft by oil pump gear. To correctly time the oil pump, make sure that lower edge of tooth (T) which is in line with drive slot aligns with pump casting parting line (PL).

free from burrs and tighten retaining cap screws to a torque of 85 ft.-lbs. on series 3000 & 4000; or 130 ft.-lbs. on Series 5000.

OIL PAN
All Models

40. To remove the engine oil pan drain oil, remove oil filter cover plate from left side of pan, withdraw filter element then remove filter body. Oil pan can now be unbolted and removed.

When installing, tighten the ⅜-inch oil pan cap screws to a torque of 35 ft.-lbs. and the ½-inch cap screws to a torque of 85 ft.-lbs. On Series 5000, tighten oil pan to clutch housing cap screws to a torque of 300 ft.-lbs.

OIL PUMP
All Models

41. **REMOVE AND REINSTALL.** To remove the engine oil pump, first remove oil pan as outlined in paragraph 40, remove the two retaining cap screws and withdraw pump.

The ignition distributor on non-diesel models; or the injection pump on series 3000 & 4000 diesel models; is driven by a slot in the oil pump drive gear. Correct engine timing, therefore, depends on proper installation of oil pump on these models; on Series 5000, oil pump is not timed. When installing oil pump on series 3000 or 4000, proceed as follows: Be-

fore installing pump, remove distributor (non-diesel models). On diesel models remove injection pump as in paragraph 64. Turn crankshaft until No. 1 piston is on compression stroke and continue turning until "TDC" mark on flywheel is aligned with mark on clutch housing. Refer to Fig. JD24 and rotate the oil pump drive gear until drive slot is approximately 45 degrees from mounting bolts and "V" mark on drive lug is toward outlet side of pump as shown. Make sure that lower end of gear tooth (T) which is in direct line with drive slot aligns with parting line (PL) of pump casting as shown by broken line, then carefully install pump in engine.

When properly installed and with No. 1 piston at top center on compression stroke, the "V" mark on drive hub should be toward engine block and drive slot approximately 15 degrees from parallel with crankshaft as shown in Fig. JD25.

NOTE: If oil pump drive gear is one tooth out of time, the position of drive slot will be advanced or retarded 24 degrees. Engine timing would be changed 48 degrees.

42. **OVERHAUL.** To overhaul the removed oil pump, first remove pump cover and examine pump gears and cover for wear or scoring. If wear is not excessive, further disassembly is not necessary.

If further disassembly or inspection is indicated, drive out the groove pin which retains drive gear or hub to shaft, and press shaft and driven gear downward out of drive gear.

NOTE: On diesel models, the drive gear is separate from hub as shown in Fig. JD26. If drive gear only is to

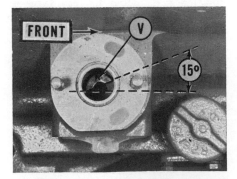

Fig. JD25—Oil pump is correctly timed if, with No. 1 piston at TDC on compression stroke, the drive slot is 15 degrees from parallel with crankshaft. "V" mark (V) on heavy drive lug must be toward engine block.

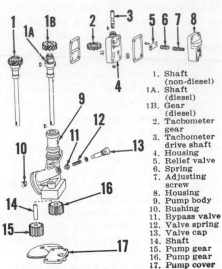

Fig. JD26—Exploded view of engine oil pump showing component parts. Drive gear (1B) is renewable on diesel models.

1. Shaft (non-diesel)
1A. Shaft (diesel)
1B. Gear (diesel)
2. Tachometer gear
3. Tachometer drive shaft
4. Housing
5. Relief valve
6. Spring
7. Adjusting screw
8. Housing
9. Pump body
10. Bushing
11. Bypass valve
12. Valve spring
13. Valve cap
14. Shaft
15. Pump gear
16. Pump gear
17. Pump cover

be renewed, it may be pressed from hub without removing groove pin or disassembling pump.

Remove plug (13—Fig. JD26) and withdraw bypass valve and spring; then clean the pump parts in a suitable solvent. Pump specifications are as follows:

Drive gear OD.........2.312 -2.316
Pump gear OD.........2.268 -2.270
Pump gear thickness
 (non-diesel) ....1.000 -1.002
Pump gear thickness
 (diesel) ....1.500 -1.502
Drive shaft OD ....0.687 -0.689
Pump idler gear
 shaft OD 0.6284-0.6290
Pump gear bore ID.....2.272 -2.274
Pump gear bore depth
 (non-diesel) ....1.003 -1.007
Pump gear bore depth
 (diesel) ....1.503 -1.507

To assemble the pump, turn pump drive shaft so that pin hole in upper end of shaft is 90 degrees from hole in gear or hub and press shaft into hub until lower end of shaft is $8\frac{11}{32}$ inches from nearest face of hub as shown in Fig. JD27. Drill a $\frac{9}{32}$-inch hole in shaft using pin hole in hub as a guide and secure hub to shaft with groove pin. Insert shaft in housing and install Woodruff key and driven gear. Support opposite end of shaft when pressing gear in place to prevent stress on groove pin. Install the assembled pump as outlined in paragraph 41.

PRESSURE RELIEF VALVE
All Models

43. **ADJUSTMENT.** The oil pressure regulator is located on right

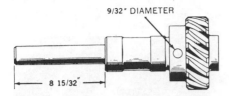

Fig. JD27—To renew pump drive shaft or gear, use the dimensions shown above to drill the pin hole. See text for details.

side of engine block immediately below the tachometer drive as shown in Fig. JD28. To adjust the pressure, install a master gage in the gallery plug (P), and with engine warm, adjust throttle control to maintain an engine speed of 1900 rpm. Remove cap nut (N), loosen jam nut and adjust the pressure to 30 psi by turning the slotted adjusting screw in or out as required.

OIL COOLER

All Diesel Models

44. Series 3000 and 4000 diesel tractors are equipped with an engine oil cooler which attaches to right side of engine block as shown in Fig. JD29. On Series 5000, the oil cooler is attached horizontally to right side of block behind right side rail. The oil is cooled by the engine coolant liquid which circulates through tubes in cooler body.

To remove the oil cooler, drain cooling system and remove coolant hoses. On Series 5000, remove right side rail. Disconnect oil return line at block, then remove the cap screws which retain cooler to bypass valve body. Remove cooler bracket cap screws and lift off cooler assembly.

The removed unit can be disassembled using Fig. JD30 as a guide.

CRANKCASE VENTILATOR PUMP

All Models

45. All models are equipped with an engine driven ventilator pump which provides forced ventilation of the crankcase. The pump is driven by the crankshaft timing gear through an idler gear, and is integral with the engine governor on non-diesel models. The air supply must first pass through the engine air cleaner and is therefore filtered of dust and dirt.

The pump impeller rotates at crankshaft speed and revolves in a casing bore which is offset in relation to impeller and partially filled with engine oil. In operation, the oil is picked up and forced to outer wall of casing by centrifugal force where it rotates with the impeller. Air is drawn into the ventilator pump from the inlet tube and is trapped between the impeller blades and the rotating stream of oil. As the impeller turns, the air and a small amount of oil is discharged into the timing gear case at front of engine. The air and combustion gases from crankcase pass upward into rocker arm cover and are exhausted through the vent tube.

To remove the ventilator pump on diesel models, disconnect the inlet air line, remove the attaching cap screws and withdraw the pump assembly. Pump can be cleaned without disassembly by immersing in solvent then using compressed air to remove the loosened foreign material. To disassemble the pump, press drive gear

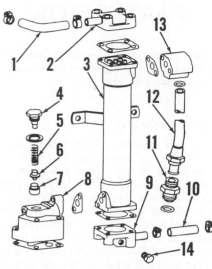

Fig. JD30 — Exploded view of engine oil cooler used on series 4000 diesel. Oil cooler used on other models is similar.

1. Hose	8. Bypass body
2. Cap	9. Cap
3. Cooler	10. Hose
4. Plug	11. Fitting
5. Spring	12. Return hose
6. Bypass valve	13. Elbow
7. Bushing	14. Drain plug

from shaft, remove Woodruff key and the cover attaching cap screws. Inspect bushings and shaft for wear or scoring and renew if indicated. Assemble and install by reversing the disassembly procedure.

Refer to paragraph 70 for service on the combined ventilator pump and governor used on non-diesel models.

Refer to paragraph 27 for service on idler gear.

Fig. JD28—Adjust the engine oil pressure to 30 psi by removing the nut (N) on regulating valve housing, and turning the exposed adjusting screw. Plug (P) is tapped into engine oil gallery. Plug may be removed to install a master gage.

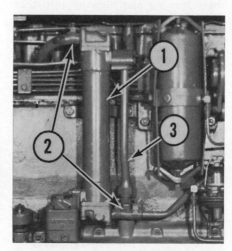

Fig. JD29 — Series 4000 diesels are equipped with an engine oil cooler (1) as shown. To remove oil cooler, disconnect coolant hoses (2) and oil return line (3). Series 3000 diesel is similar.

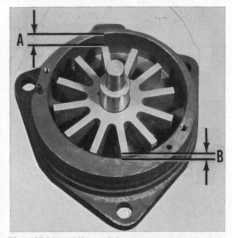

Fig. JD31—All models are equipped with a positive displacement rotor pump with an oil seal. A metered supply of engine oil is pumped to ventilator pump and is thrown to outside of pump housing by centrifugal force. Rotor is offset to bottom of housing as shown by (A) and (B). Air is drawn into pump from air cleaner inlet tube at top of rotor and is forced into timing gear cover by the rotating oil when the clearance narrows at the bottom (B).

CARBURETOR

Marvel Schebler USX or Zenith Series 69 carburetors are used. Refer to the appropriate following paragraph.

Marvel-Schebler

46. All models are equipped with a vacuum actuated, diaphragm type accelerator pump, idle and load mixture adjustment needles and an idle speed adjusting screw. Suggested initial settings are 2½ turns open from closed position for load mixture adjustment needle and 1½ turns open for idle mixture adjustment needle. Refer to Fig. JD31A for location of adjustment points. Back out the idle speed stop screw (3) until it clears stop, then turn screw in 1½ turns after it contacts stop and begins to open throttle butterfly.

A definite measurement of the float level is not given, but float should close needle valve about midway between limit stops in fuel bowl. Refer to Fig. JD31B. To check float setting after removing bowl cover, first remove the float and lift out inlet needle (2). Reinstall the float and, using a pencil and marking on wall of fuel bowl, mark position of float lever with carburetor inverted and right side up, as shown by broken lines. Reinstall inlet needle (2); reinstall float and check float position with carburetor body inverted. Float should rest midway between the two marked positions as shown. If it does not, adjust by bending float lever or installing an extra gasket (1) beneath inlet needle seat.

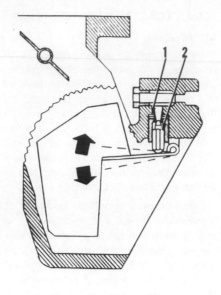

Fig. JD31B—To check the float for adjustment, remove inlet needle (2) and reinstall float. Mark limits of float travel as indicated by dotted lines. With needle installed, float should rest midway between the two marks. Float level can be lowered by adding an extra gasket (1).

Two types of accelerator pump discharge jets have been used. Refer to Figs. JD31C and JD31D. On early models (USX20, USX21, USX30 and USX31), the spring-loaded jet (E) presses into carburetor body and extends into center of venturi alongside main nozzle. On late models, the unit screws into carburetor body and is ported through side of venturi. On all models, check ball must unseat with minimum pressure but seal completely when ball is seated. On early models, the discharge jet can be checked by slipping a short piece of tubing over protruding end of jet, then drawing and blowing into tube with mouth, with fuel bowl cover removed. On late models, unscrew jet from carburetor body, working from bottom end, then check the removed jet. Late model jet contains a tapered weight instead of a spring, to seat

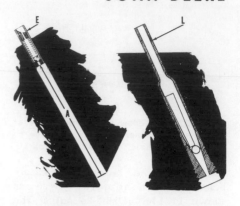

Fig. JD31D—Cross sectional views of early (E) and late (L) accelerator pump discharge jets. Early model is pressed into housing until distance (A) equals 1-17/32-inches. Late model is threaded into housing from bottom. Check ball is spring loaded in early model and seated by weight in late unit.

the check ball, so threaded end of jet must be down when checking ball seating.

To renew the early jet, insert a $\frac{5}{32}$-inch brass rod in bottom opening and drive the jet upward out of carburetor body. Install new jet from bottom with long end of discharge slot up and 5° from parallel with mounting flange stud holes as shown in Fig. JD31E. Use M-505 driver or install jet until distance (A—Fig. JD31D) measures $1\frac{17}{32}$-inches.

On all models, accelerator pump check valve is contained in diaphragm housing as shown in Fig. JD31F. Check ball (C) and spring can be inspected or renewed after removing seat (S). Distance (D) from end of diaphragm stop to gasket surface of housing should measure as follows:

USX-20, USX-30 .... 25/64-inch
USX-21, USX-31, USX-35 29/64-inch
USX-36 .... 31/64-inch

Diaphragm stop pin can be driven into or out of housing if distance (D) is not correct.

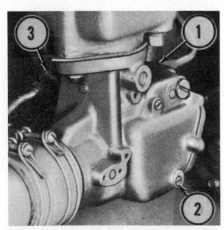

Fig. JD31A—Installed view of carburetor showing idle mixture adjustment (1), load adjustment (2) and idle speed stop screw (3).

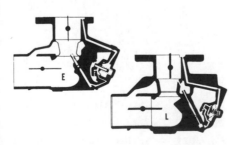

Fig. JD31C—Cross sectional views of early carburetor (E) and late carburetor (L), showing accelerator pump and passages.

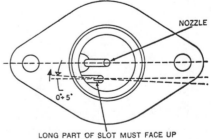

Fig. JD31E — Top view of early Marvel Schebler USX carburetor showing proper position of slot in upper end of discharge jet.

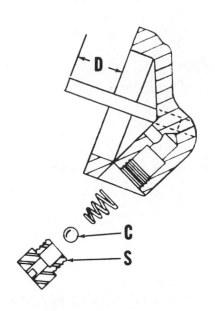

Fig. JD31F—Cross sectional view of diaphragm housing showing inlet check valve (C) and seat (S). Refer to text for recommended measurement of distance (D).

Zenith

46A. The Zenith, Model 69 carburetor used on some engines is equipped with a vacuum controlled accelerator pump. Initial setting of idle mixture adjusting needle is 1½ turns open; load mixture adjusting needle, 1¾ turns open. Adjust speed and throttle linkage as outlined in paragraph 68. Make final mixture adjustments with engine at operating temperature.

Float setting is 1¹⁹⁄₃₂-inches when measured from gasket surface of carburetor body to farthest edge of float

with needle valve closed. To adjust the float, bend levers next to float body.

The accelerator pump piston and spring assembly can be withdrawn from top of bowl chamber after removing upper body. Both pistons of accelerator pump assembly should fit cylinder with a minimum of 0.001 or a maximum of 0.003 clearance in cylinder. Examine cylinder for scoring or deep scratches, and pistons for wear, scoring or other damage. A

check valve is located in bottom of accelerator pump cylinder bore. Check valve is a drive fit and seats on a stepped shoulder in bore. Removal of check valve requires the use of a special tool (Zenith Part No. C161-15) which screws into check valve enabling valve to be withdrawn from top of cylinder bore. Check valve will be damaged in removal, and must be renewed if removed. Install check valve until it bottoms, using Zenith Tool C161-197.

LP-GAS SYSTEM

The pressure fuel tank is equipped with filler, vapor return, pressure relief, bleed and liquid and vapor withdrawal valves which can only be serviced as complete assemblies. Before renewal is attempted on any of these units, the fuel tank must be completely exhausted of fuel. If little fuel remains in the tank, drive tractor to an open area and allow engine to run until fuel is exhausted; then open bleed valve and allow any remaining pressure to escape.

If a considerable quantity of fuel is in the tank, consult an LP-Gas dealer about pumping out and saving the fuel. LP-Gas fuel is heavier than air and tends to settle in low spots, presenting the danger of explosion or fire even if released outdoors, therefore any released fuel must be burned.

A suggested method of disposing of fuel which is perfectly safe and meets fire regulations in most places is as follows:

Drive tractor to an open, well ventilated area and attach a hose 20 feet or more in length to the filler valve of the tank. Place

the open end of hose downwind from tractor to prevent damage to the unit. Prior to attaching hose to tank, place some burning waste at open end of hose so the escaping gas will be ignited immediately. Allow the fuel to completely burn, then tow tractor to the shop.

The fuel gage sending unit consists of a magnetic sender unit which can be renewed at any time, and a float unit which can only be renewed if tank is completely empty.

The safety relief valve is set to open at 312 psi to protect the fuel tank against excessive pressures. This pressure should never be adjusted. If the relief valve is faulty or inoperative, renew the unit.

UL regulations prohibit any welding or repair on LP-Gas tanks. In the event of defect or damage, the tank must be renewed rather than repaired.

Fuel lines and components may be removed at any time without emptying tank if liquid and vapor withdrawal valves are closed and engine allowed to run until the fuel is exhausted in lines and filter.

SYSTEM ADJUSTMENT
LP-Gas Models

47. Before starting engine for the first time, or after installing a new carburetor, make the following preliminary adjustments:

Refer to Fig. JD32 and turn the load adjusting needle (3) approximately 6 turns open. Turn the idle adjusting needle (1) one turn open and the throttle stop screw (2) ½ turn from closed position.

Start engine and bring to operating temperature. Place hand throttle in the slow idle position and adjust throttle stop screw to obtain a slow idle engine speed of 450 rpm. Turn idle adjusting screw either way as required, to point of smoothest idle.

1. Check ball
2. Main nozzle
3. Discharge jet (late)
4. Idle jet
5. Minimum fuel jet
6. Idle speed stop screw
7. Economizer jet
8. Discharge jet (early)
9. Needle seat
10. Load adjusting needle
11. Idle mixture needle

Fig. JD31G—Exploded view of typical Marvel-Schebler USX carburetor. Accelerator pump discharge jet (3) is used on Models USX35 and USX36. Jet (8) is used on other models.

The load adjustment should properly be set under load. If a dynamometer is available, adjust load needle for best performance under rated load. If a dynamometer is not available, a reasonably satisfactory shop adjustment can be obtained as follows: Disconnect three spark plug wires (four-cylinder models) or four spark plug wires (six-cylinder models). Be sure to ground the removed wires. Start engine and open hand throttle to first stop (1900 rpm) position. Adjust carburetor by opening load needle until highest engine speed is obtained, then close needle until speed just begins to drop. Shut off ignition, reinstall the removed spark plug wires, then recheck engine low idle. Make minor adjustments as necessary.

TROUBLE SHOOTING

LP-Gas Models

48. If the engine fails to start and if trouble is determined to be in the fuel system, set the carburetor adjustments as outlined in paragraph 47 and check to make sure that carburetor choke operates properly. Check air cleaner for plugging and make sure air cleaner is properly serviced.

The following paragraphs list the more common troubles that can be attributed directly to the fuel system. Many of the same troubles can be caused by malfunction of the ignition system or of the valves and rings; therefore, check ignition system and compression also, when diagnosing trouble.

HARD STARTING could be caused by:

a. Improperly blended fuel.

b. Excess-flow valve in withdrawal

Fig. JD32—LP-Gas models use the single barrel carburetor shown. Refer to text for carburetor adjustment.

1. Idle adjusting needle
2. Idle speed stop screw
3. Load adjusting needle

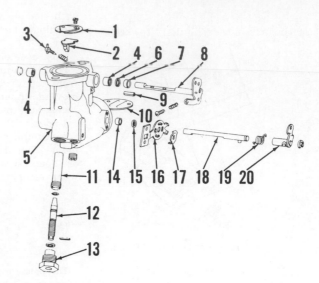

Fig. JD32A — Exploded view of LP-Gas carburetor.

1. Throttle disc
2. Nozzle valve
3. Idle needle
4. Needle bearing
5. Body
6. Packing
7. Retainer
8. Throttle shaft
9. Stop pin
10. Choke disc
11. Nozzle
12. Load needle
13. Nut
14. Packing
15. Retainer
16. Choke bracket
17. Clip
18. Choke shaft
19. Choke spring
20. Choke lever

valve closed. Close withdrawal valve to reset, then open slowly.

c. Incorrect starting procedure.

d. Plugged fuel filter or lines.

e. Liquid fuel in lines.

f. Automatic fuel shut-off not operating properly. Check solenoid and valve.

g. Plugged vent hole on converter.

h. Defective low pressure diaphragm in converter.

i. High pressure valve stuck or valve spring broken in converter.

LACK OF POWER could be caused by:

a. Throttle not opening properly due to maladjusted, bent or broken linkage.

b. Plugged vent hole in converter.

c. Clogged fuel strainer or lines.

d. Excess-flow valves closed.

e. Sticking high pressure valve in converter.

f. Restricted low pressure valve in converter.

g. Defective diaphragms or converter adjustment.

h. Engine not up to operating temperature.

i. Improperly adjusted carburetor.

j. Air leaks in carburetor fuel line or carburetor or manifold gaskets.

k. Clogged air filter.

POOR FUEL ECONOMY could be caused by:

a. Improperly adjusted carburetor or converter.

b. Leaks in fuel lines or tank.

c. Sticking converter valves.

d. Lack of power from any of the causes outlined above.

CONVERTER FREEZING UP could be caused by:

a. Running on liquid fuel before

engine is warm.

b. Water circulating backwards through converter.

Leaks or restrictions in the fuel system can sometimes be detected by frost forming at the point of the restriction. Check for frost at the fuel filter and the withdrawal valves on all complaints of lack of power or hard starting.

CARBURETOR

49. LP-Gas models use the single barrel updraft carburetor shown in Fig. JD32 and Fig. JD32A. Throttle shaft (8) operates in needle bearings

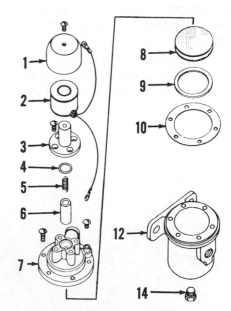

Fig. JD33—Exploded view of fuel strainer and shut-off valve.

1. Cover
2. Coil
3. Plunger housing
4. "O" ring
5. Spring
6. Plunger
7. Housing
8. Filter
9. Retaining ring
10. Gasket
12. Housing
14. Plug

(4) which should be checked when carburetor is overhauled. To remove the nozzle (11), first remove the load adjusting screw (12) and cap (13). Note depth to which nozzle is screwed into body, then remove nozzle with a screwdriver. When reassembling, re-install nozzle in the approximate position it occupied before removal.

FUEL STRAINER AND SHUT-OFF VALVE

50. All of the fuel must pass through the strainer before reaching the converter. The strainer contains a filter element, consisting of a felt pad, chamois disc, and brass screens which remove all solids and gum from the fuel. A solenoid operated, automatic shut-off valve is located on top of the strainer. A spring, plus system pressure keeps this valve closed when the ignition switch is turned off.

If strainer is excessively cold or shows frost, it is probably clogged. To clean the strainer, close both withdrawal valves and run engine until fuel is exhausted. Remove plug (14 — Fig. JD33) from bottom of strainer and open the liquid withdrawal valve momentarily. Pressure from the fuel tank will blow out any accumulation of dirt.

To remove the strainer assembly, close both withdrawal valves and run engine to exhaust any fuel, disconnect fuel lines and the lead-in wire, then unbolt and remove the unit.

Remove the screws retaining cover (7) to strainer body (12) and lift off cover assembly. Remove filter pack

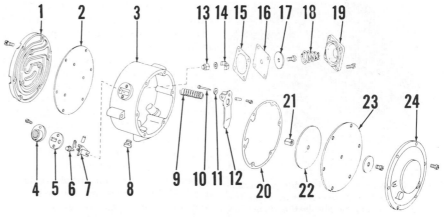

Fig. JD35—Exploded view of LP-Gas converter (regulator).

1. Rear cover	10. Valve seat pin	17. Plate
2. Gasket	11. Low pressure	18. Spring
3. Body	valve	19. Cover
4. Inlet cover	12. Lever	20. Gasket
5. Gasket	13. Link	21. Diaphragm button
6. Valve seat	14. Damper	22. Plate
7. Lever	15. Gasket	23. Low presure
8. Plug	16. High pressure	diaphragm
9. Spring	diaphragm	24. Front cover

(8) by prying out retaining ring (9). Filter pack can be cleaned in a volatile solvent and air dried. Reinstall filter pack with chamois disc toward the top.

Disconnect the wire and lift off case (1) from shut-off valve; then lift off coil (2). Remove plunger housing (3) and lift out plunger and spring. See Fig. JD34. Inspect and renew any damaged parts.

Assemble plunger housing by reversing the disassembly procedure, then test by connecting the unit to a battery. An audible "click" will be heard when the solenoid opens the valve.

After strainer is installed, turn on the vapor withdrawal valve and use

a soap solution to test for leaks at the connections.

CONVERTER (REGULATOR)

51. OPERATION. Fuel enters converter as a liquid at tank pressure, through high pressure valve (6—Fig. JD35). The converter is connected to the engine cooling system which supplies the heat required for vaporization. Fuel pressure is reduced in the converter by means of the regulating valves, to approximately atmospheric pressure required for engine operation. The plug (8) in bottom of converter is provided to drain out the coolant compartment if the radiator is drained in cold weather.

52. R&R AND OVERHAUL. To remove the converter, first close both withdrawal valves and run engine until fuel is exhausted from converter and lines. Drain cooling system, remove plug (8—Fig. JD35) and allow coolant to drain from converter. Disconnect the coolant lines and fuel lines from converter, remove the attaching bolts and lift off the converter assembly.

The high pressure valve (4 through 7) can be removed for service without removing converter from tractor. To remove, close both withdrawal valves and exhaust fuel by running engine. Disconnect inlet fuel line, then remove the screws retaining inlet cover (4). The entire valve assembly will be removed with cover. Discard the gasket (5), and examine the seating surface of valve (6). To remove valve from lever (7), remove

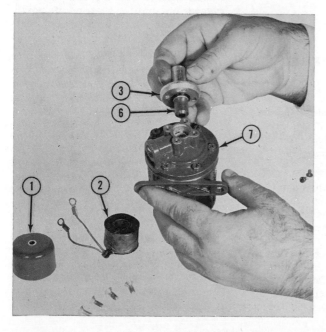

Fig. JD34—Removing the fuel shut-off plunger and plunger housing from fuel strainer. Refer to Fig. JD33 for legend.

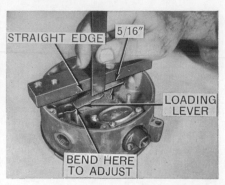

Fig. JD36—Use a straight edge and scale to adjust the low pressure lever. Bend the lever, if necessary to obtain the dimension shown.

spring lock and withdraw valve. Renew valve (6) if seat is worn, cut or ridged. Clean the metal parts in a suitable solvent and reassemble by reversing the disassembly procedure, using a new gasket.

To disassemble the converter after it is removed from tractor, remove the end covers (1 and 24), and high pressure diaphragm cover (19). Examine the diaphragms (16 and 23) for cracks or pin holes. Spring (18) is calibrated to maintain a converter pressure of 6 psi. Renew spring if it is bent, rusty, or has taken a permanent set. Remove the pin retaining low pressure valve lever (12) and remove the lever and spring. Examine the seating surface of valve seat (11) and examine spring (9) for rust, broken coils or loss of tension.

Reassemble by reversing the disassembly procedure, using Fig. JD35 as a guide. Always use new gaskets and make sure that diaphragms are not wrinkled in assembly. When assembling low pressure valves, use a straight edge and rule to measure the assembled distance between free end of low pressure lever (12) and gasket surface of converter body (3) as shown in Fig. JD36. Bend lever if necessary, to obtain the specified $\frac{5}{16}$-inch distance below gasket surface of body.

Install the assembled converter, turn on the vapor withdrawal valve and check for leaks, using a soap and water solution. Adjust the carburetor as outlined in paragraph 47.

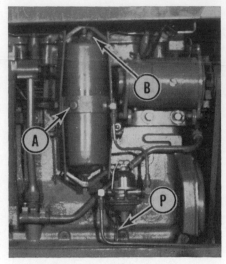

Fig. JD37—To bleed the diesel fuel system, open bleed screw (A) to first stage filter and use lift pump lever (P) to pump air from filter. Bleed second stage (upper) filter by opening bleed screw (B). Make sure lever (P) is in lower position before starting engine.

DIESEL FUEL SYSTEM

FUEL FILTER, PRIMARY PUMP AND LINES
All Models

53. The first stage (lower) fuel filter is fitted with a renewable element which should be renewed every 600 hours of operation. Element in second stage (upper) filter should be renewed at major overhaul. The primary fuel pump is equipped with a sediment bowl which should be checked daily.

The fuel tank is mounted vertically in front of the radiator, and shut-off valve is accessible from underneath tractor at front.

54. **BLEEDING.** To bleed the system refer to Fig. JD37. Open bleed screw (A) and actuate lever on lift pump (P) until air-free fuel flows. Open bleed screw (B) and continue pumping until air is ejected from second stage filter. Make sure lever (P) is in down position before starting engine. Loosen the pressure line connections at injector assemblies and, with throttle open, turn engine over with starter until fuel is being pumped from injector lines. Tighten the connections and start the engine. If engine will not start, or misses, repeat the above procedure until system is free of trapped air.

TROUBLE SHOOTING
All Models

55. If the engine does not run properly and the fuel system is suspected as the source of trouble, refer to paragraph 56 if engine runs but misses or paragraph 64A or 65A if engine fails to start. Refer also to paragraph 54 for bleeding procedures.

INJECTOR NOZZLES
All Models

NOTE: Early models are equipped with either Bosch or Scintilla injector nozzles of the type shown in Fig. JD38. Some late models may use 9.5 mm Roosa (Fig. JD39B) or Scintilla (Fig. JD39A) nozzles.

56. **TESTING AND LOCATING A FAULTY NOZZLE.** If the engine does not run properly and the quick checks outlined in paragraph 55 point to a faulty injector, locate the faulty unit as follows:

If one engine cylinder is misfiring, it is reasonable to suspect a faulty injector. Generally, a faulty injector can be located by loosening the high pressure line fitting to each injector in turn, while the engine is running at a slow idle. As in checking spark plugs in a spark ignition engine, the faulty unit is the one which, when its line is loosened, least affects the running of the engine.

Remove the suspected injector unit from the engine as outlined in paragraph 57 or 57A. If a suitable nozzle tester is available, check the unit as in paragraph 58. If a tester is not available, reconnect the fuel line to the injector, and with the nozzle tip directed where it will do no harm, crank the engine and observe the spray pattern. Five or six evenly spaced, finely atomized sprays should emerge. If the spray pattern is ragged or unduly wet or if nozzle dribbles, the nozzle valve is not seating properly and the injector assembly should be renewed or overhauled.

CAUTION: Fuel leaves the nozzle tips with sufficient force to penetrate the skin. Keep unprotected parts of body clear of nozzle spray when testing.

57. **REMOVE AND REINSTALL (EARLY MODELS).** To remove an injector, wash the injector, lines and surrounding area with clean diesel fuel to remove any accumulation of dirt or foreign material.

Remove bleed-back line and disconnect the pressure line at the injector; then remove the attaching cap screw and withdraw the injector assembly. When installing, tighten the retaining cap screw to a torque of 36 ft.-lbs.

57A. REMOVE AND REINSTALL (LATE 9.5 mm TYPE). Wash the injector, lines and surrounding area with clean diesel fuel to remove any accumulation of dirt or foreign material. Remove cowl and hood. Disconnect leak-off pipe at fitting adjacent to injection pump. Expand lower clamp on each leak-off boot and move clamp upward next to top clamp; then remove leak-off pipe and all leak-off boots as a unit.

Disconnect high-pressure line, remove nozzle clamp cap screw, clamp and spacer; then withdraw the injector assembly.

NOTE: If injector cannot be easily withdrawn by hand, the special OTC puller, JDE-38 will be required. DO NOT attempt to pry nozzle from its bore.

Before reinstalling injector nozzle, clean nozzle bore in cylinder head, using OTC Tool JDE-39, then blow out foreign material with compressed air. Turn tool clockwise only, when cleaning nozzle bore. Reverse rotation will dull tool.

Renew carbon seal at tip of injector body and seal washer at upper seat, whenever injector has been removed.

NOTE: Nozzle tip may be cleaned of loose or flaky carbon using a brass wire brush. DO NOT use a brush, scraper or other abrasive on Teflon coated surface of nozzle body between the seals. The coating may become discolored by use, but discoloration is not harmful.

Insert the dry injector nozzle in its bore using a twisting motion. Tighten pressure line connection finger tight; then install hold-down clamp, spacer and cap screw. Tighten cap screw to a torque of 20 ft.-lbs. Bleed the injector if necessary, as outlined in paragraph 54, then tighten pressure line connection securely. Complete the assembly by reversing disassembly procedure.

58. NOZZLE TESTER. A complete job of testing and adjusting an injector requires the use of special test equipment. Only clean, approved testing oil should be used in the tester tank. The nozzle should be tested for opening pressure, seat leakage, back leakage and spray pattern. When tested, the nozzle should open with a sharp popping or buzzing sound and cut off quickly at end of injection, with a minimum of seat leakage and a controlled amount of back leakage.

Use the tester to check the injector as outlined in the following paragraphs:

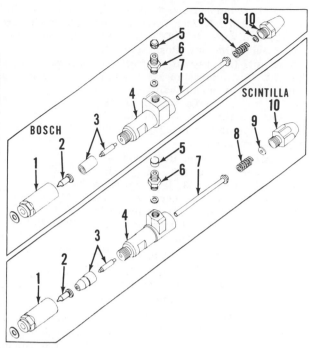

Fig. JD38 — Exploded view of the two diesel injector types used on early diesel engines. Complete injectors are interchangeable but parts are not. Opening pressure is adjusted by means of shims (9).

1. Nut
2. Spray tip
3. Nozzle valve
4. Body
5. Cap
6. Inlet nipple
7. Pressure pin
8. Pressure spring
9. Adjusting shim
10. Spring cap

CAUTION: Fuel leaves the nozzle tips with sufficient force to penetrate the skin. Keep unprotected parts of body clear of nozzle spray when testing.

59. SPRAY PATTERN. In early models each nozzle tip contains five equally spaced nozzle holes. Late models have six holes, equally spaced. Pump the tester at a steady rate of approximately 30 strokes per minute and observe the spray pattern from the nozzle holes. A finely atomized, conical spray should emerge from each of the nozzle holes. If the spray pattern is not satisfactory, disassemble and overhaul the injector as outlined in paragraph 63, 63A or 63B.

60. SEAT LEAKAGE. Pump the tester handle slowly to maintain a gage pressure 100-200 psi below nozzle opening pressure while examining the nozzle tip for fuel accumulation. If the nozzle is in good condition, there should be no noticeable accumulation for a period of at least 10 seconds. If a drop or undue wetness appears on the nozzle tip, renew the injector or overhaul as outlined in paragraph 63, 63A or 63B.

61. BACK LEAKAGE. After performing the seat leakage test as outlined in paragraph 60, release the tester handle and observe the gage needle to determine the condition of the nozzle valve piston and the lapped sealing surfaces of the injector assembly. Gage needle should drop slowly and steadily from the test reading. Excessive speed of drop indicates a worn or scored valve piston or leakage at one or more of the lapped joints.

NOTE: Leakage of injector tester connections or of tester check valve will show up as excessive back leakage. If all injectors tested fail to meet this test, the tester rather than the injector should be suspected.

62. OPENING PRESSURE. The correct opening pressure on early models should be between 2400 and 2600 psi, and should not vary more than 100 psi for all of the injectors in any one engine. To adjust the pressure on early models, remove cap (10—Fig. JD38) and add or remove shims (9) to increase or decrease spring pressure. Note: if a used injector or spring (8) is installed, adjust pressure to 2400 psi. If a new spring is used, adjust the pressure to 2600 psi to allow for pressure drop as the spring takes a set. On late models, refer to paragraph 63A or 63C, for procedure to adjust the pressure.

63. OVERHAUL EARLY MODELS. Before disassembling the removed injector assembly, wash the unit in clean diesel fuel and blow off with clean, dry compressed air.

Carbon accumulation from exterior of spray tip can be removed with a brass wire brush. Set up a pan with a clean cloth in the bottom and fill pan with clean diesel fuel. As parts are removed, place them in pan but do not allow them to touch one another. Clamp nozzle in a soft jawed vise and remove tip retaining nut (1—Fig. JD38). Nozzle valve assembly (3) and spray tip (2) will be removed with the nut. Push spray tip out of nut. If spray tip is frozen in the nut, use a piece of steel tubing which will fit about halfway down on the tip as shown in Fig. JD38A and tap the tip free. Do not tap directly on point of tip or spray holes may be distorted.

Reverse the injector in the vise and remove spring cap (10—Fig. JD38), shims (9), spring (8) and pressure pin (7). Be sure to keep the shim pack (9) together for convenience in opening pressure adjustment when injector is reassembled.

Clean all parts thoroughly in the clean fuel, using a brass wire brush to remove any gum or varnish deposits. Parts which cannot be thoroughly cleaned may be soaked in a non-corrosive carbon solvent.

NOZZLE TIP. Clean the nozzle tip holes, first with a 0.010 broach mounted in a pin vise; then with a 0.011 broach. Make sure that all holes are fully open and clean. Test the holes with a 0.013 cleaning wire. If the wire can be inserted in the holes, renew the tip. Clean the center passage of the nozzle tip using a 0.040 nozzle hole reamer.

NOTE: Series 3010, 4010 and early Model 5010 nozzles contained five evenly spaced tip holes. All later models have six holes. Orifice size is 0.011 for all nozzles except 5-hole model used in early 5010, in which orifice size was 0.012.

NOZZLE VALVE AND SEAT. Inspect needle valve and seat with a magnifying glass. Renew valve and seat if seating surfaces are scored or worn, or if the lapped piston surface is scratched or scored. The needle valve and seat is a matched unit and must be renewed as an assembly.

Test the fit of nozzle valve in its seat as follows: Hold seat in a vertical position and start the valve into seat. Valve should slide slowly into seat under its own weight. Note: Dirt particles, too small to be seen with the naked eye, will restrict the valve action. If valve sticks and is known to be clean, free up the valve using a mixture of tallow and clean diesel fuel as a solvent, and rotating the valve in its seat.

Machined surfaces of nozzle valve and tip may be lapped, if necessary by using a lapping block as shown in Fig. JD39, to provide a metal to metal seal.

Clean all parts thoroughly in clean diesel fuel and reassemble while wet. Locating marks are etched on outside of nozzle tip and on nozzle holder under inlet fitting. Make sure these marks are aligned, then tighten the nozzle retaining nut to a torque of 60-65 ft.-lbs.

63A. OVERHAUL 9.5 mm SCINTILLA. First wash the unit in clean diesel fuel and blow off with clean, dry compressed air. Remove carbon stop seal and sealing washer. Clean carbon from spray tip using a brass wire brush, then polish tip using an approved lapping compound on a piece of hard felt. Clean Teflon coating on injector body with a soft cloth after soaking in solvent. DO NOT use a wire brush or abrasive on Teflon coating.

Clamp nozzle in a soft jawed vise and remove cap (8—Fig. JD39A), shim pack (7), spring (6) and valve stop (5). If nozzle valve (4) is stuck and cannot be jarred from nozzle body, reinstall valve stop (5) and cap (8), turning cap 3 or 4 turns only, into valve body. Connect injector assembly to nozzle tester and use hydraulic pressure to force valve from body.

NOTE: Bendix Corporation supplies a fixture (Bendix Part No. 11-5736) as part of nozzle service tool kit, which can be used in conjuction with a nozzle tester to expel a stuck valve. If a pressure of 5000 psi does not free the valve, renew the injector unit.

Keep the parts of each nozzle separate, and immerse in clean diesel fuel in a compartmented pan as they are removed. Nozzle body and valve are individually fitted and matched, and cannot be mixed or interchanged. Adjusting washers (7) and spring (6) are individually selected, and control opening pressure. These parts, together with the valve stop (5) and cap (8) may be interchanged or renewed separately, but adjustment is simplified if parts are kept together.

Clean all parts thoroughly in clean diesel fuel, using a brass wire brush and lint-free wiping towels. Hard carbon or varnish can be loosened with Bendix "Speediclene" or "Metalclene", "Lakeseal BAS-TU" or other non-corrosive carbon solvent.

Clean carbon from the spray-tip orifices using a 0.010 cleaning needle held in a pin vise; then change to a 0.011 needle to make sure all carbon is removed. Check nozzle wear

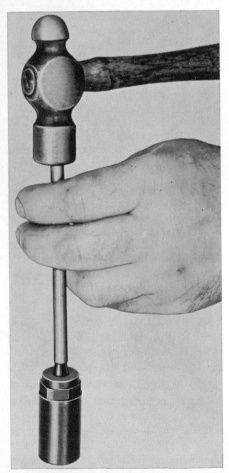

Fig. JD38A—Use a piece of steel tube as a driver to dislodge a frozen tip. Do not drive on end of tip.

Fig. JD39 — When lapping flat surfaces use the figure eight motion as shown.

Fig. JD39A — Exploded view of Pencil-Type Scintilla Nozzle, showing component parts. Shims (7) adjust opening pressure.

1. Carbon seal
2. Seal washer
3. Nozzle body
4. Nozzle valve
5. Valve stop
6. Pressure spring
7. Adjusting shims
8. Spring cap

using a 0.012 needle. Renew the complete injector nozzle unit if the 0.012 needle can be inserted in any orifice; or if orifice is mis-shaped or has rounded edges when viewed through a magnifying glass. Clean the sac-hole in spray tip with Bendix 11-5749 sac-hole drill. Flush or blow all loose carbon from nozzle body after carbon is loosened.

Carefully check large end of nozzle valve for burrs or belling from contact with valve stop. Remove any defects with a fine grit hand stone, until nozzle valve can be inverted and started in piston area of valve body. With a wax pencil, scribe an easily visible mark 1⅜-inches from large end of valve, then polish piston area of valve and body using Bendix Green Rouge, Part No. 11-5307.

CAUTION Apply a small quantity of Green Rouge to valve piston, invert the valve and insert, large-end first, into valve body. Polish by hand only, with an oscillating motion, working up to (but not beyond) the previously scribed pencil mark. If valve is inserted in body beyond the mark, it may pass honed surface of body, causing it to stick.

The Bendix Nozzle Tool Kit contains tools for reconditioning and lapping the nozzle valve and seat. Use a high-speed (1750-3500 rpm), ¼-inch electric drill, light pressure, and remove a minimum amount of metal in the reconditioning operation.

Examine valve contact surface of stop (5). If wear is evident, invert the stop when reassembling. If both sides are worn, renew the stop. NOTE: The valve stop limits the lift of nozzle valve and lift increases as parts become worn.

When reassembling, tighten spring cap (8) to a torque of 20 - 25 ft.-lbs. Check the injector as outlined in paragraphs 58 through 61. If opening pressure is not within the limits of 2500 - 2550 psi, remove cap (8) and add or remove shims (7) as required. Install injector nozzles as outlined in paragraph 57A.

63B. OVERHAUL 9.5 mm ROOSA-MASTER. First wash the unit in clean diesel fuel and blow off with clean, dry compressed air. Remove carbon stop seal and sealing washer. Clean carbon from spray tip using a brass wire brush. Also, clean carbon or other deposits from carbon seal groove in injector body. DO NOT use wire brush or other abrasive on Teflon coating on sides of nozzle body between the seals. Teflon coating can be cleaned with a soft cloth and solvent, coating may discolor from use, but discoloration is not harmful.

Clamp the nozzle in a soft jawed vise, loosen lock nut (13 — Fig. JD39B) and remove pressure adjusting screw (14), ball washer (10), upper spring washer (7), spring (8) and lower spring seat (7).

If nozzle valve (5) will not slide from body when body is inverted, use a suitable valve extractor; or re-install on nozzle tester with spring and lift adjusting screw removed, and use hydraulic pressure to remove the valve.

Nozzle valve and body are a matched set and should never be intermixed. Keep parts for one injector separate and immerse in clean diesel fuel in a compartmented pan, as unit is disassembled.

Clean all parts thoroughly in clean diesel fuel using a brass wire brush and lint-free wiping towels. Hard carbon or varnish can be loosened with a suitable, non-corrosive solvent.

Clean the spray tip orifices first with a 0.008 cleaning needle held in a pin vise; then with a 0.011 needle.

Clean the valve seat using a Valve Tip Scraper and light pressure while rotating scraper. Use a Sac Hole Drill to remove carbon from sac hole.

Piston area of valve and guide can be lightly polished by hand, if necessary, using Roosa Master No. 16489 lapping compound. Use the valve retractor to turn valve. NOTE: Move valve in and out slightly while rotating, but do not apply down pressure while valve tip and seat are in contact.

Valve and seat are ground to a slight interference angle. Seating areas may be cleaned up, if necessary, using a small amount of 16489 lapping compound, very light pressure and no more than 3 - 5 turns of valve on seat. Thoroughly flush all compound from valve body after polishing.

When assembling the nozzle, back lift adjusting screw (11) several turns out of pressure adjusting screw (14), and reverse disassembly procedure, using Fig. JD39B as a guide.

Fig. JD39B — Cross sectional view of Pencil-Type Roosa-Master Nozzle. Nozzle tip (1) and valve guide (6) are parts of finished nozzle body and not serviced separately.

1. Nozzle tip
2. Carbon seal
3. Nozzle body
4. Seal washer
5. Nozzle valve
6. Valve guide
7. Spring seat
8. Pressure spring
9. Boot clamp
10. Ball washer
11. Lift adjusting screw
12. Boot
13. Lock nut
14. Pressure adjusting screw

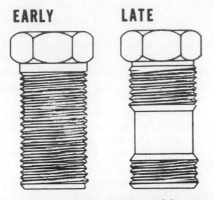

Fig. JD39C—Views of early and late type pressure adjusting screws used in Roosa-Master nozzle.

63C. Connect nozzle to tester and turn pressure adjusting screw (14) into nozzle body until opening pressure is 2550 psi with a used spring; or 2750 psi with a new spring.

After spring pressure has been adjusted and before tightening locknut (13), valve lift must be adjusted as follows: Hold pressure adjusting screw from turning and, using a small screwdriver, thread lift adjusting screw (11) into nozzle until it bottoms on lower spring seat. To be sure screw is bottomed, actuate tester handle until a pressure 250 psi above opening pressure is obtained. Nozzle valve should not open.

With lift adjusting screw bottomed, turn screw counter-clockwise as follows: On Series 3000 and 4000, back screw out ½-turn to establish the recommended 0.009 lift. On Series 5000, back screw out ¾-turn to establish the recommended 0.0135 lift. On all models refer to Fig. JD39C. While holding pressure adjusting screw from turning, back off the locknut if necessary, and check to see whether early or late type pressure adjusting screw is used. On early screw tighten locknut to a torque of 110-115 inch-pounds while holding adjusting screw from turning. On late models, tighten locknut to 70-75 inch-pounds.

Recheck the opening pressure; then recheck spray pattern, seat leakage and back leakage as outlined in paragraphs 59 through 61. Install the unit as outlined in paragraph 57A.

NOTE: When adjusting opening pressure and injector has not been disassembled, back out the lift adjusting screw at least one full turn before moving pressure adjusting screw; to prevent accidental bottoming while attempting to set the pressure.

INJECTION PUMP
Series 3000-4000

A Roosa-Master Model DBG injection pump is used on all models. Pump is vertically mounted on right side of engine and driven by engine camshaft. Proper pump timing depends on correct installation of oil pump as outlined in paragraph 41.

Injection pump service requires the use of specialized skill, training and equipment. This section therefore will cover only the information required for removal, installation and field adjustment of the pump.

64. REMOVE AND REINSTALL. To remove the injection pump, first shut off the fuel supply and thoroughly clean dirt from pump, lines and connections. Cap all fittings as they are disconnected to prevent dirt entry. Disconnect the fuel lines and controls, remove nuts and washers from mounting studs and withdraw pump from engine.

Before installing pump, crank the engine until number one piston is coming up on the compression stroke, and continue cranking until, on series 3010, the 10° before TDC timing mark is aligned with scribe mark in flywheel timing window (W—JD40). On series 4010, correct pump timing is 14° before TDC. On series 3020 and 4020 correct pump timing is at TDC. Injection pump drive slot in oil pump shaft should be approximately 15° from parallel with crankshaft as shown in Fig. JD25. The "V" mark on heavy lug should be toward engine block. Remove timing hole cover from injection pump body and turn pump rotor until scribe lines on governor cage and cam ring are aligned as

shown in Inset—Fig. JD40. Be sure thrust spring is in place and insert the pump so that drive tang on shaft enters slot in oil pump shaft and tighten the retaining stud nuts slightly. Turn the crankshaft two full turns in the normal direction of rotation and recheck timing. The normal backlash in timing gears is enough to permit a slight error in pump timing. After timing has been rechecked and adjusted, tighten the stud nuts securely, reinstall timing hole cover and lines, then bleed the system as outlined in paragraph 54.

64A. ADJUSTMENT AND TIMING. To check injection timing without removing pump, proceed as outlined in paragraph 64, but without removing pump. Loosen the two stud nuts (A—Fig. JD40) and shift pump on studs if retiming is required. If injection timing is off several degrees, check installation of oil pump as outlined in paragraph 41, and the condition of camshaft and oil pump drive gears.

64B. SPEED AND LINKAGE ADJUSTMENT. All speed adjustments are made on the throttle linkage located underneath the dash. To make the adjustments, remove the cowling and instrument panel rear access door and proceed as follows:

With engine at operating temperature, move hand throttle (1—Fig. JD41) down to it's stop without pulling out on throttle knob (3). Adjust the length of throttle rod (7—Fig. JD40A) if necessary, to provide a no-load engine speed of 2130-2170 rpm (or a loaded engine speed of 1900 rpm). Move hand throttle up to slow idle position and adjust stop screw

Fig. JD40—Right side of series 3000 and 4000 engine showing method of timing injection pump. See text for details.

W. Flywheel timing window.

A. Attaching stud nuts.

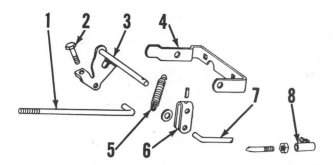

1. Link
2. Stop screw
3. Shaft
4. Bracket
5. Spring
6. Arm
7. Throttle rod
8. Socket

(9—Fig. JD41) to provide the recommended slow idle speed of 600 rpm. Pull out on throttle knob (3) and move hand throttle all the way down into over-travel and adjust stop screw (7) to provide a high idle, no-load speed of 2380-2420 rpm (or a loaded engine speed of 2200 rpm). Depress the foot throttle pedal and adjust the stop screw (2—Fig. JD40A) located on speed control cross shaft to provide a no-load transport speed of 2650 rpm.

Series 5000

65. **REMOVE AND REINSTALL.** To remove the injection pump, first shut off the fuel and thoroughly clean dirt from pump, lines and connections. Cap all fittings as they are disconnected, to prevent dirt entry. Loosen injector line clamp and pump-to-injector lines at both ends to prevent bending or springing the lines.

Remove timing cover (T—Fig. JD-40B) and turn engine crankshaft until No. 1 piston is coming up on compression stroke and injection pump timing marks are aligned as shown in Fig. JD40.

Remove injection pump flange nuts (A—Fig. JD40B) and plate (P) on front of timing gear cover. Remove the nut (N—Fig. JD40C) and lock plate securing injection pump drive

gear (G) to shaft and, using a short brass drift and hammer, jar the injection pump drive shaft rearward from its keyed position in injection pump drive gear. Lift the pump from engine.

When installing the injection pump, make sure timing marks on injection pump drive gear (G—Fig. JD40C) and the camshaft gear are aligned as shown at (T). Be sure Woodruff key is in injection pump drive shaft slot, and install pump in drive gear. Install locking plate and nut (N). Install the two pump mounting flange stud nuts loosely, and with No. 1 piston on compression stroke and "TDC" flywheel timing mark aligned with pointer, rotate pump body slightly, if necessary, until timing scribe marks on injection pump cam ring and governor cage are aligned as shown in Fig. JD40. Securely tighten the mounting flange stud nuts when marks are aligned. Reinstall timing hole cover and lines, and bleed the system as outlined in paragraph 54.

65A. **ADJUSTMENT AND TIMING.** The injection pump is correctly timed when "TDC" flywheel timing

mark is aligned with timing pointer; No. 1 piston is on compression stroke (both valves closed); and scribe marks on cam and governor cage are aligned when viewed through injection pump timing port. Refer to Fig. JD40D. NOTE: Shut off the fuel at fuel tank before removing injection pump timing window.

If the scribe mark on governor cage (front mark) is visible in timing port but does not align with rear mark, loosen the two mounting flange stud nuts (A) and shift pump body slightly until marks are aligned. If marks cannot be aligned, remove the injection pump as outlined in paragraph 65, and reposition the injection pump drive gear.

66. **SPEED AND LINKAGE ADJUSTMENT.** To check and adjust the speed control linkage, remove cowl and right rear side shield. Refer to Fig. JD40E.

With engine at operating temperature and running, pull down (open) throttle lever (1) without pulling out on knob (3). When throttle lever contacts stop, engine no-load speed should be 2150 rpm when checked with a master tachometer; if it is not, shorten or lengthen forward throttle link (26) until proper speed is obtained.

Pull out on knob (3) and pull down (open) throttle until it contacts override stop. Engine no-load speed should be 2400 rpm; if it is not, turn high idle stop screw (7) in or out until correct speed is obtained.

Close (push up) throttle lever without pulling out on knob (3) until throttle contacts stop; and adjust slow idle stop screw (S—Fig. JD40F) if necessary, to obtain the suggested slow-idle speed of 580-620 rpm.

Fully open foot throttle (20—Fig. JD40E) and adjust stop screw (25)

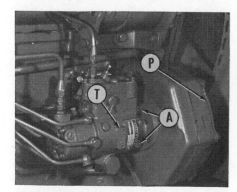

Fig. JD40B — Right side of Series 5000
A. Attaching flange nuts
P. Front plate
T. Pump timing cover

Fig. JD40C—Timing gears on Series 5000, showing timing marks (T), pump drive gear (G) and injection pump shaft nut (N). Injection pump may be removed and installed without removing timing gear cover, by removing cover plate (P—Fig. JD40B).

Fig. JD40D — To check injection timing on Series 5000, remove timing covers (C and P). Loosen flange nuts (A) for minor adjustment.

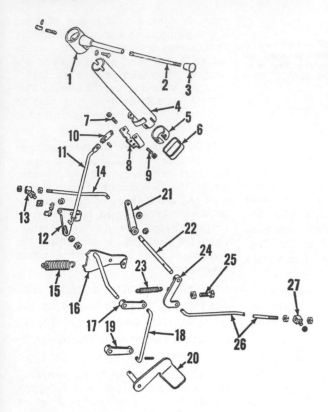

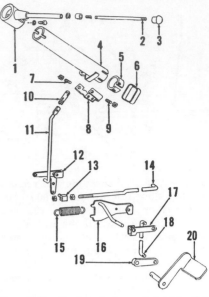

21. Arm
22. Shaft
24. Arm
25. Stop screw
26. Control rod
27. Swivel

Fig. JD41—Exploded view of throttle linkage of the type used on all models. Minor differences in construction may be noted on examination.

1. Throttle hub	11. Rod
2. Lever shaft	12. Bell crank
3. Knob	13. Swivel
4. Tube	14. Link
5. Facing	15. Spring
6. Clamp	16. Arm
7. Stop screw	17. Arm
8. Bracket	18. Link
9. Stop screw	19. Arm
10. Link	20. Foot pedal

to obtain the suggested no-load engine speed of 2650 rpm.

Pull out on knob (3) and move throttle lever counter-clockwise past slow idle position. Engine should stop; if it does not, back out the slow idle stop screw (9) until fuel delivery is just cut off. CAUTION: Fuel shut-off (upper) screw on injection pump governor arm should not quite touch pump body in cut-off position; if it does, pump governor shaft seal may be damaged. Adjust as follows: Back out the slow idle stop screw (9) and carefully move lever counter-clockwise until shut-off screw on pump governor arm just contacts pump body. Turn stop screw (9) clockwise to contact throttle tube stop, plus an additional ½-turn.

Fig. JD40F — Slow idle stop screw (S) on Series 5000 is located on injection pump governor housing cover as shown.

NON-DIESEL GOVERNOR

SPEED AND LINKAGE ADJUSTMENTS

All Non-Diesel Models

68. All models are equipped with variable speed linkage. Adjustment is made with positive stops at five different engine speeds as follows:

Slow Idle	450 rpm
Normal Idle	600 rpm
@ Rated PTO Speed (3020)	
No-load	2360 rpm
Loaded	2100 rpm
@ Rated PTO Speed (3010—All 4000)	
No-load	2170 rpm
Loaded	1900 rpm
@ Rated Drawbar (3020)	
No-load	2690 rpm
Loaded	2500 rpm
@ Rated Drawbar (3010—All 4000)	
No-load	2440 rpm
Loaded	2200 rpm
@ Transport (Foot Throttle)	
No-load	2690 rpm
Loaded	2500 rpm

To adjust the linkage, proceed as follows: Refer to Fig. JD42. With engine not running, disconnect governor

to carburetor control rod at governor. Adjust rod so that it is ½-hole short when governor lever and carburetor throttle are both in wide-open position. Reconnect the rod.

Start engine and warm to operating temperature. Pull out on control knob (3—Fig. JD41) and push control lever

Fig. JD42—Left side of engine block showing governor installation.

1. Idle speed stop screw
2. High idle stop screw

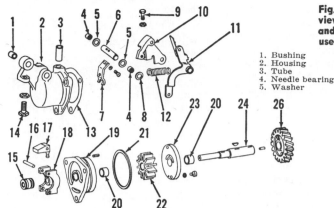

Fig. JD43 — Exploded view of ventilator pump and governor assembly used on non-diesel models.

1. Bushing	6. Shaft
2. Housing	7. Fork
3. Tube	8. Oil seal
4. Needle bearing	9. Stop screw
5. Washer	10. Arm
	11. Arm
	12. Spring
	13. Gasket
	14. Stop screw
	15. Thrust bearing
	16. Weight pin
	17. Flyweight
	18. Carrier
	19. Pump body
	20. Bushing
	21. Packing
	22. Rotor
	23. Cover
	24. Shaft
	26. Gear

(1) up as far as it will go. Engine speed should be 450 rpm. To adjust, back off stop screw (9) if necessary, turn idle speed adjusting screw on carburetor in or out until proper speed is obtained, then adjust stop screw (9) to just contact the stop on speed control tube (4). Move hand throttle down until knob (3) moves inward, then push control lever (1) up to normal slow idle position. Do not pull out on knob (3). Engine speed should be 600 rpm. Adjust by means of governor arm stop screw (1—Fig. JD42).

NOTE: High governed speeds should be adjusted under load. The no-load speeds given are approximate only. Do not hesitate to vary the no-load speeds to obtain correct loaded engine speed. If a dynamometer is available, use dynamometer to apply approximately 80% of rated load and adjust governed speeds to the given loaded speed. If a dynamometer is not available, adjust to the suggested no-load speeds.

Move control lever down to the first stop (rated pto speed). Do not pull out on knob (3—Fig. JD41). Engine speed should be as given in table for rated pto speed. If it is not, adjust by moving swivel (13) on governor control rod (14) by means of the adjusting nuts.

Pull out on control knob (3) and move control lever down to final hand throttle stop (rated drawbar speed). Engine speed should be as indicated in preceding tables. Adjust by means of high speed stop screw (7).

Depress foot control (20) and check transport speed. Adjust by means of control arm stop screw (2—Fig. JD42) located on governor housing.

VENTILATOR PUMP & GOVERNOR
All Non-Diesel Models

69. **REMOVE & REINSTALL.** The governor unit is combined with the crankcase ventilator pump on non-diesel models. To remove the assembly, disconnect the air inlet line and governor linkage. Remove the three attaching cap screws and withdraw pump and governor assembly.

When reinstalling, check the governed speeds as outlined in paragraph 68.

70. **OVERHAUL.** To disassemble the removed unit, tap gently on pump mounting flange to free the gasket, then withdraw the governor housing. Remove the two screws retaining governor fork (7—Fig. JD43) to shaft (6) and withdraw shaft and levers. Needle bearings (4) and oil seal (8) can be renewed at this time. Check governor weights (17) and carrier (18) for wear. To inspect ventilator pump bushings (20) and shaft (24), pull drive gear (26) with a puller or press, remove Woodruff key, the two screws retaining the cover (23) and remove cover. See paragraph 45 for a description of ventilator pump operation.

COOLING SYSTEM

RADIATOR
All Models

71. To remove radiator, drain cooling system and remove muffler, side panels, grilles, screens, cowl and hood. Remove the tie rods attaching fuel tank to radiator, disconnect hydraulic fluid line clamps and remove the screws retaining the fan shroud. Remove air cleaner hose, disconnect radiator hoses. Remove radiator retaining cap screws and slide radiator out left side of tractor.

NOTE: On LP-Gas models it will be necessary to disconnect fuel shut-off control rods and slide them rearward to clear radiator. On Series 5000 it is necessary to remove air cleaner and hydraulic oil cooler.

FAN AND WATER PUMP
All Models

72. **REMOVE & REINSTALL.** To remove fan and/or water pump, drain cooling system and remove muffler, cowl, hood, side panels and grilles. Remove the screws attaching fan shroud to radiator and cap screws attaching fan to pump hub; then, slide fan and shroud together out left side of tractor. Loosen fan belt, disconnect by-pass line and lower radia-

tor hose. Where interference exists, disconnect oil cooler supply line. Unbolt and remove pump.

To install, reverse removal procedure.

73. **OVERHAUL.** To disassemble the pump, remove the spring type bearing retainer from slot in pump housing. On diesel models remove pump cover. Press shaft and bearing assembly (with fan hub attached) out of impeller and housing with an arbor press. Pump seal can be renewed at this time. Shaft and bearings are

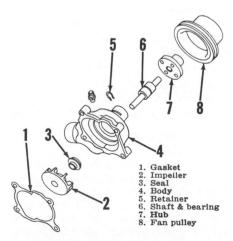

1. Gasket	
2. Impeller	
3. Seal	
4. Body	
5. Retainer	
6. Shaft & bearing	
7. Hub	
8. Fan pulley	

Fig. JD44—Exploded view of water pump.

available as an assembly only, other pump parts are available individually.

When reassembling, press shaft and bearing assembly in housing until groove in outer race aligns with slot in housing, then install retainer. Support hub end of shaft and press impeller on pump shaft until a clearance of 0.015-0.035 exists between impeller blades and housing. Tighten fan retaining bolts to a torque of 36 ft.-lbs.

THERMOSTAT AND WATER MANIFOLD

Series 3000

74. The 160° F. thermostat is con-tained in a thermostat housing at center of cylinder head. To renew the thermostat, drain cooling system and disconnect the bypass line at upper end. Remove the cap screws retaining elbow to thermostat housing, raise the elbow and extract the thermostat.

Series 4000—5000

75. Two 160° F. thermostats are contained in housings on top of cylinder head. To renew either or both thermostats, drain cooling system and disconnect bypass line at upper end. Remove the cap screws retaining water manifold to the two thermostat housings, lift up the manifold and extract thermostats.

IGNITION AND ELECTRICAL SYSTEM

DISTRIBUTOR

Non-Diesel Models

76. Delco-Remy ignition distributors are used on all non-diesel models. Specifications are as follows:

Model 1112577 & 1112623 (Series 3000)
Breaker contact gap 0.022
Breaker arm spring
pressure ....17-21 oz.
Cam angle (degrees):
Model 1112577 25-34
Model 1112623 31-34
Model 1112581 & 1112624 (Series 4000)
Breaker contact gap 0.022
Breaker arm spring
pressure ....17-21 oz.
Cam angle (degrees):
Model 1112581 31-37
Model 1112624 31-34

The advance data given applies to all models and is given in distributor degrees and distributor rpm.
Start advance ...... 0-2 @ 225 rpm
Intermediate
advance .... 5-7 @ 350 rpm
Intermediate
advance ....10-12 @ 825 rpm
Maximum advance..14-16 @ 1200 rpm

77. INSTALLATION AND TIMING. Proper timing of ignition distributor depends on correct installation of engine oil pump as outlined in paragraph 41. With oil pump properly installed, crank engine until No. 1 piston is coming up on compression stroke and continue cranking until, on gasoline models, the "TDC" mark

on flywheel is 23/32-inch past the mark on timing window. On LP-Gas models, align "TDC" mark on flywheel with scribed mark on timing window. The drive slot in oil pump drive gear should be approximately 15 degrees from parallel with crankshaft and "V" mark on drive lug should point toward block as shown in Fig. JD45. Install distributor so that rotor arm is in No. 1 firing position. Install distributor clamps loosely. Rotate distributor housing counter-clockwise until points are closed, remove coil wire from center of distributor cap and turn ignition switch to "ON" position. Hold coil wire 1/8-inch from engine block and slowly rotate distributor housing clockwise until a spark jumps the gap between coil wire and engine block. Tighten the distributor clamps at this point. Start engine and open the throttle until an engine speed of 1900 rpm is obtained, then check the timing with a light. Make minor adjustments as necessary to obtain the recommended timing of 20 degrees BTC for gasoline models, or 25 degrees BTC for LP-Gas models. Firing order is; Series 3000...1-3-4-2, Series 4000...1-5-3-6-2-4.

GENERATOR AND REGULATOR

All Models

78. Non-diesel models are equipped with a 12-volt electrical system. Diesel models use a 24-volt system. Delco-Remy units are used on all models except 5010, which uses Prestolite. Specifications are as follows:

Delco-Remy

Generator Model 1100380, 12-Volt
Brush spring tension ......28 oz.
Field draw
Volts 12.0
Amperes 1.58-1.67
Output (cold)
Volts 14.0
Amperes 20.0
RPM 2300

Generator Model 1103026, 24-Volt
Brush spring tension ......28 oz.
Field draw
Volts 24.0
Amperes 0.75-0.85
Output (cold)
Volts 28.5
Amperes 10.0
RPM 2500

Regulator Model 1119135, 12-Volt
Cutout Relay
Air gap 0.020
Point gap 0.020
Closing range 11.8-13.5
Voltage Regulator
Air gap 0.075
Voltage range 13.8-14.5
Current Reguator
Air gap 0.075
Setting range (amperes).18.5-21.5

Regulator Model 1119219, 24-Volt
Cutout Relay
Air gap 0.017
Point gap 0.032
Closing range 24.0-27.0
Voltage Regulator
Air gap 0.075
Voltage range 27.5-29.5
Current Regulator
Air gap 0.075
Setting range (amperes).8.5-11.5

Prestolite

Generator Model GKF-7101B, 24-Volt
Brush spring tension ....18-36 oz.
Field draw
Volts 30.0
Amperes 1.02-1.12

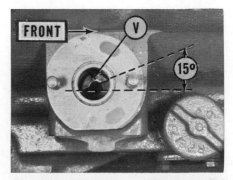

Fig. JD45—When No. 1 piston is at TDC on compression stroke for distributor installation, drive slot in oil pump drive gear should be in the position shown. If it is not, oil pump must be retimed.

Output (cold)
    Volts 30
    Amperes 10
    RPM 1650
Regulator Model VBO-6401AJ-3
Cutout Relay
    Air gap 0.025-0.027
    Point gap 0.015
    Closing range 25.0-26.5
    Opening amperes 1.5-3.0
Voltage Regulator
    Air gap 0.057-0.060
    Voltage range (at
        80° F.) 27.8-28.8
Current regulator
    Air gap 0.057-0.060
    Setting range (amperes) .9.0-11.0

ALTERNATOR AND REGULATOR
Series 5000 So Equipped

78A. A Motorola Alternator and transistorized regulator are used. Because the charging circuit is in the 24 volt system, all circuit components must be insulated from the tractor (and alternator) frame, which is the return path from either of the 12 volt circuits. Refer to paragraph 79B for circuit description.

Alternators are basically the same on Model 5010 so equipped and Model 5020, but important differences exist. Refer to Fig. JD45A for a view of rear end frame on Model 5010. Brush

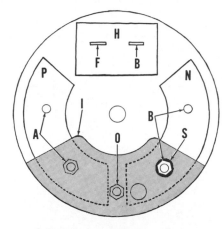

Fig. JD45B—Schematic view of rear end frame of alternator used on Model 5020. Isolation diode (I) is mounted on one accessory terminal (A) and one battery return terminal (B), but is insulated from (B) terminal by sleeve (S) and washer. The regulator mounts on alternator frame at brush holder (H). Refer to Fig. JD45A for parts identification and for view of earlier alternator.

holder (H) is located on left side of frame and isolation diode (I) is attached to the two accessory terminals (A) on the positive diode heat sink (P). On Model 5020, rear end frame is turned clockwise 90° (Fig. JD45B) and brush holder (H) is at top of frame. Isolation diode (I) is mounted

on one accessory terminal (A) and one battery return terminal (B), therefore the return terminal must be insulated from the isolation diode as well as alternator frame.

The primary purpose of the isolation diode is to permit use of charging indicator lamp. The isolation diode and brush holder can be renewed without removal or disassembly of alternator. All other service requires removal and disassembly.

The regulator is fully transistorized and consists of only a voltage regulator. Current regulation and reverse current protection are provided in the design of the alternator unit.

Failure of the isolation diode is usually indicated by the charging indicator lamp which glows with key switch off and engine stopped if shorted; or with engine running if diode is open.

Failure of a rectifying diode is usually indicated by a humming noise when alternator is running and diode is shorted; or by a steady flicker of charge indicator light at slow idle speed when diode is open. Either fault will reduce generator output.

To check the charging system, refer to Fig. JD45A or JD45B and proceed as follows:

(1). With key switch and all accessories off and engine not running,

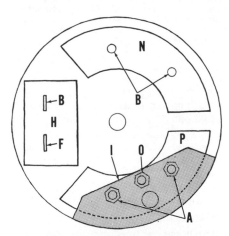

Fig. JD45A—Schematic view of rear end frame of Motorola Alternator used on Model 5010 so equipped. Brush holder (H) is positioned on left (engine) side of frame at approximately 9 O'Clock as shown, and isolation diode (I) is attached to the two accessory (Positive) terminals (A). Refer to Fig. JD45B for view of alternator used on Model 5020.

A. Accessory   I. Isolation diode
    terminals   N. Negative heat
B. Battery return   sink
    terminals   O. Output terminal
F. Field terminal   P. Positive heat sink
H. Brush holder

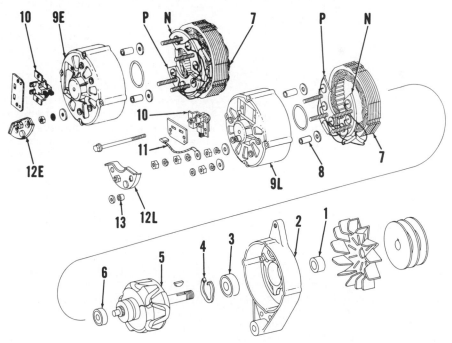

Fig. JD45C—Exploded view of Motorola Alternator of the type used. Parts which differ are marked (E) (early) for Model 5010 or (L) (late) for Model 5020. Most of the other parts are interchangeable.

1. Spacer   6. Bearing   11. Brush return wire
2. Front end frame   7. Stator   12. Isolation diode
3. Bearing   8. Insulated sleeve N. Negative heat
4. Snap ring   9. Rear end frame   sink
5. Rotor   10. Brush holder   P. Positive heat sink

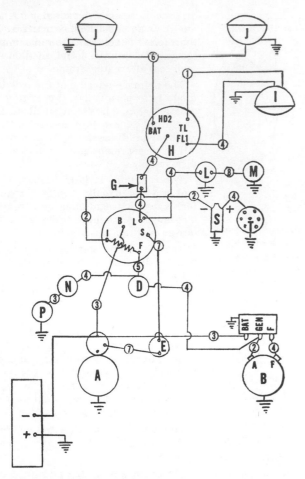

Fig. JD45D—Wiring diagram typical of that used on non-diesel models. Letters indicate system components; numbers indicate color code. Refer to Fig. JD45F for color code and parts identification.

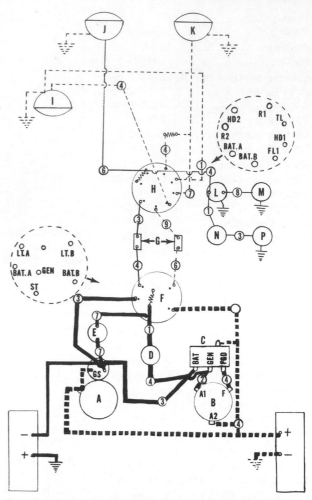

Fig. JD45E—Series 3000-4000 diesel wiring diagram. Heavy markings indicate circuits which carry the 24-volt starting and generating current as well as the grounded 12-volt accessory circuits. Letters indicate system components; numbers indicate color code. Refer to Fig. JD45F for color code and parts identification.

connect a low reading voltmeter to terminals A-B. Reading should be 0.1 volt or less. A higher reading would indicate a short in isolation diode (I), key switch or wiring. Find and correct the trouble then proceed to test 2.

(2). With a suitable voltmeter connected to terminals A-B, start engine and gradually increase engine speed to approximately 1300 rpm. Reading should be 29 volts. If a lower reading is obtained, proceed as outlined in test 4; if reading is 29 volts, proceed to test 3.

(3). If a 29 volt reading was obtained in test 2, move voltmeter lead from Terminal A to Terminal O. Reading should drop to 28 volts, reflecting the 1-volt resistance built into isolation diode. If battery voltage (24 volts) is obtained, isolation diode is open and must be renewed.

(4). If a reading lower than the recommended 29 volts was obtained when checked as in test 2, stop the engine and disconnct (or remove) the

voltage regulator. Connect a jumper wire from terminal (A) on positive heat sink to terminal (F) on brush holder (H). Connect a suitable voltmeter to terminals (A-B). Start engine and slowly increase engine speed while watching voltmeter. If a reading of 29 volts can now be obtained at 1300 engine rpm or less, renew the regulator. If a reading of 29 volts cannot be obtained, renew or overhaul the alternator.

CAUTION: Do not allow voltage to rise to above 30 volts when making this test.

Refer to Fig. JD45C for an exploded view of Motorola Alternator of the type used.

STARTING MOTOR
All Models

79. Delco-Remy starting motors are used on all models except 5000, which uses Prestolite. Specifications are as follows.

Delco Remy

Model 1107522, 12-Volt

Brush spring tension .... 35 oz.
No-Load Test
    Volts 11.8
    Amperes 75-100
    RPM 6450-8750
Lock Test
    Volts 5.0
    Amperes 720-870

Model 1107785, 12-Volt

Brush spring tension ...... 35 oz.
No-Load Test (includes solenoid)
    Volts 10.6
    Amperes 65-100
    RPM 3600-5100
Lock Test (includes solenoid)
    Volts 3.5
    Amperes 300-360

Model 1108665, 12-Volt

Brush spring tension ...... 24 oz.
No-Load Test
    Volts 11.8
    Amperes 40-70
    RPM 6800-9200

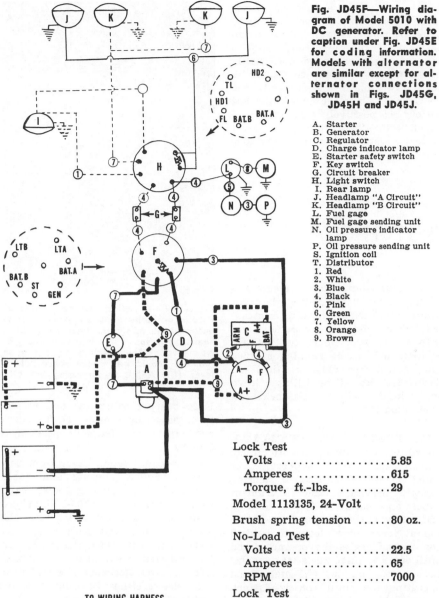

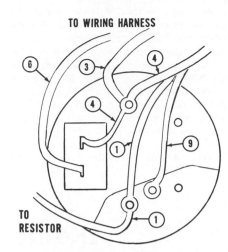

Fig. JD45F—Wiring diagram of Model 5010 with DC generator. Refer to caption under Fig. JD45E for coding information. Models with alternator are similar except for alternator connections shown in Figs. JD45G, JD45H and JD45J.

A. Starter
B. Generator
C. Regulator
D. Charge indicator lamp
E. Starter safety switch
F. Key switch
G. Circuit breaker
H. Light switch
I. Rear lamp
J. Headlamp "A Circuit"
K. Headlamp "B Circuit"
L. Fuel gage
M. Fuel gage sending unit
N. Oil pressure indicator lamp
P. Oil pressure sending unit
S. Ignition coil
T. Distributor
1. Red
2. White
3. Blue
4. Black
5. Pink
6. Green
7. Yellow
8. Orange
9. Brown

Fig. JD45H—Alternator wiring connections for late Model 5010 tractors with regulator wiring built into main wiring harness. Refer to Fig. JD45F for color code.

Fig. JD45J—Alternator wiring for Model 5020 with regulator attached to alternator frame. Refer to Fig. JD45F for color code.

Lock Test
  Volts 5.85
  Amperes 615
  Torque, ft.-lbs. 29
Model 1113135, 24-Volt
Brush spring tension ......80 oz.
No-Load Test
  Volts 22.5
  Amperes 65
  RPM 7000
Lock Test
  Volts 6.5
  Amperes 500
  Torque, ft.-lbs. 20

Prestolite
Model MEW-8001B, 24-Volt
Brush spring tension ....52-65 oz.
No-Load Test (includes solenoid)
  Volts 20
  Amperes 65
  RPM 5000
Lock Test (includes solenoid)
  Volts 4.0
  Amperes 420
  Torque, ft.-lbs. 22

circuit. Refer to Figs. JD45D through JD45J for schematic wiring diagrams, and to the following paragraphs for a short circuit description.

Non-Diesel Models

79A. All electrical accessories are grounded through the tractor frame in the conventional manner. Refer to Fig. JD45D. One 12 volt, 78 plate, 75 ampere-hour battery is supplied as original equipment.

Diesel Models

79B. In the 24 volt, split-load system the 12 volt "A Circuit" battery is positively grounded to tractor frame. The 12 volt "B Circuit" battery is negatively grounded. In the 12 volt accessory circuits, the tractor frame completes the circuit. The electrical starting and generating circuits are iso-

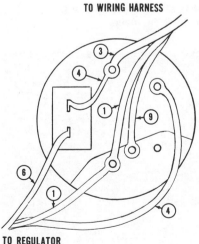

CIRCUIT DESCRIPTION

Non-Diesel Models are equipped with a 12 volt electrical system. All diesel models use a 24 volt, split-load system in which the full current is supplied only to the starting and generator circuits; the other electrical units receiving 12 volt current from half the

Fig. JD45G—Alternator wiring connections for early Model 5010 tractors where regulator wiring is separate from main wiring harness. Refer to Fig. JD45F for wiring color code.

lated, and the tractor frame serves as the series connection between "A Circuit" and "B Circuit" batteries when the 24 volt units are operating. Fig. JD45E shows wiring diagram for Series 3000 and 4000 diesel. Fig. JD45F shows Model 5010 with DC generator. Series 5000 tractors with alternator are similar except for alternator connections shown in Figs. JD45G, JD45H and JD45J.

Four 6 volt, 51 plate, 115 ampere-hour batteries are used on Series 5000; other units use two 12 volt, 78 plate, 75 ampere-hour batteries. On Series 5000, the two front batteries are negatively grounded and supply the "B Circuit"; the two rear batteries positively grounded and supply the "A Circuit." On other models the left battery is positively grounded and supplies the "A Circuit"; the right battery negatively grounded and connected to "B Circuit."

Fig. JD47—To adjust the power take-off linkage on series 3000 and 4000, remove access doors as shown. Place lever (L) in neutral position and disconnect rod (R) from lever. When clutch starts to engage, rod (R) should be ⅜-inch above hole in lever (L). Adjust by loosening nut (N).

ENGINE CLUTCH

NOTE: This section covers tractors equipped with Syncro-Range, manual shift transmission only. For models equipped with power shift transmission, refer to paragraphs 91 through 98.

LINKAGE ADJUSTMENT
Series 3000-4000

The transmission and power take-off are disengaged from engine by flywheel mounted single disc clutches which are independently controlled. The transmission clutch is controlled by a foot pedal, the power take-off clutch by a hand lever.

80. TRANSMISSION CLUTCH. To check the transmission clutch pedal free play, place transmission lever in "PARK" position and have engine running at 1900 rpm. Linkage should be readjusted if free play (P—Fig. JD46) measures less than ½ inch on Series 3000 Row-Crop Utility, Grove or Orchard tractors or ¾-inch on other models.

To adjust the linkage, release the latch and swing left battery box out where interference exists. Loosen castellated nut and clamp bolt (A), and with transmission lever in "PARK" position and engine running at 1900 rpm, adjust free play to 1-inch on Series 3000 Row-Crop Utility, Grove or Orchard tractor or 1½-inches on other models.

80A. POWER TAKE-OFF CLUTCH. To adjust the power take-off clutch, remove the control support cover which is located at rear of steering support housing below instrument panel and controls. Refer to Fig. JD47. Disconnect the clutch operating rod at lever, place lever in neutral position and raise the operating rod until clutch just starts to engage. Loosen top jam nut (N) and adjust operating rod until rod is ⅜-inch above hole in lever. Reconnect rod to lever and tighten jam nut.

When PTO clutch operating lever is held in rearmost position, PTO clutch is disengaged and a brake is applied which stops shaft rotation.

Series 5000

The transmission clutch consists of twin, flywheel mounted clutch discs controlled by a foot operated clutch pedal.

81. TRANSMISSION CLUTCH. Total clutch pedal travel should be 6⅞-inches when measured at pad as shown at (T—Fig. JD47A). Adjust by shortening or lengthening the exposed portion of pedal stop screw (A). Clutch pedal free travel should be approximately 1½-inches. Adjustment of clutch pedal free travel is made internally by adjusting the clearance (C—Fig. JD47B) between clutch re-

lease bearing collar (2) and release lever adjusting screws (1). To check and adjust free clearance, proceed as follows:

Remove inspection cover from bottom of clutch housing and check clearance (C) using a feeler gage. Clearance should be 0.105-0.112; if less than 0.060, back out lock screw (L) about ¼-turn, and with a spanner wrench, turn threaded sleeve (S) until specified clearance is obtained. After correct clearance is established, turn sleeve either way until nearest slot is aligned with locking screw (L), and lock in place.

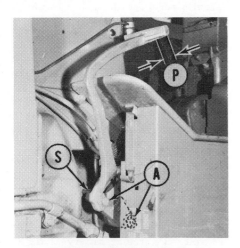

Fig. JD46—On series 3000-4000 adjust the transmission clutch pedal to obtain a free play (P) of one inch on Series 3000 Row-Crop Utility, Grove or Orchard tractors or 1½inches on other models.

Fig. JD47A—On Series 5000, adjust total clutch pedal travel (T) to 6⅞-inches by turning adjusting stop screw (A).

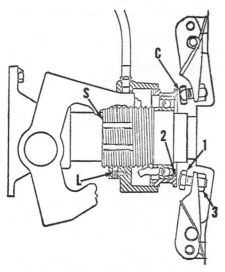

Fig. JD47B — Clutch pedal free travel on Series 5000 is adjusted internally by turning threaded sleeve (S) after loosening lock screw (L). Refer to text.

C. Clearance	2. Clutch release
1. Release lever	collar
screw	3. Lock nut

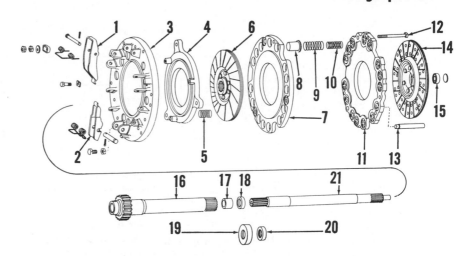

Fig. JD48—Exploded view of Series 3000-4000 dual plate clutch assembly. Parts 19 and 20 are used only on tractors not equipped with power take-off.

1. Transmission	8. Retainer	16. PTO clutch
clutch finger	9. Spring	shaft
2. PTO clutch finger	10. Spring	17. Bushing
3. Drive plate	11. Pressure plate	18. Oil seal
4. Pressure plate	12. Bolt	19. Quill
5. Release spring	13. Flywheel pin	20. Oil seal
6. PTO disc	14. Clutch disc	21. Transmission
7. Pressure plate	15. Pilot bearing	clutch shaft

81A. POWER TAKE-OFF CLUTCH. The hydraulically controlled, multiple disc pto clutch requires no adjustment. Refer to paragraph 112G for pressure adjustment of control valve.

TRACTOR SPLIT

Series 3000-4000

82. To detach (split) engine from clutch housing to obtain access to engine clutch and flywheel, proceed as follows:

Remove muffler, cowl and hood. Remove crankcase vent pipe and disconnect oil pressure switch and tachometer cable. On diesel models, disconnect throttle rod and fuel shutoff cable and disconnect oil cooler to reservoir line at cooler hose. Cap the line and cooler to prevent dirt entry. On non-diesel models, disconnect throttle rod and coil wire. On LP-Gas models disconnect fuel control rods at front end and pull them rearward through instrument panel.

On all models, relieve any pressure in hydraulic components by actuating the control levers, remove the hydraulic fluid line clamps and spacers, then disconnect the main hydraulic line at firewall and clutch housing. Disconnect steering motor fluid pipes. Be sure to cap all disconnected hydraulic fittings to prevent dirt entry. Disconnect wiring harness and battery cables from starter. If tractor is equipped with front end weights, remove the weights. Support engine and transmission separately, remove

the connecting cap screws and roll transmission assembly rearward away from engine.

To attach, reverse the above procedure and tighten the connecting cap screws as follows:

¾-inch cap screws......275 ft.-lbs.
½-inch cap screws...... 85 ft.-lbs.

NOTE: The power take-off clutch plate is held engaged by the clutch fingers. It is therefore free in flywheel when tractor is split. When re-attaching clutch housing to engine, use an alignment tool (Fig. JD50) to align the plate, then secure in place by installing a soft wood wedge under outer end of one pto clutch finger. After tractor is reattached, engage the pto clutch and wedge will fall to bottom of clutch housing where it will cause no trouble. If a metal wedge is used, remove clutch housing lower plate and remove the wedge before starting tractor.

Series 5000

82A. To detach (split) engine from clutch housing, first remove muffler, cowl and hood. Relieve any pressure in main hydraulic system by actuating the control levers or steering. Remove ether tube from cold starting unit and manifold. Disconnect main hydraulic pump inlet line at oil filter base on left side of transmission, and allow pump and oil cooler to drain. Disconnect both battery ground straps. Disconnect left hand steering tube at coupling in front of firewall and main

hydraulic return line at rear. Disconnect wires and cables from starter. Drain cooling system and disconnect heat indicator sending unit from cylinder head.

Disconnect wiring harness from fuel tank sending unit and generator, detach regulator bracket from cylinder block and lay wiring harness and regulator rearward over firewall. Disconnect throttle linkage and tachometer cable. Disconnect right rear steering tube at both ends and remove the tube. Disconnect main hydraulic pressure tube at firewall.

Support engine from a hoist and transmission housing from below. Remove cap screws securing side rails to clutch housing and all cap screws securing engine to clutch housing and separate the units.

Attach by reversing the outlined procedure; tighten clutch housing and side frame cap screws to a torque of 300 ft.-lbs.

R&R AND OVERHAUL

Series 3000-4000

83. To remove the clutch assembly after engine has been separated from clutch housing, remove the six retaining cap screws and withdraw the assembly. Transmission clutch disc (14—Fig. JD48) will be freed by the removal, and may be renewed without disassembly of clutch unit. Check friction faces of flywheel and transmission clutch pressure plate (11) for heat checks, wear or scoring. Pressure surface of flywheel must be true to 0.006 when measured with a

straight edge and feeler gage. Guide dowels (13) are a press fit in flywheel and are renewable.

Disassembly and reassembly of the removed unit will be facilitated by removal of flywheel or use of a spare flywheel. To disassemble, place flywheel, front side down, on a bench. Place clutch assembly in operating position in flywheel and secure with three alternately spaced jack screws, jam nuts and flat washers as shown in Fig. JD49. Tighten jam nuts evenly to take pressure from transmission clutch operating fingers (1); then remove the jam nuts, washers and operating bars from transmission clutch operating bolts (12). Back off the jack screw jam nuts evenly until spring pressure is released, remove jack screws, then remove the clutch components. PTO clutch disc (6—Fig. JD48), transmission clutch pressure plate (11), PTO front pressure plate (7) or clutch springs may be renewed at this time. Drive cover (3) may be disassembled for renewal of components, all of which are available as service parts.

Check the removed parts against the specifications which follow:

Guide dowels OD 0.624-0.625
Dowel holes in pressure
    plates ID 0.645-0.655

Spring Test Data

Outer Spring
    Approximate free length.3¼ in.
    Lbs. pressure @ 1¾-in. .105-129

Inner Spring
    Approximate free length.3 9/16 in.
    Lbs. pressure @ 1¾-in. .50.5-61.5

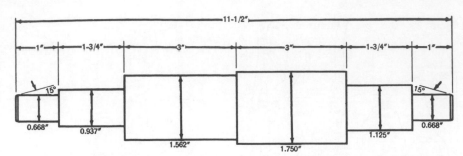

Fig. JD50—A clutch pilot can be turned out of a 12-inch piece of round stock, using the dimensions shown. Tool is used on series 3000 and 4000.

PTO Release Spring
    Approximate free length.2¼ in.
    Lbs. pressure @ ⅝-in. .13.5-17.5

Transmission Clutch Disc

Diameter
    Series 3000 . . . . .11 inches
    Series 4000 . . . . .12 inches
Thickness (New)
    Series 3000 . . . . .0.334-0.356
    Series 4000 . . . . .0.375-0.392

PTO Clutch Disc

Diameter
    Series 3000 . . . . .10 inches
    Series 4000 . . . . .11 inches
Thickness (New)
    All Series 0.317-0.319

Tightening Torques
    Flywheel attaching
        cap screws . . . . .85 ft.-lbs.
    Clutch cover
        attaching screws . . . . .35 ft.-lbs.

To assemble the removed unit, position transmission clutch disc in flywheel with long end of clutch hub down. Insert the three clutch operating bolts (12—Fig. JD48) in pressure plate (11) making sure bolt heads fit in recesses of pressure plate. Install plate over guide dowel and install clutch springs on plate; then install PTO front pressure plate (7). Position PTO clutch disc (6) with long end of hub up, and place the three PTO clutch springs on pressure plate. Assemble the drive cover, if disassembled; then position it over flywheel making sure guide dowels and operating bolts are aligned. Install the three jack screws. Use the clutch drive shafts to align discs, or fashion an aligning tool using the dimensions shown in Fig. JD50; then tighten the jam nuts on jack screws. Install the operating bars, washers and jam nuts on operating bolts (12—Fig. JD48). Reinstall flywheel and clutch unit using the aligning tool or clutch shafts and adjust as outlined in paragraph 83A.

83A. CLUTCH ADJUSTMENT. Whenever clutch has been disassembled, or when flywheel or transmission clutch disc has been renewed, both sets of operating fingers must be adjusted as follows:

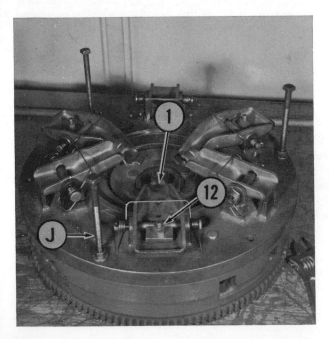

Fig. JD49 — To disassemble series 3000 and 4000 clutch, attach unit to a flywheel with three jack screws (J), then remove the nuts and washers from bolts (12) on clutch fingers (1). Back off the nuts on jack screws evenly, until spring pressure is released.

Fig. JD51 — On series 3000 and 4000, use JDE-19 clutch finger adjusting gage and adjust each transmission clutch finger individually, to 0.003 clearance using a feeler gage.

When using the special gage provided by the manufacturer, adjust transmission clutch fingers to clear center leg of gage by 0.003 as shown in Fig. JD51. Adjust PTO clutch fingers to clear upper gage surface by 0.003 as shown in Fig. JD52. NOTE: PTO clutch must be ENGAGED while making adjustment by forcing a screwdriver (or other wedge) under outer edge of one finger while adjustment is made on other fingers. Move wedge before adjusting third finger.

Series 5000

84. To remove the clutch after engine has been detached from clutch housing, proceed as follows: Remove three alternate drive plate retaining screws and install full thread jack screws (S—Fig. JD52B) which must be provided with nuts and flat washers. Screws must be 3 inches or longer, and should be bottomed in flywheel. Tighten the nuts finger tight against drive plate (14—Fig. JD52A) and remove the three remaining cap screws. Slide the three release levers (7) toward outer rim of clutch and push out the pins connecting levers to links. Back off jack screw nuts evenly, allowing clutch unit to separate. Remove drive plate (14), springs (15), pressure plate (5), rear clutch disc (3), separator plate (4) and front clutch disc from flywheel.

Examine friction surfaces of flywheel (2), separator plate (4) and pressure plate (5). Surfaces must not be scored, grooved or out-of-true

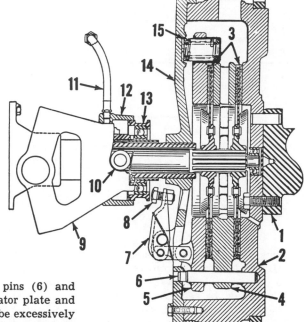

Fig. JD52A — Cross sectional view of transmission clutch used on Series 5000.

1. Crankshaft flange
2. Flywheel
3. Clutch discs
4. Separator plate
5. Pressure plate
6. Drive pin
7. Release lever
8. Adjusting screw
9. Operating cam
10. Cam roller
11. Lube line
12. Carrier
13. Release collar
14. Clutch cover
15. Clutch spring

more than 0.006. Drive pins (6) and drive pin holes in separator plate and pressure plate must not be excessively worn or damaged. If clutch face of flywheel is remachined, do not remove more than 0.035 of the surface to true the flywheel.

Clutch springs (15) consist of an inner and outer spring. Examine springs for rust, pitting or distortion, and check springs against the specifications which follow:

Free length (approx.)
Outer spring .... $3\frac{1}{4}$ in.
Inner spring .... $2\frac{27}{32}$ in.
Test length (all) .... $1\frac{3}{4}$ in.
Pressure at test length (lbs.)
Outer spring 105–129
Inner spring 79.2–96.8

To assemble the removed unit, position rear clutch disc on aligning tool (Fig. JD52D) with long hub to rear. Position separator plate (4—Fig. JD52A) in front of rear disc, then place front disc on aligning tool with long hub forward. Install the assembled driving unit in flywheel, fitting separator plate over drive pins and alignment tool in pilot bearing bore.

Place the pressure plate on a bench, friction surface down, install the 12 outer and 12 inner clutch springs in spring cups; then position drive plate and connect release lever links. Install the assembled drive plate and pressure plate over end of aligning tool and on drive pins. Install jack screws in three alternate holes and tighten nuts evenly. Tighten the clutch retaining cap screws to a torque of 35 ft.-lbs.

84A. CLUTCH ADJUSTMENT.

Whenever clutch has been disassembled, release levers must be adjusted as follows: Use an adjusting gage of

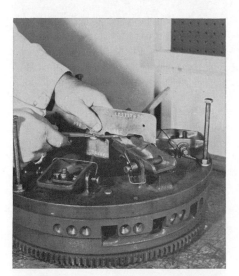

Fig. JD52—When adjusting the pto clutch fingers on series 3000 and 4000, use a chisel, screwdriver or other wedge under outer end of one finger to hold pto clutch pressure plate engaged. Adjust other fingers to 0.003 using special gage and feeler gage as shown. Move wedge to adjust third finger.

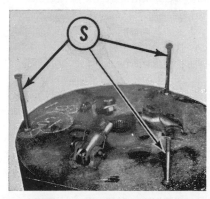

Fig. JD52B — Use three full-thread jack screws (S) as shown, to remove clutch unit from flywheel on Series 5000. Refer to text.

Fig. JD52C — Use special gage to adjust release lever height on Series 5000. Refer to text.

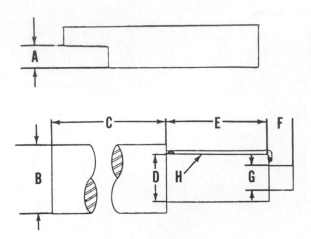

Fig. JD52D—Series 5000 adjusting gage (upper view) and clutch pilot (lower view) can be made locally, using the dimensions shown. Dimensions are in inches.

A. 0.810
B. 1.937
C. 6.0
D. 1.247
E. 2.750
F. 0.625
G. 0.667
H. 1/16 wire (tack welded)

the dimensions shown in upper view, Fig. JD52D. Measuring from rear face of pto drive hub in clutch cover, adjust the screw in each release lever to clear gage by 0.005. If gage is not available, adjust levers equally within 0.005 to extend 0.810 beyond rear surface of pto drive hub in clutch cover. Check and adjust operating linkage as outlined in paragraph 81, after tractor is assembled.

CLUTCH SHAFT

Series 3000-4000

85. To renew either the transmission or PTO clutch shaft, it is first necessary to detach (split) tractor between the clutch housing and transmission case. To split the tractor, refer to the following paragraph.

85A. **TRACTOR SPLIT.** First drain the transmission and hydraulic system fluid and remove batteries and battery boxes. Disconnect transmission shifter linkage and clutch pedal return spring.

Disconnect and cap the main hydraulic pump inlet line, steering valve return line, rockshaft cylinder inlet line and the hydraulic brake lines. Move tractor seat to extreme rearward position. Loosen the control support cover knob, raise support cover to clear platform, then remove operators' platform.

Disconnect the rockshaft push-pull cable. Remove the remote cylinder hydraulic lines, hydraulic brake lines and wiring shields. Disconnect and pull wiring forward to clear transmission case and remove transmission top cover. Remove front PTO guard and bearing quill being careful not to damage seal.

Support both ends of tractor separately, remove the connecting cap screws and separate the two units.

When reconnecting, tighten the retaining cap screws to a torque of 150 ft.-lbs.

85B. **R & R AND OVERHAUL SHAFTS.** After splitting tractor as outlined in paragraph 85A, clutch shafts may be withdrawn rearward out of clutch housing. To avoid damage to inner seal, remove both shafts as a unit. Withdraw transmission clutch shaft carefully to avoid damaging seal.

Both oil seals can be renewed at this time. Install seals with lips to rear. Check the bushing inside PTO clutch shaft for wear. Inside diameter of a new bushing is 1.3795-1.381. When renewing, press bushing into shaft the full depth of counterbore.

Series 5000

86. To renew either the transmission or pto clutch shaft, first detach (split) tractor between clutch housing and transmission case. Refer to the following paragraph.

86A. **TRACTOR SPLIT.** First relieve all pressure in main hydraulic system and brakes, drain hydaulic system and remove batteries.

NOTE: To relieve pressure in brake system accumulator, open right hand brake bleed screw and depress right hand brake pedal with engine not running.

Remove right, front platform extension, disconnect hydraulic pipe from fitting at bottom of accumulator and remove accumulator.

Remove operators' platform and disconnect light wires. Remove both hydraulic brake lines and remote cylinder hydraulic lines on tractors so equipped. Disconnect main hydraulic pump inlet and return lines. Remove pto clutch control valve.

Remove transmission top cover and the two, top flange cap screws. Disconnect front shifter rods from both cam arms. Support the two tractor halves separately, remove the securing cap screws and separate the units.

When reconnecting, tighten ¾-inch cap screws to a torque of 300 ft.-lbs. and ⅞-inch cap screws to a torque of 445 ft.-lbs.

86B. **R & R AND OVERHAUL SHAFTS.** To remove the clutch shafts after clutch housing is separated from transmission, drive roll pin (2—Fig. JD52E) out of pto shifter cam (3). Pull pto shift shaft (1) upward out of shift cam and remove the cam, shift fork (4) and shift collar (6); then carefully withdraw transmission clutch shaft (8), pto drive shaft (7) and pto drive gear (5) as a unit from housing. Avoid damaging clutch shaft outer seal as assembly is withdrawn. Support the removed shafts to prevent damage to inner oil seal, drive out the roll pin which retains transmission clutch shaft in pto drive shaft; and withdraw clutch shaft forward out of pto drive shaft.

Both seals can be renewed at this time. Seals should be installed with lips to rear. Assemble by reversing the disassembly procedure. Pto disconnect assembly is correctly assembled when beveled side of lever is up and pointed to right when pto is engaged.

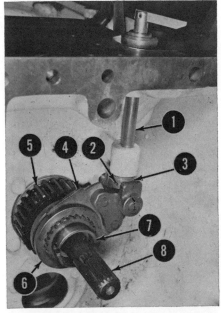

Fig. JD52E — Rear view of Series 5000 clutch housing showing clutch shafts, pto drive gear, pto disconnect mechanism and associated parts.

1. PTO shift shaft
2. Roll pin
3. Shift cam
4. Shift fork
5. PTO drive gear
6. Shift collar
7. PTO drive shaft
8. Clutch shaft

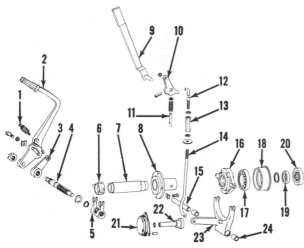

Fig. JD53 — Exploded view of clutch linkage and associated parts used on series 3000 and 4000.

1. Spring
2. Pedal
3. Arm
4. Shaft
5. Fork
6. Collar
7. Tube
8. Support
9. PTO lever
10. Lever arm
11. Detent pawl
12. Rod
13. Coupling
14. Rod
15. Shaft
16. PTO operating sleeve
17. Bearing
18. PTO operating collar
19. Collar
20. Throwout bearing
21. PTO brake
22. Brake shaft
23. Operating fork
24. Bushing

Fig. JD53A — Exploded view of clutch release mechanism and associated parts used on Series 5000.

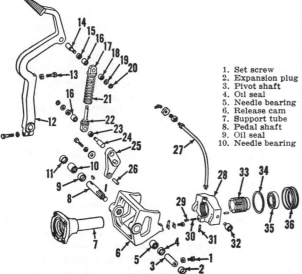

1. Set screw
2. Expansion plug
3. Pivot shaft
4. Oil seal
5. Needle bearing
6. Release cam
7. Support tube
8. Pedal shaft
9. Oil seal
10. Needle bearing
11. Oil seal
12. Clutch pedal
13. Stop screw
14. Dowel
15. Oil seal
16. Needle bearing
17. Spring end
18. Oil seal
19. Washer
20. Snap ring
21. Spring
22. Spring end
23. Oil seal
24. Dowel
25. Hub
26. Dowel pin
27. Lube tube
28. Carrier
29. Plunger
30. Spring
31. Set screw
32. Cam follower
33. Sleeve
34. Sealing ring
35. Release bearing
36. Collar

CONTROL LINKAGE
Series 3000-4000

87. To overhaul the clutch control linkage it is necessary to remove the clutch housing from tractor. To remove the housing first detach clutch housing from engine as outlined in paragraph 82, attach a chain hoist to clutch housing, then detach from transmission housing as outlined in paragraph 85A. Refer to Fig. JD53 for an exploded view of linkage.

Clutch release bearings (17 and 20) and PTO operating fork and linkage can be removed from front of housing without making rear split. Shaft (15) is a tight press fit in right side of housing and must be driven out toward the left. If service is required on detent pin (11) or its spring, drive out the roll pin and lift out the detent assembly. To install, compress the spring in a vise and wire securely in the compressed position. Install the

parts, insert roll pin, then cut and remove the wires.

Transmission clutch fork (5) and arm (3) are splined to operating shaft (4). Correct position of installation is indicated by index marks on shaft, fork and arm.

When reassembling, adjust linkage as outlined in paragraphs 80 and 80A.

Series 5000

87A. To overhaul the clutch control linkage, first detach (split) engine from clutch housing as outlined in paragraph 82A.

Before any disassembly is attempted, depress clutch pedal until the holes in spring end (17—Fig. JD53A) and pin (22) are aligned, and insert an ⅛-inch steel pin or nail through the parts to retain the spring. Remove retaining ring and washer securing spring pin (22) to pivot (24); loosen clamp bolt in clutch pedal (12) and remove pedal and spring as an assembly. Remove grease fitting and nut from release bearing lube line (27) and free the line from housing. Rotate actuating cam (6) downward and withdraw release bearing and carrier as a unit. Remove expansion plug (2) from right side of housing and loosen set screw (1). Pull cam pivot (3), using a ⅜-inch bolt as a puller. Outer end of cam pivot is threaded to receive puller bolt.

Remove cotter pin and nut from inner end of actuating shaft (8) and remove shaft and actuating cam (6).

Needle bearings (5, 10 and 16) should be packed with Lubriplate or similar high-temperature grease when linkage is overhauled or bearings are renewed. Lips of seals should face outward, away from bearings. Cam follower rollers (32) are retained in release bearing carrier (28) by roll pins which must be removed if disassembly is indicated. Detent plunger (29) fits against a flat on set screw (31) to prevent screw from loosening. Plunger and spring are free when set screw is removed.

Disassemble and/or assemble return spring (21) and spring ends in a suitable press. Spring free length should be approximately $6\frac{1}{32}$-inches, and spring should test 231-281 lbs. when compressed to a length of 4 inches.

Assemble by reversing the disassembly procedure and adjust linkage as outlined in paragraph 81.

SYNCRO-RANGE TRANSMISSION
(For Power Shift Models, Refer to Paragraph 91)

The "Syncro Range" transmission is a mechanically engaged transmission consisting of three transmission shafts and a single, mechanically connected, remote mounted control lever as shown in Fig. JD54. The four basic gear speeds are selected by coupling one of the shaft idler gears to the

splined main drive bevel pinion shaft, and can only be accomplished by disengaging the engine clutch and bringing the tractor to a stop. The high, low and reverse speed ranges within the four basic speeds are selected by shifting the couplers on the transmission drive shaft and, because of the

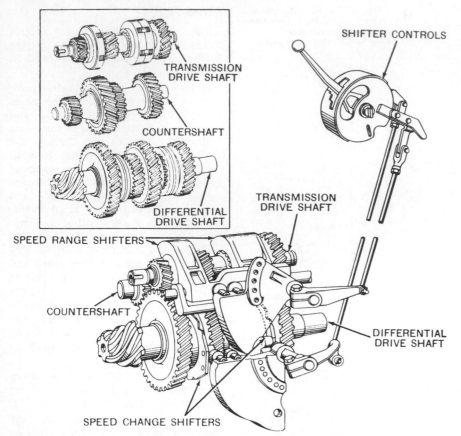

SHIFTER CONTROLS

TRANSMISSION DRIVE SHAFT

COUNTERSHAFT

DIFFERENTIAL DRIVE SHAFT

SPEED RANGE SHIFTERS

TRANSMISSION DRIVE SHAFT

COUNTERSHAFT

DIFFERENTIAL DRIVE SHAFT

SPEED CHANGE SHIFTERS

Fig. JD54—Schematic view of the "Syncro Range" transmission components. The single shift lever moves either shifter cam to change to the selected gear. Series 3000-4000 transmission is pictured; Series 5000 is similar in major details.

design of the couplers, can be accomplished by disengaging the engine clutch and moving the control lever, without bringing the tractor to a halt.

NOTE: The rotating speeds of the transmission drive shaft and its idler gears are automatically equalized by the synchronizing clutches. All other phases of shifting are under the direct control of the operator. The fact that clashing of gears is eliminated by the synchronizing clutches does not relieve him of the responsibility of using care and judgment in re-engaging the clutch after the gears have been shifted.

The idler gears and bearings on the main shaft and bevel pinion shaft are pressure lubricated by a separate transmission oil pump.

INSPECTION

All Models

88. To inspect the transmission gears, shafts and shifters, first drain the transmission and hydraulic fluid and remove the operator's platform, then remove the transmission top cover. Examine the shaft gears for worn or broken teeth and the shifter linkage and cam slots for wear.

CONTROL QUADRANT

This section covers disassembly and overhaul of the shifter controls mounted in tractor steering support. Removal, inspection, overhaul and adjustment of shift mechanism inside the transmission housing is included with transmission gears and shafts.

All Models

88A. **R&R AND OVERHAUL.** To overhaul the control quadrant, remove the cowl and raise the dash enough to clear quadrant. Right mounting handle may be removed for convenience. Disconnect the shifter rods (6 and 13—Fig. JD55) at upper end and remove service card access door and cards. Remove snap ring (17), steel washer (16) and bronze washer (15) from outer end of quadrant shaft (19) and, while holding shaft with a wrench, remove the nut and washer securing shaft to bracket (1). Quadrant assembly may then be completely disassembled.

If lever (7) or pivot (9) are damaged and need to be renewed, proceed as follows: Clamp the lower curved portion of lever (7) in a soft-jawed vise and slip a $\frac{5}{32}$-inch cotter pin inside the two roll pins (8). Grasp roll

pins with a good vise-grip plier and extract with a twisting motion. When reassembling, leave at least $\frac{3}{16}$-inch of roll pin protruding from lever (7). To drive pin farther will damage bushing (10).

When reassembling, install shaft (19) loosely in bracket (1) and start the retaining nut. Install the remainder of parts in proper order, omitting the bronze washer (15), steel washer (16) and snap ring (17). Make sure that thrust washer (2) has not slipped over the shoulder of shaft (19), then tighten the shaft retaining nut. Steel washer (16) is available in thicknesses of 0.018, 0.036 and 0.060. Install washers of sufficient number and thickness to limit the end play of quadrant components to 0.015 when snap ring is installed.

After quadrant is reassembled, check the action of the neutral-start switch as follows: Move control lever to neutral or park position and turn key switch to "Start." If the starter fails to operate, loosen the two cap screws retaining starter arm (18) to speed range quadrant (12) and shift starter arm on quadrant until neutral-start switch is depressed.

TRANSMISSION DISASSEMBLY AND ASSEMBLY

Paragraphs 88B through 88E outline the general procedure for removal and installation of the main transmission components. Disassembly, inspection and overhaul of the removed assemblies is covered in overhaul section beginning with paragraph 89, which also outlines those adjustment procedures which are not an exclusive part of assembly.

Series 3000-4000

88B. **DISASSEMBLY.** Any disassembly of transmission gears, shafts and controls requires that tractor first be separated (split) between transmission and clutch housing as outlined in paragraph 85A. Transmission must be disassembled in the approximate sequence outlined in the following paragraphs; however, disassembly need not be completed once defective or damaged parts are removed.

Remove the transmission top cover and rockshaft housing or transmission rear cover. Remove the detent spring caps from right side of housing, raise upper shifter arm to its highest position and remove cotter pin and slotted nut from inner end of shaft. Be careful not to drop the parts in housing, as they will be difficult to remove. Withdraw shifter arm and shaft from

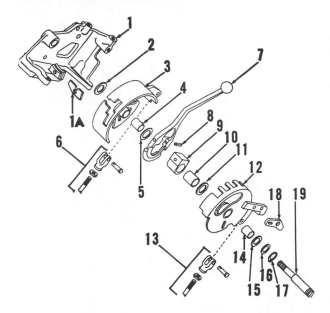

Fig. JD55 — Exploded view of the dash mounted shifter controls. The notch in latch (1A) fits around the lower rocker of lever (7) and moves to positively lock the opposite quadrant (3 or 12) when the other is shifted.

1. Bracket
2. Washer
3. Quadrant
4. Bushing
5. Washer
6. Link
7. Lever
8. Roll pin
9. Pivot
10. Bushing
11. Washer
12. Quadrant
13. Link
14. Bushing
15. Washer
16. Washer
17. Snap ring
18. Arm
19. Shaft

race (23—Fig. JD62) and use a gear puller to remove race from front end of pinion shaft.

Use a soft drift and heavy hammer and bump the pinion shaft rearward as far as possible. Install a "C" clamp through the hole provided in front face of housing as shown at (C—Fig. JD58) and tighten the clamp over one edge of the front gear (18—Fig. JD62); then continue to bump the shaft rearward until the three snap rings (8, 14 and 16) are exposed. NOTE: On series 3020 and 4020 tractors a fourth snap ring is used. The additional snap ring is located between shifter gear (10) and gear (12). Using a pair of snap ring pliers, expand and unseat the snap rings forward on the shaft. After snap rings are unseated, use a long brass drift to drive the bevel pinion shaft rearward out of shaft and gears.

NOTE: The snap rings are the same diameter but differ in thickness, the thicker snap ring being to the rear so that no snap ring can fall into another groove as the shaft is removed.

Bushing (6), in gear (5), is presized and may be renewed if worn or scored.

Remove the countershaft front bearing retainer and shim pack. Using a hammer and soft drift and working from rear of countershaft, drive the countershaft front bearing cup from transmission case; then lift out the countershaft.

Overhaul the transmission main components as outlined in paragraph 89 through 90D; assemble as outlined in paragraph 88C.

88C. ASSEMBLY. Install countershaft and front bearing cup; install front bearing retainer and removed shim pack and tighten retaining cap screws to a torque of 35 ft.-lbs.; then,

housing. Oil seal may be renewed at this time. Slide the upper shift rail (4—Fig. JD63) from housing, rotate shifters (1 and 2) upward, then lift them from housing. Shifter cam (5) may now be withdrawn.

Jack up one rear wheel of tractor and turn bevel ring gear until one of the flat surfaces of differential housing is toward transmission oil pump as shown in Fig. JD56; then unbolt and remove pump, together with the inlet and outlet tubes.

Working through the front bearing retainer and using a brass drift, drive the transmission drive shaft rearward to force the rear bearing cup part of the way out of housing. Tape the synchronizer clutches together to keep them from separating while shaft is being removed. Remove PTO idler gear from front of transmission, remove front bearing retainer and front bearing cup, using care not to damage or lose the shims, then lift out the transmission drive shaft assembly.

Remove the cotter pin and nut from inner end of lower (speed change)

shifter cam shaft and remove shaft and arm. Withdraw the shifter rail forward out of transmission housing and lift out the speed change shifter cam and shifter forks.

Remove the differential assembly as outlined in paragraph 96. Remove the cap screws retaining the pto idler thrust washer (1—Fig. JD57) and remove the washer and gear (2). Be careful not to lose any of the loose needle bearings located inside the idler gear. On Series 3010 and 4010 remove the shifter pin (3) retaining shift collar (4), then pull the pto front bearing (5) and remove shifter collar and gears.

Tighten a bearing puller in the center groove of pto idler gear inner

Fig. JD56 — To remove series 3000 and 4000 transmission pump after rockshaft housing is removed, turn differential until flat (F) is nearest pump, then remove the attaching cap screws (C).

Fig. JD57 — Front face of series 3010 and 4010 transmission housing showing the attached pto drive gears. On late models, a single pto drive gear is used. Gears must be removed before transmission can be disassembled.

1. Idler plate
2. Idler gear
3. Shifter pin
4. Shift collar
5. Bearing
6. PTO shaft
7. 510 rpm gear
8. 1000 rpm gear

Fig. JD58—To remove the output (bevel pinion) shaft on series 3000 and 4000, install a "C" clamp (C) over front gear and housing, then drift shaft rearward to remove bearing (B). See text for details.

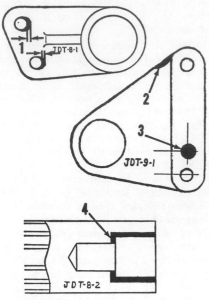

Fig. JD59 — To use the special 3010 and 4010 John Deere service tools on 3020 and 4020 tractors, alter them as shown.

1. Elongate ⅛-inch
2. Relief notch ¼-inch deep
3. ⅞-inch hole centered as follows: 1¼ inch above C/L of lower hole, ⅞-inch from edge
4. Enlarge to 2.000-2.004, 1½-inch deep

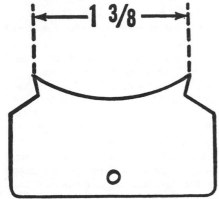

Fig. JD60—If bevel pinion shaft installation tools are not available, make three snap ring retainers of 5/16 inch material using the pattern shown. Spread the snap rings and insert the retainers in open end while installing shaft. Tool can be used on all models.

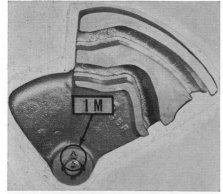

Fig. JD61—When installing the shifter cam on series 3000 and 4000, make sure that the index marks (IM) on shaft and cam are aligned as shown.

1. Pinion shaft
2. Bearing cone
3. Bearing cup
4. Shim
5. Gear
6. Bushing
7A. Thrust washer (3020 only)
7B. Thrust washer (3010 & 4010 only)
8. Anap ring
9. Dowel
10. Shifter gear
11. Collar
12. Gear
13. Thrust washer
14. Snap ring
14A. Snap ring (3020 & 4020 only)

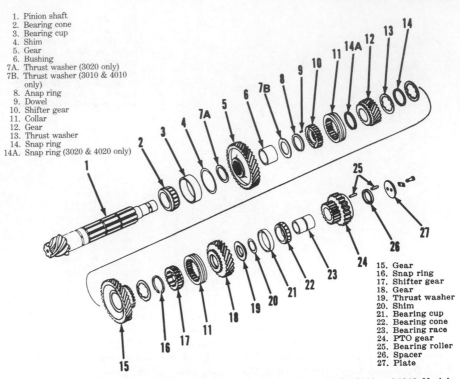

15. Gear
16. Snap ring
17. Shifter gear
18. Gear
19. Thrust washer
20. Shim
21. Bearing cup
22. Bearing cone
23. Bearing race
24. PTO gear
25. Bearing roller
26. Spacer
27. Plate

Fig. JD62 — Transmission output (bevel pinion) shaft showing components for 3010 and 4010. Models 3020 and 4020 are similar except for the following: Notched thrust washer (7B) and dowels (9) are used on 3010 and 4010 only. Thrust washer (7A) is used only on 3020. Model 4020 does not use either thrust washer. Snap ring (14A) is used between gears (10 and 12) on 3020 and 4020 only.

check countershaft end play using a dial indicator. Adjust end play, if necessary, to the recommended 0.001-0.004 by varying shim pack thickness. Shims are available in thicknesses of 0.006, 0.010 and 0.018.

If bevel pinion shaft or transmission housing were renewed; or if shaft bearings are not properly preloaded, adjust shaft and bearings as outlined in paragraph 90E.

Special John Deere service tools are almost essential in installing the bevel pinion shaft and gears. The tools required are as follows:

JDT-2 Snap Ring Expanding Cone.

JDT-3 Snap Ring Retainers (3)

JDT-8 (1, 2 & 3) Installing Arbor For Series 4000.

JDT-9 (1 & 2) Installing Arbor For Series 3000.

NOTE: The early service tools must be modified as shown in Fig. JD59, for use on 3020 and 4020 tractors.

To install the shaft and gears, refer to Fig. JD62 for order of assembly. All components must be installed on shaft as shaft is inserted. Note on Models 3010 and 4010 that the rear shifter gear (10) contains two roll pins (9) which fit into notches in thrust washers (7B) to keep washer from turning independently of shaft. On all models, snap rings must be spread

and held in the expanded position to allow the shaft to be easily inserted. To expand the snap rings use snap ring pliers and expanding cone JDT-2. Hold the ring on the cone and insert the snap ring retainers JDT-3. NOTE: If special tools are not available, the retainers can be fashioned of 5/16 inch material using the pattern shown in Fig. JD60. Attach the arbor of the installing tool (JDT-8 or JDT-9) to front of transmission case, then insert the dummy shaft from front. Install the shaft components on dummy shaft in proper order beginning with gear (18—Fig. JD62) and working back to gear (5). When all units are in place on dummy shaft, install pinion shaft (1) from rear through the shaft gears, replacing the dummy shaft. After shaft is in place, seat the snap rings in their grooves, then install spacer (19), the previously determined thickness of shim pack (20) bearing cone (22) and bearing inner race (23). Installation of bearings can be facilitated by heating them to a temperature not to exceed 300°F.

Install speed change shifter forks and cam, and insert shift rail through forks. Install cam shaft and arm, making sure that index marks are aligned as shown in Fig. JD61. Tighten the retaining castellated nut

Fig. JD62A—Bevel pinion shaft and associated parts used on Series 5000.

1. Shaft
2. Bearing cone
3. Bearing cup
4. Shims
5. 1st-3rd speed gear
7. Thrust washer
8. Snap ring
9. Dowel
10. Shifter gear
11. Collar
12. 6th-8th speed gear
13. Thrust washer
14. Snap ring
15. 2nd-5th speed gear
16. Snap ring
17. Shifter gear
18. 4th-7th speed gear
19. Thrust washer
20. Shim
21. Bearing cup
22. Bearing cone
27. Retainer
28. Locking plate

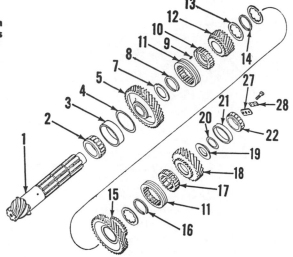

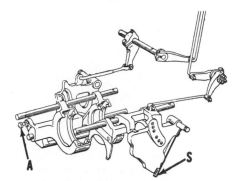

Fig. JD26B — Assembled view of Series 5000 shifter mechanism and associated parts, showing parking lock spring (S) and reverse shifter stop screw (A).

to provide a shaft end play of 0.001-0.005, and install the cotter pin.

To install the transmission drive shaft and shifter mechanism, place shaft in transmission housing and install front bearing cup, shims and retainer. Tighten cap screws securely but do not lock in place. Install rear bearing cup and transmission oil pump. Check transmission drive shaft end play, using a dial indicator and adjust to 0.004-0.006 by means of shims (19—Fig. JD63E). Shims are available in thicknesses of 0.006, 0.010 and 0.018.

Place shifter cam (5—Fig. JD63) in housing and install shifter (1 and 2) and rail (4). Make sure rollers (3) are in place and engage the slots in cam (5). Install shaft (6) carefully to keep from damaging oil seal. Make sure "V" mark on shaft indexes with corresponding mark on cam, install slotted nut and adjust end play to 0.002-0.005 and install cotter pin. Refer to Fig. JD61.

Series 5000

88D. DISASSEMBLY. To disassemble the shafts, gears and controls located in transmission housing, detach (split) transmission from clutch housing as outlined in paragraph 86A. Transmission must be disassembled in the approximate sequence outlined, but may stop at the point where all damaged parts have been removed.

Remove rockshaft housing or transmission rear cover. Remove transmission oil pipes and countershaft front bearing retainer and adjusting shims. Slide countershaft forward and lift from transmission.

NOTE: It may be necessary on some models, to turn bevel pinion shaft until blank spline on reverse shifter gear (17—Fig. 62A) points toward countershaft, turn in stop screw (A—Fig. JD62B), then shift the reverse fork rearward to provide removal clearance.

Remove transmission oil filter and oil filter relief valve housing from left side of transmission housing. Remove the snap ring, shifter pawl retainer, spring and pawl from each side of transmission. Remove the retaining snap rings, thrust washers and roll pins; then remove both shifter cam shafts from transmission housing. Drive out the roll pins retaining the two upper shifter rails; remove the three rails and lift out shifter forks and cams. Remove transmission oil pump from front of transmission housing and rear bearing quill from transmission drive shaft, then lift out the shaft.

Remove the differential assembly as outlined in paragraph 96. Remove bevel pinion shaft front bearing plate (27—Fig. JD62A), bearing cone (22) and cup (21) from front of shaft. Remove shim pack (20) and retaining washer (19), keeping shim back together for reinstallation when unit is assembled. Slide front idler gear (18), shift collar (11) and shifter gear (17) forward, unseat and expand snap ring (16) and move it forward on shaft. Expose and unseat snap rings (8 and 14) and move them forward out of their grooves; then drift the shaft rearward, removing the gears as they are free.

Overhaul the transmission main components as outlined in paragraphs 89B through 90D; assemble as outlined in paragraph 88E.

88E. ASSEMBLY. If bevel pinion shaft or transmission housing were renewed; or if shaft bearings are not properly preloaded, adjust shaft and bearings as outlined in paragraph 90E.

The special John Deere Snap Ring Expanding Cone (JDT-2) and Snap Ring Retainers (JDT-3) (See Fig. JD60) should be used when installing the bevel pinion shaft. Slide the snap rings (8, 14 & 16—Fig. JD62A) down the cone and insert the retainer (Fig. JD60) to hold rings open, and leave retainers in place while shaft is being installed.

Insert the shaft (1—Fig. JD62A) and bearing cone (2) through bearing cup (3) in transmission housing and install the larger gear (5) and thrust washer (7) on shaft. Place the thicker snap ring (8) on shaft and continue to insert the shaft with gears and washers in proper order until the front gear (18) is installed. Remove the snap ring retainers (Fig. JD60) and seat the snap rings in their grooves, starting with rear snap ring. Installation of bearings can be facilitated by heating them to a temperature of 300° F. Tighten front bearing plate retaining cap screws to a torque of 35 ft.-lbs.

Install the transmission drive shaft front bearing cup and spacer, if removed, then install transmission drive shaft and transmission oil pump. Tighten oil pump retaining cap screws to a torque of 35 ft.-lbs. Install rear bearing quill and shims and check drive shaft end play using a dial indicator. Adjust end play to 0.004-0.006, if necessary, by adding or removing shims behind rear bearing quill. Tighten rear bearing retaining cap screws to a torque of 35 ft.-lbs. when bearings are properly adjusted.

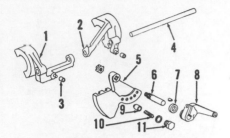

Fig. JD63—Exploded view of speed range shifter mechanism used on all models except Series 5000.

1. Reverse fork	7. Oil seal
2. High-Low fork	8. Shift arm
3. Cam roller	9. Detent plunger
4. Shift rail	10. Spring
5. Shift cam	11. Plug
6. Shaft	

Install shifters in transmission by reversing the disassembly procedure, using Fig. JD62B as a guide. Place speed range shifter cam in any detent position except reverse, and adjust reverse shifter stop screw (A) to provide 0.005-0.015 clearance between screw and case. Tighten the lock nut.

Install countershaft and adjust countershaft end play to 0.001-0.004 by varying the thickness of shim pack behind front bearing retainer. Tighten cap screws to a torque of 35 ft.-lbs.

Complete the assembly by reversing disassembly procedure.

OVERHAUL

To overhaul the transmission, first disassemble the unit as outlined in paragraph 88B or 88D. Disassemble, overhaul and inspect the components as outlined in the appropriate following paragraphs; then reassemble as in paragraph 88C or 88E.

Series 3000-4000

89. SHIFTER CAMS AND FORKS. Refer to Fig. JD63 for an exploded view of speed range shifter mechanism and Fig. JD63A for speed change shifters. Examine shifting grooves in cams (5—Fig. JD63 and 6—Fig. JD63A) for wear or other damage. Parking lock spring (8—Fig. JD63A) is retained in cam by roll pin (9). The spring must have sufficient tension to shift the front shift coupling into engagement. Shift forks (F) are riveted to fork carriers (1 and 2) and are renewable.

89A. TRANSMISSION PUMP. The removed transmission pump may be disassembled by removing the cover and lifting out the gears. Check the pump parts for wear, scoring or other damage, and against the specifications which follow:

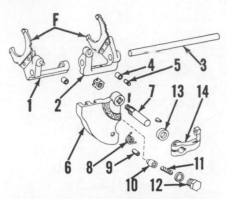

Fig. JD63A — Exploded view of series 3000 and 4000 speed change shifter mechanism which operates on the bevel pinion shaft. The forks (F) are riveted to carriers (1 and 2) and are renewable separately.

1. Shift fork	7. Shaft
2. Shift fork	8. Spring
3. Shift rail	9. Roll pin
4. Cam roller	10. Detent plunger
5. Roller pin	11. Spring
6. Shift cam	12. Plug

Pump gear diameter OD . 2.082 -2.084
Pump gear thickness 0.6235-0.6245
Idler gear ID . . . . 0.626 -0.628
Idler shaft OD . . . . 0.624 -0.625
Gear bore in housing ID . 2.090 -2.092
Gear bore depth 0.6262-0.6278
Pump gear radial
 clearance . . . . 0.003 -0.005
Transmission drive
 shaft OD 1.0010-1.0016
Pump drive bore ID 1.0035-1.0045
Cover cap screw torque . . . . 21 ft.-lbs.

Series 5000

89B. SHIFTER CAMS AND FORKS. Refer to Fig. JD62B for an assembled view of shifter mechanism and to Fig. JD63B for an exploded view of park pawl and associated parts. Examine shift grooves and detents in shifter cams for wear or other damage. Parking lock spring (S—Fig. JD62B) should require a pull of 17-21 lbs. to deflect free end of spring 2-inches.

Speed range detent spring should have a free length of approximately 1⅛-inches and test 12-14 lbs. when compressed to a height of ¹³⁄₁₆-inch. Speed change detent spring should have a free length of 1¹¹⁄₃₂-inches and test 27-33 lbs. when compressed to ¾-inch. Approximate free length of park pawl spring is 2⅛-inches and spring should test 9-13 lbs. when compressed to a height of 1¹¹⁄₁₆-inches.

89C. TRANSMISSION PUMP. The transmission pump is driven by the pto drive shaft at engine speed, and operates whenever engine is running.

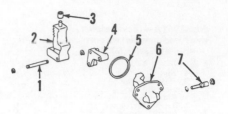

Fig. JD63B — Exploded view of parking lock used on Series 5000.

1. Shaft	5. Packing ring
2. Lock	6. Operating cover
3. Roller	7. Arm shaft
4. Operating arm	

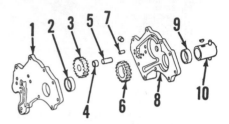

Fig. JD63C — Exploded view of transmission pump used on Series 5000.

1. Cover	6. Drive gear
2. Bushing	7. Dowel
3. Idler gear	8. Body
4. Bushing	9. Bushing
5. Shaft	10. Drive shaft

The pump cover is retained to housing only by dowels, after pump is removed.

To disassemble the removed pump, tap the cover (1—Fig. JD63C) from its doweled position on housing (8) and remove the pump gears and drive sleeve (10). The drive gear (6) is keyed to drive sleeve but is a slip fit on sleeve. Drive sleeve bushings (2 & 9) in cover and housing; and idler gear bushing (4) are renewable. Check the pump parts for wear, scoring or other damage, being sure to check pto clutch valve bore in transmission pump housing and the removed pto clutch valve. Specifications are as follows:

Pump gear diameter OD . . 2.853 -2.854
Pump gear thickness . . . . . 0.625 -0.626
Gear bore in housing 2.8565-2.8585
Gear bore depth 0.6272-0.6296
Pump gear radial
 clearance 0.0012-0.0016
Pump gear end
 clearance 0.0012-0.0036
Drive sleeve OD . . . . 1.937 -1.938
Drive sleeve
 bushing ID 1.9395-1.9415
Idler gear bushing ID . . . 0.7517-0.7527
Idler gear shaft OD 0.7497-0.7503
PTO clutch valve OD . . . 0.6861-0.6865
PTO clutch valve
 bore ID . . . . 0.687 -0.688

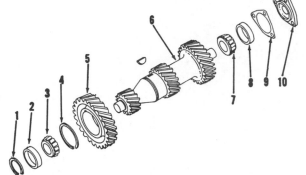

1. Snap ring
2. Bearing cup
3. Bearing cone
4. Snap ring
5. Gear
6. Countershaft
7. Bearing cone
8. Bearing cup
9. Shim
10. Bearing quill

All Models

90. **COUNTERSHAFT.** Refer to Fig. JD63D. The countershaft is a one-piece unit except for the bearings and high-speed gear (5). The high-speed gear is keyed to shaft and retained by snap ring (4); and may be removed with a press after removing the snap ring. Countershaft bearings should have 0.001-0.004 end play when shaft is properly installed.

90A. **TRANSMISSION DRIVE SHAFT.** To disassemble the removed transmission drive shaft proceed as follows:

Remove snap ring (1—Fig. JD63E), remove bearing cone (3) with a press or bearing puller, then remove reverse range pinion (4).

Withdraw synchronizer drum (5), plates (6 and 7) and blocker (8) from shaft.

CAUTION: The four detent balls and springs (9) will be released when blocker is withdrawn. Use care not to lose these parts.

Remove snap ring (10) and use a press or bearing puller to remove drive collar (11), then withdraw low range pinion (12) from shaft. NOTE: On most models, low range pinion is retained to shaft by a snap ring which must first be removed.

Remove low range synchronizer drum and plates and the low and high range blocker rearward from shaft being careful not to lose the detent balls and springs.

Remove snap ring (18), bearing cone (16) and high range pinion (14) from front of shaft, then withdraw the remaining synchronizer clutch parts.

The high and low range drive collar can be pressed from shaft after removing snap ring (13), if renewal is indicated.

90B. **SYNCHRONIZER CLUTCHES.** The purpose of the synchronizer clutches is to equalize the speeds of the transmission drive shaft and the selected range pinion for easy shifting without stopping the tractor. The synchronizer clutches operate as follows:

The range drive collars (11) are keyed to the shaft. Synchronizer clutch drums (5) are splined to the range pinions. The blocker ring (8) is centered in drive collar slots by the detent assemblies (9). Synchronizer clutch discs (6 and 7) are connected alternately (by drive tangs) to the clutch drum and blocker ring. When the engine clutch is disengaged and the control lever moved to change gear speeds, the first movement of shifter linkage moves clutch discs (6 and 7) into contact. The difference in rotational speeds of shaft and pinion causes the blocker to try to rotate on the drive collar. The drive lugs inside the blocker ring ride up the ramps in the drive collar causing the blocker to be temporarily locked in the center of the drive collar. The clutch discs are thus compressed and the speeds of pinion and shaft are equalized. When the speeds become equal, the thrust force on blocker is relieved and shifting pressure causes the drive lugs to move back down the ramps to a center position in drive collar slot. The synchronizer drum is now permitted to move toward drive collar until the splines are engaged and the range pinion securely coupled to shaft.

90C. **INSPECTION AND ASSEMBLY.** Inspect the transmission drive shaft for scoring or wear in areas of range pinion rotation and make sure oil passages are open and clean. Carefully inspect the blocker rings for damage to the drive lugs and inspect friction faces of blocker rings and synchronizer drums. Check synchronizer discs for wear, using a micrometer. The thickness of a new disc on Series 5000 is 0.123; on other models, 0.078. Renew any disc which measures less than 0.103 for Series 5000 or less than 0.060 for other models. Check drive tangs on discs for thickening due to peening. If thickness of drive tang is twice that of friction surface, renew discs.

Reassemble the transmission drive shaft by reversing the disassembly procedure. A special installing cone is required to install detent assemblies in blocker rings and to install blocker assemblies on drive collar. See Fig. JD63F.

1. Snap ring
2. Bearing cup
3. Bearing cone
4. Reverse range pinion
5. Synchronizer drum
6. Clutch discs
7. Clutch plates
8. Blocker
9. Detent assembly
10. Snap ring
11. Drive collar
12. Low range pinion
13. Snap ring
14. High range pinion
15. Input shaft
16. Bearing cone
17. Bearing cup
18. Snap ring
19. Shim
20. Bearing housing

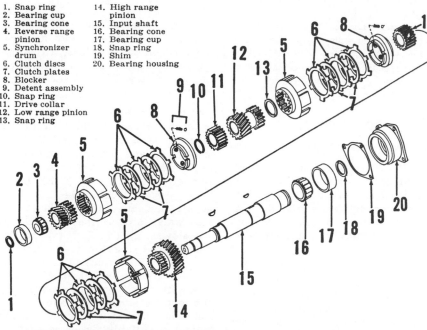

Fig. JD63E — Exploded view of transmission input shaft showing component parts. An early model 4010 is shown. Most other models use an additional snap ring located between drive collar (11) and low range pinion (12).

90D. BEVEL PINION SHAFT. Except for bearing cups in housing, and rear bearing cone on shaft, the bevel pinion shaft is disassembled during removal. Refer to Fig. JD62 or Fig. JD62A for exploded views.

The low speed gear (5) contains a renewable bushing; all other gears ride directly on the shaft. Check inside diameter of gear (or bushing), and diameter of shaft, against the specifications which follow:

Shaft OD at gears:

    Series 3000 .... 2.3615-2.3625
    Series 4000 .... 2.6121-2.6131
    Series 5000 .... 2.7996-2.8006

Gear (or bushing) ID:

    Series 3000 .... 2.3665-2.3675
    Series 4000 .... 2.6171-2.6181
    Series 5000 .... 2.8046-2.8056

The bevel pinion shaft is available only as a matched set with the bevel ring gear. Refer to paragraph 98 for information on renewal of ring gear. Refer to paragraph 90E for mesh position adjustment procedure if bevel gears and/or housing are renewed.

90E. PINION SHAFT ADJUSTMENT. The cone point (mesh position) of the main drive bevel gear

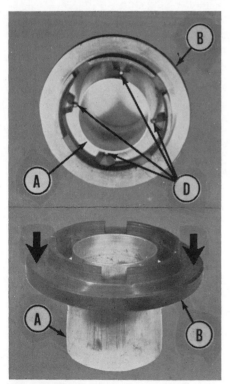

Fig. JD63F — Using the installing cone, JDT4, to install the detent assemblies in blocker. Use a heavy grease to hold the springs and balls (D) in the blocker drive lugs, then carefully start the assembly over narrow edge of tool (A). Push blocker (B) down to lower edge of tool, then use the tool to transfer the blocker to the drive collar located on the shaft.

Fig. JD63G—Cone point (mesh) adjustment of bevel ring gear and pinion is factory determined. Installation numbers (CP) are stamped on transmission housing and rear end of pinion as shown. See text for method of determining shim pack thickness.

and pinion is adjustable by means of shims (4—Fig. JD62 or JD62A) which are available in thicknesses of 0.003, 0.005 and 0.010. The cone point will only need to be checked if the transmission housing or ring gear and pinion assembly are renewed. To make the adjustment, proceed as follows:

The correct cone point of housing and pinion are factory determined and assembly numbers are etched on left upper housing flange and rear face of pinion as shown at (CP—Fig. JD63G). To determine the shim pack thickness, add the appropriate guide number to the number stamped on rear face of pinion, then subtract the sum from number etched on housing.

Guide numbers are as follows:

    Series 3010-4010 .... 1.443
    Series 3020-4020 .... 1.755
    Model 5010 .... 1.627
    Model 5020 .... 1.814

The result is the correct shim pack thickness. To add or remove shims, or to check shim pack thickness, use a punch and drive out rear bearing cup (3—Fig. JD62 or JD62A).

The bevel pinion bearings are adjusted to a pre-load 0.004-0.006 by means of shims (20—Fig. JD62 or JD62A). If adjustment is required, it should be made before installing the gears as follows:

First make sure that cone point is correctly adjusted as outlined above, then install shaft (1), cone (2), washer (19), the removed shim pack (20) plus one 0.010 shim, and bearing cone (22). On series 3000 and 4000 if a suitable 2½-inch pipe spacer is available, install it instead of the press fit inner bearing race (23). On all models, install plate (27) and the retaining cap screws. Measure the shaft end play using a dial indicator, then when disassembling, remove shims equal to the observed end play plus 0.005. Assemble the shaft and gears as outlined in paragraph 88C or 88E.

POWER SHIFT TRANSMISSION

Series 3020 and 4020 tractors are optionally available with a power shift transmission which provides 8 forward and 4 reverse speeds by moving a shift lever, without stopping the tractor or touching the foot-operated feathering valve.

OPERATION

All Power Shift Models

91. POWER TRAIN. The power shift transmission is a manually controlled, hydraulically actuated planetary transmission consisting essentially of a clutch pack and planetary pack shown schematically in Fig. JD64.

Hydraulic control units consist of three clutch packs (C1, C2 & C3) and four disc brakes (B1 through B4). In addition, a multiple disc clutch (PTO) is used in the pto train. All units are hydraulically engaged, and mechanically disengaged when hydraulic pressure to that unit is released. The power train also contains

a hand operated single disc transmission disconnect clutch (DC) mounted on the flywheel, a foot operated inching pedal, a mechanical disconnect for towing, and a park pawl.

Three hydraulic control units are engaged for each of the forward and reverse speeds. In 1st speed, Clutch 1 is engaged and power is transmitted to the front planetary unit by the smaller input sun gear (C1S); Brake 1 is engaged, locking the front ring gear to housing, and the planet carrier walks around the ring gear at it's slowest speed. Clutch 3 is also engaged, locking the rear planetary unit, and output shaft turns with the planet carrier. Second speed differs from 1st speed only by disengaging Brake 1 and engaging Brake 2, causing planet carrier and output shaft to rotate at a slightly faster speed.

Third speed and 4th speed are identical to 1st & 2nd except that Clutch

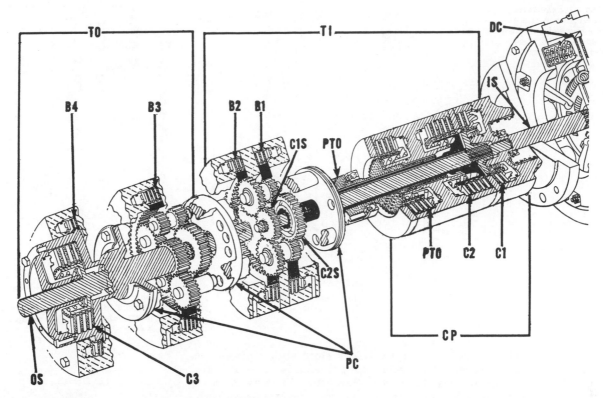

Fig. JD64 — Schematic view of Power Shift transmission showing primary function of units. Disconnect Clutch (DC) is for cold weather starting only, and is not to be used for starting or stopping tractor motion. The Power Take-Off (PTO) clutch and gear are located in, but not a part of transmission power train.

B1. Brake 1—Low Input
B2. Brake 2—High Input
B3. Brake 3—Reverse Output
B4. Brake 4—High Output
C1. Clutch 1—Low Input
C2. Clutch 2—High Input

C3. Clutch 3—Low Output
CP. Clutch Pack
C1S. C1 Sun Gear
C2S. C2 Sun Gear
DC. Disconnect Clutch (Starting Only)
IS. Input Shaft
OS. Output Shaft

PC. Planet Carrier
PTO. Power Take-Off drive units
TI. Transmission Input (Consisitng of clutch pack and front half of planetary pack)
TO. Transmission Output (Consisting of rear half of planetary pack and output shaft)

1 is disengaged and Clutch 2 engaged, and power enters the front planetary unit through the larger input sun gear (C2S).

Fifth speed and 6th speed differ from 3rd and 4th speeds in the rear planetary unit. Clutch 3 is disengaged and Brake 4 engaged, and the output shaft turns faster than the planet carrier through the action of the rear planet pinions and output sun gear.

In 7th and 8th speeds, both Clutch 1 and Clutch 2 are engaged, locking the input planetary unit, and planet carrier turns with input shaft at engine speed. Engaging the three clutch units locks both planetary units, therefore 7th speed is a direct drive, with transmission output shaft turning with, and at the same speed as, the engine. Eighth speed is an overdrive, with transmission output shaft turning faster than engine speed.

Reverse speeds are obtained by engaging Brake 3, which locks the output planetary ring gear to housing, and the output shaft turns in reverse rotation through the action of the two sets of output planetary pinions.

It will be noted that the front planetary unit is an input unit (TI)

controlled by the two front clutch units in clutch pack, and two front brake units in planetary pack. Two input control units must be engaged to transmit power, and five input speeds are obtained by selectively engaging the input brakes and clutch units.

The rear planetary unit is an output unit (TO) controlled by the two rear brakes and rear clutch. One of the rear control units must be engaged to complete the power train. Two forward ranges and one reverse output range are provided, depending on which rear control unit is engaged.

The accompanying table lists the control units actuated to complete the power flow in each shift position:

	Front (Input) Control Units		Rear (Output) Control Unit
Forward Speeds			
1st	C1	B1	C3
2nd	C1	B2	C3
3rd	C2	B1	C3
4th	C2	B2	C3
5th	C2	B1	B4
6th	C2	B2	B4
7th	C1	C2	C3
8th	C1	C2	B4

Reverse Speeds			
1st	C1	B1	B3
2nd	C1	B2	B3
3rd	C2	B1	B3
4th	C2	B2	B3

91A. CONTROL SYSTEM. The control valve unit consists of manually actuated speed selector and direction selector valves which operate through four hydraulically controlled shift valves to engage the desired clutch and brake units. The valve arrangement prevents the engagement of any two opposing control units which might cause transmission damage or lockup.

Power to operate the transmission system is supplied by an internal gear hydraulic pump mounted on the transmission input shaft, which also supplies the charging fluid for the tractor main hydraulic system. Fluid from the hydraulic pump first passes through a full flow oil filter to the main transmission oil gallery, where the pressure is regulated at 150 psi for the transmission control functions. Excess oil passes through the regulating valve to the oil cooler and main hydraulic pump.

Fig. JD64A — Schematic view of the two manual and four hydraulic valves which control the Power Shift transmission.

A. Shift valve housing
B. Dump valve
C. Shift valve
D. Speed selector valve
E. Direction selector valve
1. Shift valve 1 & pressure port
2. Shift valve 2 & pressure port
3. Shift valve 3 & pressure port
4. Shift valve 4 & pressure port
B1. Brake 1 & pressure port
B2. Brake 2 & pressure port
B3. Brake 3 & pressure port
B4. Brake 4 & pressure plate
C1. Clutch 1 & pressure port
C2. Clutch 2 & pressure port
C3. Clutch 3 & pressure port

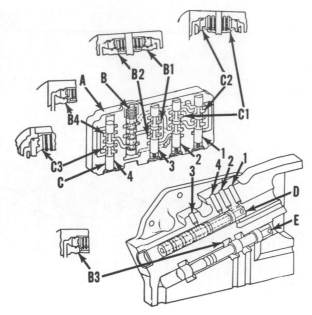

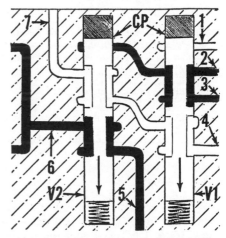

Fig. JD64B — Hydraulic pressure and flow of interlocking Shift Valves (V1 & V2) in 1st, 2nd & reverse speeds. Valve (V2) is moved downward by control pressure (CP) and valve (V1) upward by spring pressure. Pressurized fluid (dark lines) is directed to shift valve 3 through passages (5) and (6), and to Clutch 1 through passages (7) and (3). Refer also to Figs. JD64C and JD64D. Refer to Fig. JD64C for legend.

Fluid from the transmission main oil gallery is routed through the inching pedal valve to Clutches 1 and 2; and through a spring-loaded accumulator to the brake actuating pistons and to Clutch 3.

The shifting engagement rate can be adjusted by opening or closing the accumulator charging orifice to accommodate varying operating conditions and load. Closing the orifice slows the rate of pressure rise and smooths the shifting action under light load. Opening the orifice causes more abrupt shifting under no load, but reduces slippage during shift, under heavy load.

Refer to Fig. JD64A for a schematic view of control valves. The direction selector valve (E) and Shift Valve (4) controls the routing of pressure to the output control units (Clutch 3, Brake 3 and Brake 4). The speed selector valve contains four pressure ports (1, 2, 3 & 4) which control the movement of the four shift valves (C) by pressurizing the closed end (opposite the return spring) when port is open to pressure. Neutral position is provided by the selector valves or by depressing the inching pedal.

When the direction selector valve is moved to the forward detent position, system pressure is routed to Shift Valve (4). In the low range positions, the speed selector valve charging port (4) is open to pressure, Shift Valve (4) moves downward against spring pressure, and Clutch 3 is actuated. In the high range positions, charging pressure is cut off to Shift Valve (4),

shift valve return spring moves the valve upward and Brake 4 is actuated. When the direction selector valve is moved to reverse detent position, system pressure is cut off from Shift Valve (4) and mounted directly to Brake 3, without passing through a shift valve. In the neutral detent position, system pressure is cut off from all three output control units.

Shift Valves 1, 2 & 3 direct the system pressure to the input control units (Clutches 1 & 2 and Brakes 1 & 2):

Shift Valve (1) directs pressure to Clutch 2 when hydraulically actuated; and to Clutch 1 when charging port (1) is closed. Refer to Fig. JD64B and JD64C.

Shift Valve (2) routes pressure to Shift Valve (3) when hydraulically actuated, and permits the simultaneous engagement of Clutches 1 and 2 (7th & 8th speeds) when charging port (2) is closed. Refer to Figs. JD64B, JD64C and JD64D.

Shift Valve (3) directs pressure to Brake 2 when hydraulically actuated; to Brake 1 when charging port (3) is closed.

ADJUSTMENT

The multiple disc clutches and brakes require no adjustment. Linkage adjustments do not change materially because of wear, but should be checked and readjusted if necessary (if changes or malfunction occur), as part of a regular troubleshooting procedure. A change in operating conditions may require readjustment of shifting engagement rate as outlined in paragraph 92.

Fig. JD64C — Hydraulic pressure and flow of interlocking Shift Valves (V1 & V2) in 3rd, 4th, 5th and 6th speeds. Both valves are moved downward by control pressure as shown. Pressurized fluid (dark lines) is directed to shift valve 3 through passages (5) and (6), and to Clutch 2 through passages (2) and (3). Refer also to Figs. JD64B and JD64D.

V1. Shift valve 1
V2. Shift valve 2
CP. Control pressure
1. Return passage
2. Pressure port, C2
3. Inlet pressure port
4. Return passage
5. Inlet pressure port
6. Pressure port, shift valve 3
7. Pressure port, C1

92. **SHIFT RATE.** To adjust the shifting engagement rate, remove the adjustable orifice plug (Fig. JD64G) and turn the exposed slotted-head adjustment screw clockwise to slow the shift rate; or counter-clockwise for faster shifting. Turn screw ½-turn at a time, then recheck after each adjustment.

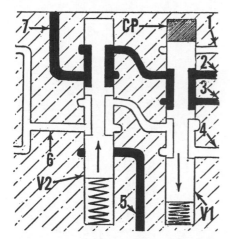

Fig. JD64D — Hydraulic pressure and flow of interlocking Shift Valves (V1 & V2) in 7th and 8th speeds. Valve (V1) is moved downward by control pressure and valve (V2) upward by spring pressure. Pressurized fluid enters through pressure port (3) and flows to both Clutch 1 and Clutch 2. Pressurized fluid from port (5) is cut off by shift valve (V2) and cannot enter passage port (6) to pressurize Brake 1 or Brake 2. Refer also to Figs. JD64B and JD64C.

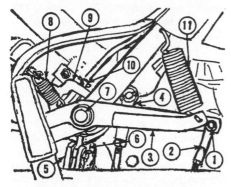

Fig. JD64E — Power shift linkage and associated parts showing points of adjustment.

1. Yoke pin
2. Park cable yoke
3. Actuating arm
4. Speed control rod
5. Direction control rod
6. Bellcrank
7. Pivot pin
8. Spring
9. Control lever
10. Pivot pin
11. Spring

92A. LINKAGE ADJUSTMENT. To adjust the control linkage, refer to Fig. JD64E and proceed as follows: Remove pin (1) securing park cable (2) to actuating arm. Move shift lever to "Park" position, pull up on park cable until park pawl is fully engaged; and adjust cable yoke until pin (1) can just be inserted through yoke and actuating arm.

Disconnect direction control rod yoke (5) from bell crank. Move shift lever to "Neutral" position. With direction control valve in center (Neutral) detent, adjust length of control rod until pin hole is aligned in bell crank and yoke.

Fig. JD64F—Pedal valve linkage and associated parts used on power shift transmission. The PTO brake and clutch pressure should be 140-160 psi with engine running at 1900 rpm. PTO brake pressure is checked with gage installed in place of plug (B) and clutch pressure is checked at opening for plug (C).

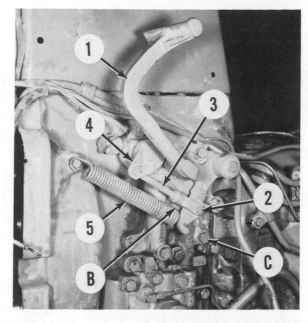

With speed control rod yoke (4) disconnected from arm on lever pivot, move shift lever to neutral position. Place speed control rod in uppermost detent position and adjust control rod yoke until connecting pin can just be inserted.

To adjust pedal valve linkage, refer to Fig. JD64F. Remove the cotter pin and disconnect operating rod (3) from valve operating arm. Fully depress pedal, turn valve operating arm (4) fully counter-clockwise and thread operating rod in or out of yoke until hooked end of rod aligns with hole in operating arm. Lengthen the rod ½-turn and reconnect.

92B. PRESSURE TEST AND ADJUSTMENT. Before checking the transmission operating pressure, first be sure that transmission oil filter is in good condition and that oil level is at top of "SAFE" mark on dip-stick. Place towing disconnect lever in "TOW" position, start engine and operate at 1900 rpm until transmission oil is at operating temperature.

Stop engine and install a 0-300 psi pressure gage in "CLUTCH" plug hole (Test Plug—Fig. JD64G). Gage should register 140-160 psi with engine operating at 1900 rpm and speed control lever in any position. If pressure is not as indicated, remove plug and add or remove shims (10—Fig. JD66B) located between plug and spring, until correct pressure is obtained.

If pressure cannot be adjusted, other possible causes are:

1. Incorrect pedal valve linkage adjustment; adjust as outlined in paragraph 92A.

2. Malfunctioning regulator valve, pedal valve or oil filter relief valve; overhaul as outlined in paragraphs 94 and 94A.

If adjustment of operating pressure does not correct the malfunction, leave pressure gage installed and completely check pressures as outlined in paragraph 92C.

Fig. JD64G — To check the transmission operating pressure, remove "Test Plug" in pedal valve housing. Engagement rate can be adjusted by removing "Adjustable Orifice Plug" in regulating valve housing and turning the exposed adjusting screw. Refer to text.

TROUBLE SHOOTING

92C. PRESSURE TEST. To make a complete check of the transmission hydraulic system pressures, first check and adjust operating pressure as outlined in paragraph 92B; then proceed as follows:

With engine speed at 1900 rpm, depress the inching pedal while noting the pressure gage reading. Gage pressure should drop to zero with pedal fully depressed.

Release the pedal slowly; gage pressure should rise at a smooth, even rate until approximately 80 psi is registered with pedal ½ to 1-inch from top; then move quickly to operating pressure with further pedal movement.

Failure to perform as outlined could indicate maladjustment of pedal valve linkage (see paragraph 92A) or malfunction of the valve (overhaul as outlined in paragraph 94).

System leakage or malfunctioning shift valves can be determined by installing 0-300 psi gage (or gages) in clutch and brake passage ports in control valve housing and transmission housing. Plug ports are marked on castings and are indicated in Fig. JD64H. Install one or more gages in passage ports and check at 1900 rpm by shifting through complete speed range for each test.

The gage should register at approximately system pressure when that control unit is actuated as shown in table accompanying paragraph 91. If the pressure registered in one or more control passage port is more than 15 psi below system pressure, leakage of that unit is indicated. If pressure is observed on any unit when that unit should not be engaged, check for sticking shift valves or leakage within control valve housing.

92D. BEHAVIOR PATTERNS. Erratic behavior patterns can be used to pinpoint some system malfunctions.

ODD SHIFT PATTERN. If tractor slows down when shifted to a faster speed; speeds up when shifted to a slower speed; or fails to shift when selector lever is moved; a sticking shift valve is indicated. Refer to paragraph 92C and table accompanying paragraph 91. Overhaul the control valve as outlined in paragraph 94B.

SLOW SHIFT. First check and adjust the shifting engagement rate as outlined in paragraph 92. Other possible causes are; improper regulating valve adjustment; improper pedal linkage adjustment; plugged fluid filter; malfunctioning regulating valve, pedal valve or oil filter relief valve; broken accumulator spring; or sticking accumulator piston.

FAST SHIFT. Shifting engagement rate improperly adjusted; see paragraph 92. Other causes could be sticking accumulator piston or high system pressure.

ROUGH PEDAL ENGAGEMENT. If tractor jumps rather than starts smoothly when pedal valve is actuated, a sticking pedal valve or broken pedal valve spring is indicated.

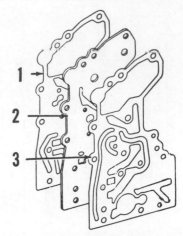

Fig. JD65—Make sure gaskets are installed on proper side of plate (2) when installing pedal and regulating valve housings. Refer to text.

1. Outer gasket
2. Gasket plate
3. Inner gasket

SLIPPAGE UNDER LOAD. If transmission slips, partially stalls or stalls under full load, first check the adjustment of pedal valve as outlined in paragraph 92A, then check transmission pressures as outlined in paragraphs 92B and 92C. If trouble is not corrected, one of the clutch or brake units; or transmission disconnect clutch, is malfunctioning. Remove and overhaul the transmission as outlined in paragraphs 93B through 94F.

TRACTOR FAILS TO MOVE. If tractor fails to move when transmission is engaged, first check to see that transmission disconnect clutch and tow disconnect are fully engaged. NOTE: If transmission disconnect clutch is disengaged or fails to hold due to malfunction, the tractor main hydraulic system will be inoperative after the supply of fluid in oil cooler and pump is exhausted.

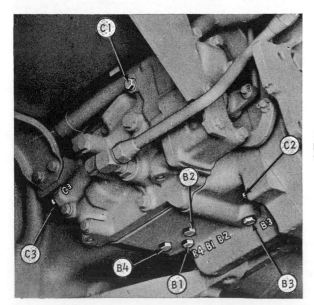

Fig. JD64H — Pressure plugs for operating brakes and clutches are marked on housings as shown.

Fig. JD65A — Left side of tractor showing approved method of disconnecting wiring and hydraulic lines for tractor split. Refer to text.

CAUTION: Do not attempt to operate tractor with disconnect clutch disengaged. Serious damage could result to main hydraulic pump.

If both disconnect units are engaged, check to see that park pawl operates properly and is correctly adjusted. Park pawl is engaged by cam action and disengaged by a return spring. If spring breaks or becomes unhooked, pawl may remain engaged even though linkage operates satisfactorily. To examine or renew the park pawl return spring, remove transmission housing cover as outlined in paragraph 95.

TRACTOR CREEPS IN NEUTRAL. A slight amount of drag is normal in the clutch and brake units, especially when transmission oil is cold. Excessive creep is usually caused by warped clutch or brake plates; observe the following:

If tractor creeps when feathering pedal is depressed and pedal properly adjusted, either Clutch 1 or Clutch 2 is malfunctioning. Check as follows: With engine speed at 1500 rpm and transmission fluid at operating temperature, shift to 2nd speed on a flat surface. Depress the inching pedal; if tractor continues to roll forward at approximately the same speed, Clutch 1 is malfunctioning, if tractor speed increases, Clutch 2 is dragging.

Disconnect speed selector control rod yoke from control arm. With selector lever in "Neutral" position, move the disconnected control rod down one detent notch. If tractor creeps forward with throttle set at 1500 rpm engine speed and transmission oil at operating temperature, Clutch 3 or Brake 4 is dragging; if tractor creeps backward, Brake 3 is malfunctioning.

NOTE: Dragging clutch or brake units, aside from causing creep, will contribute to loss of power, heat, and excessive wear. Creep is merely an indication of possible more serious trouble which needs to be corrected for best performance, or to prevent future failure.

REMOVE AND REINSTALL

92E. PEDAL AND REGULATING VALVES. Pedal valve housing and regulating valve housing attach to left side of clutch housing using common gaskets and gasket plate. Housings may be removed separately, but both should be removed, to renew the gaskets.

To remove the housings, first remove left battery and battery box. Remove pedal return spring (5—Fig. JD64F), and disconnect rod from valve operating arm (4). Remove retaining snap ring and withdraw pedal and rod assembly. Disconnect wiring from start-safety switch, and lube pipe and inlet and outlet pipes from housings. Remove control support cover and disconnect pto valve operating arm and spring, then unbolt and remove the housings, gasket plate and gaskets. Accumulator and spring can now be withdrawn from clutch housing bore for service or inspection.

Overhaul the pedal valve housing as outlined in paragraph 94, and regulating valve housing and accumulator as in paragraph 94A.

When installing, use light, clean grease to position gaskets and gasket plate, making sure gaskets are installed on proper sides of plate as shown in Fig. JD65. Make sure accumulator spring and piston are in place before installing plate and gaskets. Install regulator valve housing and retaining cap screws, then install pedal valve housing. Tighten retaining cap screws evenly and securely, and complete the assembly by reversing the disassembly procedure. Adjust as outlined in paragraphs 92, 92A and 92B.

93. CONTROL AND SHIFT VALVES. To remove the control and shift valve housing, drain transmission and remove right battery and battery box. Disconnect control valve inlet pipe at both ends and remove the pipe. Remove the cotter pins which retain control rods to control arms, remove the retaining cap screws; then remove valve housing, disconnecting the linkage as housing is removed.

Overhaul the removed unit as outlined in paragraph 94B, and install by reversing the removal procedure. Make sure linkage is connected to control arms as housing is positioned. Tighten retaining cap screws evenly and securely. Adjust as outlined in paragraph 92A.

93A. TRACTOR SPLIT. To obtain access to engine flywheel, transmission disconnect clutch and linkage; or power shift transmission main components, it is first necessary to detach (split) clutch housing from engine block. Follow the general procedure outlined in paragraph 82, except for the following:

If the regular John Deere support stand is used, transmission rear oil filter cover and element must be removed for clearance to install rear stand.

Disconnect main oil supply pipe at regulating valve housing, and wires from start-safety switch as shown in Fig. JD65A.

93B. TRANSMISSION DISCONNECT CLUTCH. The transmission disconnect clutch unit can be removed from flywheel after detaching clutch housing from engine as outlined in paragraphs 82 and 93A. Overhaul clutch as outlined in paragraph 94C.

When reinstalling, use 4000 Series (large) end of clutch pilot tool shown in Fig. JD50, or other suitable alignment tool. Tighten clutch cover retaining cap screws (C—Fig. JD65C) to a torque of 35 ft.-lbs. and bend up lock plates. Adjust the three clutch

Fig. JD65B — Right side of tractor after clutch split. Remote hydraulic lines and brackets must be removed to remove clutch housing.

Fig. JD65C—Transmission disconnect clutch installed on flywheel. Tighten cover cap screws (C) to a torque of 35 ft.-lbs. and adjust finger height (A) to 59/64-inch by turning castle nut (B).

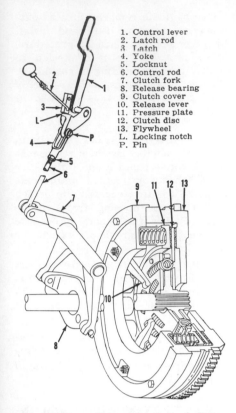

1. Control lever
2. Latch rod
3. Latch
4. Yoke
5. Locknut
6. Control rod
7. Clutch fork
8. Release bearing
9. Clutch cover
10. Release lever
11. Pressure plate
12. Clutch disc
13. Flywheel
L. Locking notch
P. Pin

Fig. JD65D — Schematic view of disconnect clutch and linkage, showing component parts.

fingers equally to a height of 59/64-inch from flat surface of clutch disc as shown at (A). Adjustment is made by removing the cotter pin and turning castellated nut (B).

93C. CLUTCH LINKAGE. The engine disconnect clutch and linkage is shown schematically in Fig. JD65D. The clutch is disconnected (for cold starting only) when control lever (1) is pulled rearward until spring-loaded latch (3) snaps into position above lock (L) on lever control arm. To re-engage the clutch, pull knob on latch rod (2) rearward and slowly allow lever (1) to move forward until fully engaged.

CAUTION: Never run tractor for more than a few seconds with engine clutch locked in disconnect position. The main hydraulic pump cannot obtain fluid unless transmission is operating, and serious damage may result. Disconnect clutch is to be used for COLD STARTING, ONLY.

Operating arm is retained to shaft of clutch lever (1) by a roll pin. Release fork pivot shaft is retained on 3020 tractors by a nut; on 4020 models by a snap ring. Pivot shaft can be withdrawn from right side of housing after removing the retaining nut

or snap ring. Release bearing carrier and bearing should be packed with high-temperature lubricant when unit is reassembled.

93D. DISCONNECT CLUTCH ADJUSTMENT. Upper end of clutch control lever (1—Fig. JD65D) should have ample movement from the time release bearing (8) contacts clutch fingers (10); until release latch (3) snaps into disconnect position. Adjust linkage as follows, with clutch housing connected to engine:

With clutch in operating position, remove pin (P) which connects control rod yoke (4) to operating arm. Move lever rearward until lockout latch snaps into position. Push down on control rod until release bearing (8) contacts clutch fingers (10); thread yoke (4) in or out on control rod until pin holes in yoke and lever arm are aligned; then lengthen rod 6 full turns of yoke. Reinsert pin (P) and tighten locknut (5), then check for complete engagement and disengagement of disconnect clutch.

93E. CLUTCH PACK AND TRANSMISSION PUMP. The transmission pump and clutch pack can be removed as a unit after detaching clutch housing from engine as outlined in paragraphs 82 and 93A; and removing disconnect clutch actuating linkage as in paragraph 93C.

Note length and location of mounting flange cap screws as they are removed. Three different lengths are used as shown in Fig. JD65F.

Connecting shafts (Fig. JD65G) may be removed with clutch pack; or may be withdrawn after clutch unit is out. Be sure that sealing ring (5) is in good condition and that shafts are in position when unit is reassembled.

Overhaul the removed clutch pack and pump as outlined in paragraphs 94D and 94E.

To assist in easier installation of clutch pack use alignment studs in the two side holes of housing and position gasket on housing using light grease. Make sure oil passages in clutch housing and in gasket are properly aligned. Insert connecting shafts with sealing ring (5) to rear. Install clutch pack with oil passages in mounting flange aligned with those of gasket and clutch housing. Install the two short retaining cap screws in upper holes (1—Fig. JD65F); long cap screw in lower hole (3). Install remainder of cap screws (2), tighten all screws evenly and securely and lock in place

by bending up corners of tab washers.

93F. CLUTCH HOUSING. The clutch housing must be removed for access to the pto drive gear train or removal of transmission planetary unit. To remove the clutch housing, first split tractor between engine and clutch housing as outlined in paragraphs 82 and 93A; and remove clutch pack as in paragraph 93E.

Remove rockshaft shields on models so equipped and remove steering support rear panel. Mark and diagram location of hydraulic tubing if necessary; then remove hydraulic tubes and system control linkage. Disconnect park cable and cable housing at rear and light wires at front connections. Remove pto front bearing quill.

Support clutch housing and steering support assembly from a hoist and remove clutch housing flange cap screws. NOTE: The two upper, center screws are accessible through inside front of clutch housing. Pry clutch housing from its doweled position on transmission case and swing housing assembly away from rear unit.

Use new gasket (1—Fig. JD65H) and "O" rings (2 & 3) when reinstalling clutch housing. "O" rings may be held in position with grease. Tighten upper center flange cap screws (from inside, front of housing) to a torque of 170 ft.-lbs. and remainder of flange screws to a torque of 300 ft.-lbs.

93G. PLANETARY PACK. To remove the transmission planetary pack, first detach (split) engine from clutch housing as outlined in paragraphs 82 and 93A; remove clutch pack as in paragraph 93E and clutch housing as in paragraph 93F. Remove rockshaft housing as in paragraph 125, and transmission top cover plate. Remove pto idler gear (4—Fig. JD65H) and drive gear and bearings (5).

Disconnect Clutch 3 pressure tube (1—Fig. JD65J) from rear bearing quill and remove quill. Using a long brass drift and working through front, center of planetary pack, jar output shaft rearward until rear bearing cup is dislodged from housing bore; then withdraw planetary output shaft rearward out of housing and output gear. Remove the four retaining cap screws (S—Fig. JD65K) and, using a hoist and ice tongs or similar tool, grasp planetary pack as indicated by arrows (Fig. JD65K) and lift the unit straight upward out of transmission housing.

Overhaul the removed planetary pack as outlined in paragraph 94F.

Before installing planetary pack, inspect or renew the four brake-passage "O" rings in bottom of transmission housing. Planetary output shaft should be installed with 0.000-0.002 bearing preload. To check bearing adjustment, make a trial installation of shaft, bearing cup and bearing quill, using one additional shim between bearing quill and housing. Tighten retaining cap screws securely and measure shaft end play using a dial indicator. Remove the shaft assembly. Deduct from shim pack, shims equal in thickness to measured end play plus 0.001. Keep remainder of shim pack together with bearing quill, for permanent installation after planetary unit is reinstalled.

Lower planetary unit straight downward, being careful not to dislodge the brake passage "O" rings. Tighten the four retaining cap screws

Fig. JD65F — Three lengths of cap screws are used to secure clutch pack mounting flange to clutch housing.

1. Short cap screws
2. Intermediate screws
3. Long cap screw

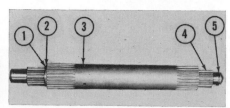

Fig. JD65G — The connecting shaft assembly splines into operating clutch hubs of clutch pack and sun gears of front planetary unit.

1. Snap ring
2. Thrust washer
3. C2 (outer) clutch shaft
4. C1 (inner) clutch shaft
5. Sealing ring

alternately and evenly, and complete the assembly by reversing the disassembly procedure.

OVERHAUL

94. **PEDAL VALVE.** Refer to Fig. JD66 for an exploded view of pedal valve housing and associated parts, and to Fig. JD66A for a partially dis-

Fig. JD65E — Cutaway view of the planetary type Power Shift transmission, showing location of component parts.

Fig. JD65H — PTO idler gear (4) and drive gear (5) must be removed to remove planetary pack. Be sure gasket (1) and "O" rings (2 & 3) are properly positioned when reinstalling clutch housing.

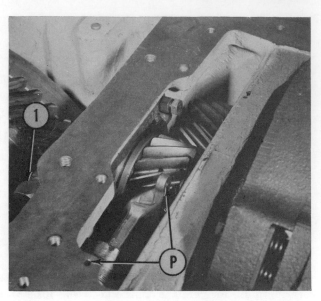

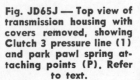

Fig. JD65J — Top view of transmission housing with covers removed, showing Clutch 3 pressure line (1) and park pawl spring attaching points (P). Refer to text.

Fig. JD65K — To remove planetary pack after cover is off, remove output shaft, pto drive gear and attaching cap screws (S). Grasp with lifting tool at points indicated and lift from transmission housing.

assembled view. Refer to paragraph 92E for removal and installation information.

Pedal shaft (7) is renewable and is a press fit in housing. Install shaft with inner edge of snap ring groove $1\frac{11}{32}$-inches from machined surface of housing. If operating shaft oil seal (9) must be renewed, install seal with lip toward inner side of housing.

Valves (25 & 29) must slide smoothly in their bores and must not be scored or excessively loose. Check the pto and pedal valve springs for distortion and against the values which follow:

PTO Valve Spring

    Free Length, Inches............$1\frac{21}{32}$
    Lbs. Test @ Inches..16.6-20.4 @ $1\frac{3}{8}$

Pedal Valve Spring (Lower)

    Free Length, Inches.............½
    Lbs. Test @ Inches.....3.3-4.1 @ $\frac{11}{32}$

Pedal Valve Spring (Center)

    Free Length, Inches............$1\frac{5}{16}$
    Lbs. Test @ Inches..15.1-18.3 @ $1\frac{3}{32}$

Pedal Valve Spring (Upper)

    Free Length, Inches.............2
    Lbs. Test @ Inches..11.4-14.3 @ $1\frac{3}{32}$

94A. REGULATING VALVE AND ACCUMULATOR. Refer to Fig. JD-66B for disassembled view of regulator valve housing and associated parts. Refer to paragraph 92E for removal and installation information.

The four valves (6, 8, 12 & 13) are interchangeable, but valve springs must be marked or tested, for proper installation.

Accumulator piston and spring can be withdrawn from clutch housing bore after regulating valve has been removed. Check the valves and piston, and their bores, for sticking or scoring. Valves and piston must move freely in bore without excessive clearance. Check the springs against the values which follow:

Oil Filter Relief Valve Spring

    Free Length, Inches............2¾
    Lbs. Test @ Inches.....27-33 @ $1\frac{3}{16}$

Oil Pressure Regulating Valve Spring

    Free Length, Inches...........4.29
    Lbs. Test @ Inches..50.4-61.6 @ 3.44

Return Pressure Relief Valve Spring

    Free Length, Inches...........$2\frac{5}{16}$
    Lbs. Test @ Inches...10.8-13.2 @ ¾

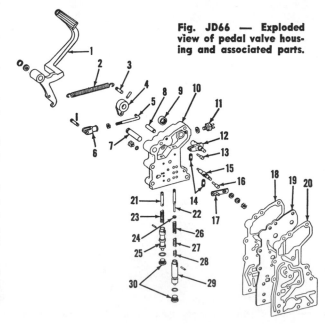

Fig. JD66 — Exploded view of pedal valve housing and associated parts.

1. Pedal
2. Return spring
3. Anchor pin
4. Operating arm
5. Operating rod
6. Yoke
7. Pivot shaft
8. Stop pin
9. Oil seal
10. Housing
11. Start safety switch
12. Operating shaft
13. Link pin
14. Link
15. PTO valve operating shaft
16. Link pin
17. PTO valve operating arm
18. Outer gasket
19. Gasket plate
20. Inner gasket
21. PTO valve rod
22. Pedal valve rod
23. Spring
24. Spring retainer
25. PTO valve
26. Upper spring
27. Center spring
28. Lower spring
29. Pedal valve
30. Plug

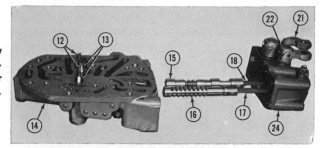

Fig. JD66C — Partially disassembled view of control valve housing. Refer to Fig. JD66E for legend.

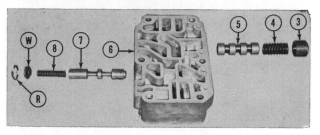

Fig. JD66D — Partially disassembled view of shift valve housing. Refer to Fig. JD66E.

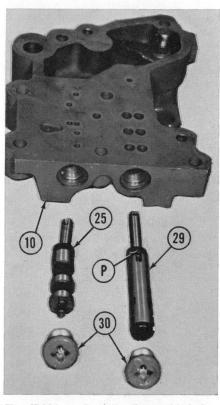

Fig. JD66A — Partially disassembled view of pedal valve housing and valves. To disassemble the pedal valve, drive out the roll pin (P). Refer to Fig. JD66 for legend.

Oil Cooler Relief Valve Spring
 Free Length, Inches..........3.80
 Lbs. Test @ Inches..33.3-40.7 @ 3.22
Accumulator Piston Return Spring
 Free Length, Inches...........3.96
 Lbs. Test @ Inches....48-57 @ 3.66
                        186-226 @ 2.78

94B. CONTROL VALVE. The shift valve housing is attached to inner face of control valve housing by a cover and six cap screws. Refer to Fig. JD-66C for a partially disassembled view of control valve housing and associated parts; and Fig. JD66D for shift valve housing. Refer to paragraph 93 for data on removal and installation of unit.

To disassemble the removed unit, proceed as follows: Remove the six cap screws retaining shift valve housing (6—Fig. JD66E) and lift off the cover, housing, gasket plate (10) and gaskets (9 & 11). Note that the two gaskets are not interchangeable. Mark the removed gaskets "Outer" and "Inner" as they are removed to aid in correct installation of new gaskets when unit is reassembled. Lift out inner detent springs (12) and plungers (13) to prevent loss. Invert control valve housing (14) and remove outer detent retaining plugs, springs and plungers. Remove the six cap screws retaining cover (24) to housing and withdraw cover, control valves and operating linkage as a unit from control valve housing. Remove plug (3) from shift valve housing and withdraw spring (4) and dump valve spool (5). Remove the four snap rings (R—Fig. JD66D) and washers (W), then withdraw shift valve springs (8) and spools (7). The four shift valve spools and springs are interchangeable.

Clean all parts in a suitable solvent and check for scoring or other damage, and for free movement of valve spools in bores. Control valve actuating mechanism need not be disassembled unless renewal of parts is indicated.

Check the valve springs for damage or distortion and against the values which follow:
Dump Valve Spring
 Free Length, Inches..........1.23
 Lbs. Test @ Inches..16.2-19.8 @ 0.73

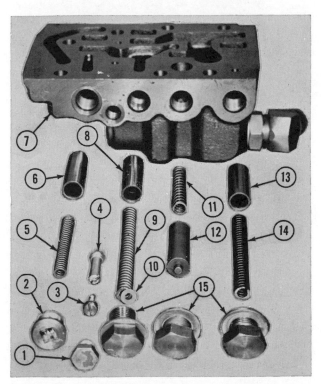

Fig. JD66B — Disassembled view of regulating valve housing and associated parts.

1. Plug
2. Plug
3. Shift rate adj. screw
4. Metering orifice
5. Spring
6. Valve
7. Housing
8. Valve
9. Spring
10. Adj. shims
11. Spring
12. Valve
13. Valve
14. Spring
15. Plugs

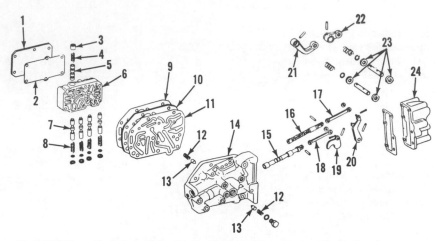

Fig. JD66E — Exploded view of control and shift valves and associated parts.

1. Cover	9. Gasket	17. Link
2. Gasket	10. Plate	18. Link
3. Plug	11. Gasket	19. Operating arm
4. Spring	12. Spring	20. Operating arm
5. Dump valve	13. Detent plunger	21. Arm
6. Shift valve	14. Control valve	22. Arm
housing	housing	23. Oil seals
7. Shift valve	15. Direction valve	24. Cover
8. Spring	16. Speed control	
	valve	

Clutch cover and pressure plate assembly must be disassembled in a suitable press; or by using a spare flywheel and three jack screws, nuts and flat washers. To disassemble clutch cover using a flywheel, install the three jack screws in alternately spaced cover mounting holes (A—Fig. JD66F). Tighten the nuts against clutch cover, then remove the release finger cotter pins and adjusting nuts (B). Back off nuts on jack screws until spring pressure is relieved; remove the jack screws, and lift out the clutch cover, spring cups, springs and pressure plate.

Check the clutch cover for scoring of friction surface and excessive wear in drive pin notches. Runout must not exceed 0.006. Check clutch springs for rust, pitting or distortion, and against the test values which follow:

Outer Spring (All Models)
Free Length, Inches............$3\frac{1}{16}$
Lbs. Test @ Inches...105-129 @ $1\frac{3}{4}$

Inner Spring (4020 Only)
Free Length, Inches............$3\frac{5}{16}$
Lbs. Test @ Inches..50.5-61.5 @ $1\frac{3}{4}$

94D. CLUTCHES. To disassemble the removed clutch pack, use a holding fixture with a 2-inch hole or drill a 2-inch hole near edge of table or

Shift Valve Springs

Free Length, Inches...........1.44
Lbs. Test @ Inches...6.8-8.4 @ 0.81

Reassemble by reversing the disassembly procedure, using Fig. JD66E as a guide. Tighten the six cap screws retaining shift valve housing to control valve evenly and alternately to a torque of 20 ft.-bs.

94C. DISCONNECT CLUTCH. Check the removed disconnect-clutch disc for loose rivets, broken springs or cracks. Facings should be smooth and free of grease or oil. Thickness of a new clutch disc is 0.422-0.447.

Friction surface of flywheel must not be heat-checked or scored, with

not more than 0.006 runout. Flywheel may be remachined provided not more than 0.060 is removed from friction surface, nor surface lowered to more than 2.150 below surface of cover mounting flange.

Fig. JD66F — Disconnect clutch can be disassembled by installing jack screws in screw holes (A) and removing adjusting nuts (B). Refer to text.

Fig. JD66G — Assembled view of removed clutch pack showing component parts.

S. Cap screws
T. Through bolts
1. PTO pressure plate
2. PTO—C2 clutch drum
3. C1—C2 pressure plate
4. C1 clutch drum
5. Manifold plate
6. Pump housing
7. Input shaft

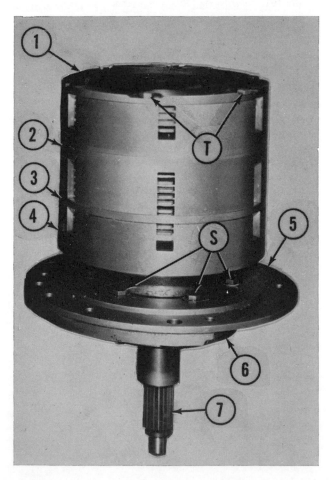

bench. Insert input shaft (7—Fig. JD-66G) and release bearing sleeve through hole, with PTO clutch pressure plate (1) up. Bend down the locking tab washers and remove through-bolts (T); then lift off pto clutch pressure plate (1), the pto clutch discs and clutch hub. PTO and C2 clutch drum (2), C1 and C2 pressure plate (3) and C1 clutch drum (4) are secured by two dowels but can be separated after jarring slightly with the heel of the hand.

C1 clutch drum can be lifted off input shaft and manifold assembly after unseating and removing the snap ring on rear splines of input shaft.

Clutch plates and discs are interchangeable for all clutches, including Clutch 3 in the planetary pack. The clutch hubs, however, will not inter-

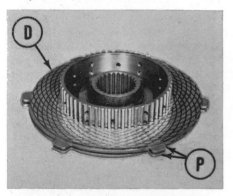

Fig. JD66H — Clutch discs (D) and plates (P) are interchangeable for all clutches and may be installed either side up.

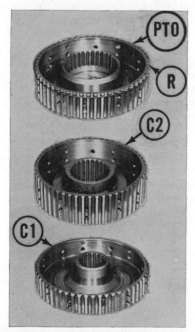

Fig. JD66J — Clutch hubs are not interchangeable. Note snap ring (R) on PTO clutch hub. PTO and C1 hubs must be inverted when installed.

change. Refer to Fig. JD66J. The internally splined clutch disc (D—Fig. JD66H) should measure 0.112-0.118 when new; the clutch plate (external lugs), 0.115-0.125. Check for warped, worn or scored clutch plates and discs, and examine lugs and splines for wear or other damage. Plates and discs should be renewed as a set for any one clutch.

Clutch C1 uses two plates and two discs; C2, five plates and five discs; the pto clutch, four plates and four discs; and C3 (in planetary pack) four plates and four discs in series 4020; or 3 plates and 3 discs in series 3020. Plates and discs for all clutches are installed alternately, either side up, beginning with an externally lugged clutch plate next to the piston and ending with an internally splined disc next to pressure plate. Examine pressure plates for scoring or wear and renew as indicated.

Clutch 1 piston uses two sets of spring (Belleville) washers (2—Fig. JD66K) retained to the clutch drum hub by a snap ring (1). The other clutches (including Clutch 3 in planetary pack) use a coil spring, retainer and snap ring. Clutch piston must be disassembled in a press, using a suitable straddle-mounted fixture and compressing spring until the retaining snap ring can be unseated.

NOTE: A fixture can be constructed using about 6 inches of 3½-inch steel pipe. Make sure both ends of pipe are square with centerline. Machine a 1x1½-inch notch in one end of pipe for working space to unseat the snap ring.

Piston can be worked out of clutch drum by grasping the strengthening ribs with pliers after spring and retainer have been removed. Refer to Fig. JD66L. Piston (1) is sealed with an expanding, cast iron ring (2) at outer edge, and a neoprene ring (3) on cylinder hub. Dowel (4) enters a hole in piston to prevent piston rotation. Renew sealing rings (2 and 3) whenever piston is disassembled, and check friction surfaces of piston and cylinder for scoring or other damage. The coil-type return spring on the pto clutch and Clutches 2 and 3 should have a free length of $2\frac{21}{32}$-inches and test 250-300 lbs. when compressed to a height of 1⅛-inches. Reassemble by reversing the disassembly procedure, making sure dowel (4) enters locating hole in piston.

Examine the oil manifold sleeve in C1 clutch drum. Renew drum assembly if sleeve is damaged. Bushing in PTO and C2 clutch drum is renewable. Inside diameter of a new bushing should be 2.0025-2.0035 after installation.

Fig. JD66K — Clutch 1 drum with piston and springs installed. Refer to text for disassembly procedure.

1. Snap ring
2. Belleville washers
3. Piston
4. Clutch drum

Fig. JD66L — C3 clutch piston housing partially disassembled. Return spring, spring retainer and snap ring are not shown.

1. Piston
2. Sealing ring
3. Sealing ring
4. Guide dowel

Disassemble, inspect and overhaul transmission pump, manifold plate and input shaft assembly before reassembling clutch pack, if service is indicated. Follow the procedures outlined in paragraph 94E.

Reassemble clutch pack by reversing the disassembly procedure, making sure clutch hubs are properly positioned and that clutch plates and discs are installed alternately, starting with an externally lugged clutch plate

next to piston and ending with an internally splined clutch disc next to pressure plate. Tighten clutch pack through bolts evenly to a torque of 28 ft.-lbs. and secure by bending lock plates.

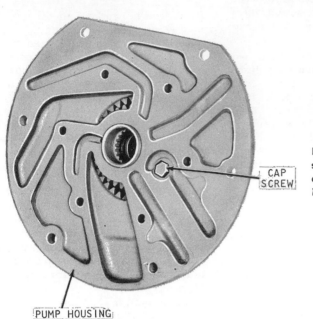

Fig. JD66M — To disassemble transmission pump after manifold and shaft is out, first remove body cap screw.

94E. TRANSMISSION PUMP, MANIFOLD PLATE AND INPUT SHAFT. To overhaul the transmission pump, manifold plate or input shaft, first disassemble clutch pack as outlined in paragraph 94D. After C1 clutch drum has been removed, bend down locking tabs on cap screws (S—Fig. JD66G) and remove the screws; then slide input shaft and manifold plate assembly out of pump housing. Remove and save the two steel check balls (Fig. JD66P) as pump and manifold units are separated. Remove the one cap screw (Fig. JD66M) and separate pump housing from body and gear assembly.

Pump gears (5 & 6—Fig. JD66N) can be lifted out of pump body (9) after housing (4) is removed. Oil seal (7) should be installed in pump body with lip of seal to rear.

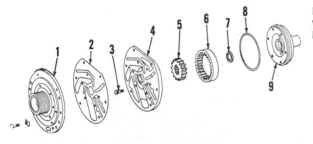

Fig. JD66N — Exploded view of transmission pump, manifold and associated parts.

1. Manifold
2. Gasket
3. Cap screw
4. Pump housing
5. Drive gear
6. Internal gear
7. Oil seal
8. "O" ring
9. Pump body

Input shaft and bearing assembly can be removed from manifold plate after unseating and removing the retaining snap ring from rear of manifold hub. Press bearing from shaft, if renewal is required, after removing shaft sealing rings and bearing retaining snap ring.

Renew gasket (2), "O" ring (8) and cast iron sealing rings on manifold plate (1) and input shaft, whenever unit is disassembled. Examine gears and housings for scoring, wear, cracks or other damage and renew as required. Assemble by reversing the disassembly procedure. Tighten cap screw (3) to a torque of 20 ft.-lbs. Place manifold (1) hub down on bench, position gasket (2) and drop the two steel check balls into manifold as shown in Fig. JD66P, before positioning pump assembly on manifold. Tighten the six cap screws retaining pump to manifold to a torque of 20 ft.-lbs. and secure by bending lock plates. Reassemble clutch pack as outlined in paragraph 94D.

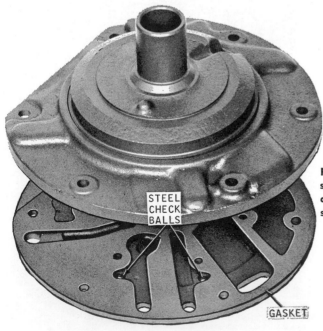

Fig. JD66P — When installing assembled pump on manifold, make sure steel check balls are in place as shown.

94F. PLANETARY PACK. To disassemble the removed planetary pack, place unit on a bench, output end up as shown in Fig. JD67. Remove the six cap screws (C) and lift off Clutch 3 piston housing (1). Remove the five through-bolts (T) and disassemble brake piston housings and associated parts until planet carrier assembly can be removed as shown in Fig. JD67A.

Each planet pinion shaft is retained to its end plate by a steel ball as shown in Fig. JD67B. The shaft is a slip fit in end plate but prevented from movement by the locking action of the steel ball and by retainer (1—Fig. JD67A). Each planet pinion contains two rows of loose needle rollers separated by a spacer. End plates are doweled to planet carrier (3). To prevent loss of any of the small parts during disassembly of the unit, proceed as follows:

Remove the cap screws retaining end plate to input (front) end of planet carrier. Lift off the pressed steel shaft retainer and remove B1 ring gear. Reinstall shaft retainer but not the cap screws. Invert the unit, making sure it is supported so that shaft retainer is held in contact with end plate; then, tap the planet carrier and output planetary assembly loose from its doweled position on input end plate. Disassemble output planetary unit using the same general procedure, making sure shafts remain in end plate and pinions remain on shafts as shown in Fig. JD67C and JD67D.

Lift off the planet pinions one at a time, keeping bearings in sets as they are removed. Catch and save the steel locking balls as shafts are removed from end plates. All planet pinion bearing rollers are interchangeable, but should be kept (and/or renewed) in sets. Front and rear bearing spacers are of different thickness, the longer spacers being used in rear (output) planetary unit.

Wash all parts in a suitable solvent. Examine gears for worn, chipped or broken teeth and inside bearing surface of planet pinions for scoring or wear. Examine pinion shafts for nicks, wear or scoring. Fiber thrust washers are used on input end plate, both sides of planet carrier, and output (rear) side of B3 piston housing. Renewable bushings are located in input planetary end plate, Brake 3 piston housing and Clutch 3 piston housing. Inspect bushings for wear or scoring and renew as required. Installed diameters of new bushings are as follows:

Input planetary end plate . 2.130 -2.132
Brake 3 piston housing . . . 3.7555-3.757
Clutch 3 piston housing . . . 3.125 -3.126

Both the internally splined brake discs and externally lugged brake plates measure 0.117-0.123 when new. The brake plate which is installed

Fig. JD67 — Assembled view of removed planetary pack.

C. Cap screw
P. Pressure ports
T. Through bolts
1. C3 piston housing
2. B4 piston housing
3. B3 piston housing
4. B2 piston housing
5. B1—B2 pressure plate
6. B1 piston housing
7. Piston return springs

Fig. JD67A — Planet carrier and gears, assembled view.

C. Cap screw
IP. Input unit
OP. Output unit
1. Retainer
2. End plate
3. Planet carrier
4. Ring gear

Fig. JD67B—Planet pinion shafts (S) are retained in end plates by steel balls (B). Refer to text.

next to piston has drilled holes in the four extended lugs to serve as pressure points for brake return springs. Brakes B1 and B2 are each equipped with two plates and two discs; brake B3 has four plates and four discs; and brake B4 has two plates and two discs on 4020 models, or one plate and one disc on 3020 tractors. Check for warped, worn or scored brake discs and plates, and for damage to lugs or splines.

Brake pistons can be removed from piston housings after unit is separated. Refer to Fig. JD67E. Using two pairs of pliers, grasp piston (2) by two opposing strengthening ribs and work piston from cylinder. Examine piston and cylinder for scoring or other damage, and renew the neoprene sealing rings (3 & 4) whenever piston is removed. Overhaul C3 piston housing and check clutch plates as outlined in paragraph 94D.

Assemble planet carrier by reversing the disassembly procedure, being sure to observe the following: Make sure fiber thrust washer is in place on front end plate (E—Fig. JD67F) before positioning Clutch 2 sun gear (2S). Install the double pinions with

Fig. JD67E—B1 piston housing with piston removed, showing component parts and details of assembly. Pistons and rings are interchangeable, but housings are not.

1. Housing
2. Piston
3. Piston ring
4. Sealing ring

Fig. JD67C — Output (rear) planetary unit removed from planet carrier. Retainer (R) must be left with unit to secure pinion shafts. Refer to text.

Fig. JD67D — Input (front) planetary unit removed from planet carrier. Retainer (R) secures pinion shafts.

Fig. JD67F — When assembling input (front) planetary unit, timing "V" marks (T) on pinions must all point to center as shown.

B. Bearing rollers
E. End plate
T. Timing marks
1S. C1 sun gear
2S. C2 sun gear

Fig. JD67G — Use four assembly studs (S) to assist in alignment, when assembling planetary unit.

B. Brake discs
D. Dowels
E. End brake plate
P. Brake plates
R. Return springs
S. Alignment studs
4. B2 piston housing
5. B1—B2 pressure plate
6. B1 Piston housing

all "V" timing marks (T) pointing to center as shown. Position a fiber thrust washer on each side of center web of planet carrier using grease, as unit is assembled. Position B1 ring gear over input planet pinions before permanently installing front pinion shaft retainer. Tighten retaining cap screws securely and fasten by bending retainer tabs, but be sure not to distort shaft retainers.

To assemble the planetary pack, place B1 piston housing (6—Fig. JD-67G) closed end down on a bench. Install four assembly studs (S) in threaded holes in return spring grooves, and alignment dowels (D) in remaining holes. Install end brake plate (E) over dowels and alignment studs, then place eight brake return springs (R) in position on plate. NOTE: Brake return springs are of three different lengths in Series 3020, and two lengths in 4020. Use intermediate length springs on 3020; or shorter springs on 4020. Place the assembled planet carrier unit in position, input (front) end down, then alternately install two brake discs (B) and one intermediate brake plate (P).

Install pressure plate (5) and B2 ring gear; then starting with a disc, alternately install two brake discs (B) and one intermediate plate (P). Install eight more brake return springs of the same length used for brake B1, install piston-end brake plate (E) and the assembled B2 piston housing, piston side down.

Install the eight long brake return springs (R) and B3 ring gear; then, beginning with a disc, alternately install four brake discs (B) and three intermediate plates (P). Install the third piston-end plate (E) and the assembled B3 piston end housing, piston side down. Install B4 sun gear, large end down, and position fiber thrust washer on hub of B3 piston housing; then install C3 clutch hub, the remaining brake disc (s) (and intermediate plate, Series 4020), the remaining eight brake return springs, and piston-end plate. Install B4 piston housing, remove the aligning studs (S) and install the five through-bolts (T—Fig. JD67). Tighten through-bolts to a torque of 35 ft.-lbs. and secure by bending tangs of lock plates.

NOTE: When properly assembled, brake passage ports (P) must be aligned as shown in Fig. JD67.

Install C3 clutch hub and the C3 clutch discs and plates alternately, beginning with an internally splined clutch disc. Install C3 piston housing (1) and retaining cap screws (C). Tighten cap screws securely and fasten by bending tangs of locking plates.

Before reinstalling planetary pack, apply 50-80 psi air pressure to each of the ports (P—Fig. JD67) in turn. Listen for air leaks and note action of brake plates. If leaks are noted or if brake return springs do not compress, recheck assembly procedure and correct the trouble before reassembling tractor.

REDUCTION GEARS, TOW DISCONNECT AND PARK PAWL

95. **TRANSMISSION TOP COVER AND PARK PAWL.** If park pawl remains engaged even though linkage is properly adjusted and operates satisfactorily, the park pawl return spring may be unhooked or broken. The unit can be inspected and spring renewed after removing transmission top cover. Proceed as follows:

Remove operator's platform, steering support rear panel and interfering hydraulic tubes and linkage. Unbolt and remove transmission top cover. Remove damaged or broken spring and install new spring by hooking spring between points (P & S—Fig. JD67J). Recheck park pawl adjustment (paragraph 92A) after new spring is installed, then reassemble by reversing the disassembly procedure.

95A. Park pawl and actuating cam can only be removed after removing planetary pack as outlined in paragraph 93G. Refer to Fig. JD67H. Remove tow-disconnect fork and coupling and output reduction gear as outlined in paragraph 95B, remove park pawl shaft retaining snap ring and slide shaft out of housing while park pawl is removed from above. Remove park pawl arm (Fig. JD67K) and Woodruff key, then remove cam assembly from inside of housing. Examine thrust bearing and washers on cam shaft and renew shaft "O" ring seal. Check the parts for wear or damage, making sure cam roller turns freely on shaft. Renew questionable parts and reassemble by reversing disassembly procedure. When installing park pawl arm (Fig. JD67K), slide arm on shaft until 0.005-0.010 end play exists, then tighten clamping screw securely.

95B. **TOW DISCONNECT.** To remove the tow-disconnect mechanism, first remove planetary output shaft as

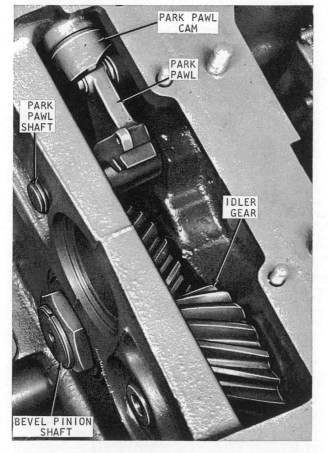

Fig. JD67H — Top view of transmission housing with planetary pack removed, showing reduction gears and park pawl.

outlined in paragraph 93G. Unscrew retaining nut behind disconnect lever (Fig. JD67L) and retaining nut on disconnect fork shaft (Fig. JD67K) and remove disconnect fork, collar (C—Fig. JD67J), reduction gear (G) and output shaft front bearing cone. Front bearing cup is retained by a snap ring and can be drifted rearward out of housing after gear has been removed.

Renew sealing washers on fork shafts and install with lever in vertical position with disconnect collar engaged. Adjust as follows, after output shaft is installed and bearings adjusted as outlined in paragraph 93G.

Block or hold reduction gear (G—Fig. JD67J) in rearmost position on shaft; then, while holding back on disconnect lever, loosen lockout (N) and turn adjusting screw (A) if necessary, until clearance (B) between rear face of gear and collar (C) is 0.005-0.010. Tighten lockout (N) and recheck the adjustment.

95C. IDLER GEAR AND SHAFT. Idler gear (Fig. JD67H) can be removed after removing planetary unit as outlined in paragraph 93G and tow disconnect and reduction gear as in paragraph 95B.

Remove cotter pin, nut and washer from rear of idler gear shaft and snap ring from front of front bearing, then drift or pull shaft forward out of housing and gear. Front end of shaft contains a ⅝ - 11 tapped hole to facilitate removal. Remove gear (7—Fig. JD67N) and spacer (6) through top opening as shaft is removed. Rear bearing can be drifted from housing if renewal is indicated. Examine bearings for wear or roughness and shaft, gear and nut for wear or other damage, and renew as indicated. Assemble by reversing the disassembly procedure, tighten shaft nut securely and install cotter pin.

95D. BEVEL PINION SHAFT. Bevel pinion shaft and bearings can be removed for service after removing planetary unit as outlined in paragraph 93G, and differential assembly as in paragraph 96. Remove idler shaft and gear by following procedures outlined in paragraph 95C.

Remove the self-locking nut (11—Fig. JD67P) from front end of bevel pinion shaft and drift the shaft rearward until front bearing cone (10) and shims (8) can be removed. Withdraw shaft and rear bearing cone from rear while lifting gear (6) and spacers (5 & 7) out through top of housing.

The bevel pinion shaft is available only as a matched set with bevel ring gear. Refer to paragraph 98 for information on renewal of ring gear. If rear bearing cup (3) must be removed, keep cone point adjusting shim pack (4) intact and reinstall the same shims or shims of equal thickness, unless gears and/or housing are renewed. If either the gears or housing are renewed, check and adjust cone point as outlined in paragraph 95E.

When reinstalling bevel pinion shaft and bearings, make a trial installation using the removed shim pack (8) plus one additional 0.010 shim. Tighten shaft nut, then measure shaft end play using a dial indicator. Note the measurement, remove nut and front bearing cone; then remove shims (8) equal in thickness to the measured end play plus 0.005, to obtain the recommended 0.004-0.006 preload of pinion shaft bearings.

NOTE: If main drive bevel gears and/or transmission housing have been renewed, cone point (mesh position) of gears must be checked and adjusted BEFORE adjusting bearing preload.

95E. CONE POINT ADJUSTMENT. The cone point (mesh position) of the main drive bevel gear and pinion is adjusted by means of shims (4—Fig. JD67P) which are available in thicknesses of 0.003, 0.005 and 0.010. The cone point will only need to be checked if the transmission housing or ring gear and pinion assembly are renewed. To make the adjustment, proceed as follows:

The correct cone point of housing and pinion are factory determined

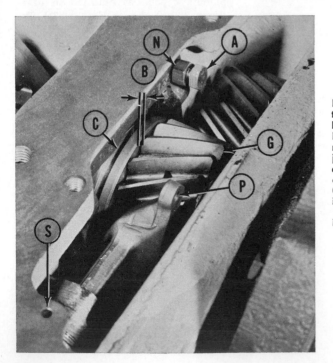

Fig. JD67J — To adjust the tow disconnect, first loosen locknut (N). Lightly pry output gear (G) rearward and turn adjusting screw (A) until clearance (B) between collar (C) and gear is 0.005 - 0.010 when lever is in engaged position. (P) and (S) are attaching points for park pawl spring.

Fig. JD67K — Right side of transmission housing showing park pawl arm and disconnect fork shaft.

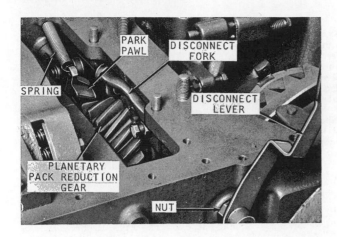

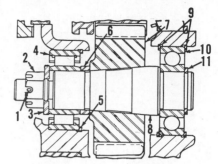

Fig. JD67L—Transmission housing with covers removed, showing reduction gears and associated parts.

Fig. JD67N — Cross sectional view of idler shaft and associated parts.

1. Cotter pin	7. Idler gear
2. Adjusting nut	8. Idler shaft
3. Washer	9. Snap ring
4. Bearing	10. Front bearing
5. Snap ring	11. Snap ring
6. Spacer	

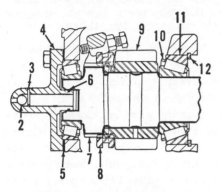

Fig. JD67M — Cross sectional view of output shaft and associated parts showing Clutch 3 pressure passage seals and shaft bearings.

2. Check ball	8. Disconnect coupling
3. Roll pin	9. Output gear
4. Bearing quill	10. Bearing cone
5. Shim pack	11. Bearing cup
6. Sealing ring	12. Snap ring
7. Output shaft	

and assembly numbers are etched on left upper housing flange and rear face of pinion as shown in Fig. JD-63G. To determine shim pack thickness, add appropriate guide number to the number stamped on rear face of pinion, then subtract the sum from number etched on housing.

Recommended guide number is 1.442 for model 3020 and 1.775 for model 4020.

The result is the correct shim pack thickness. Remove and measure the combined thickness of shim pack (4 —Fig. JD67P), then add or remove shims as required. Pinion shaft bearings must be readjusted to 0.004-0.006 preload AFTER cone point has been adjusted.

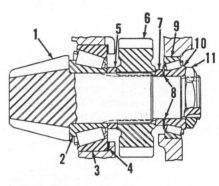

Fig. JD67P—Cross sectional view of bevel pinion shaft and associated parts.

1. Pinion	7. Spacer
2. Bearing cone	8. Shim pack
3. Bearing cup	9. Bearing cup
4. Shim pack	10. Bearing cone
5. Spacer	11. Shaft nut
6. Gear	

DIFFERENTIAL AND MAIN DRIVE BEVEL RING GEAR

REMOVE AND REINSTALL

All Models

96. To remove the differential assembly, first drain transmission and hydraulic fluid.

On models with 3-point hitch, remove rockshaft housing as outlined in paragraph 125; knock out the roll pin retaining the extension to top of hydraulic load control arm and remove extension and cam follower. Back out the set screws (A—Fig. JD-68) retaining caps (B) to each side of drawbar frame and remove the caps. Using a brass drift and working from right side of tractor, bump the load control shaft out far enough to clear yoke at lower end of load control arm. Pivot the arm forward as far as possible and; when removing

differential unit, rotate unit to provide maximum clearance between flat side of differential and load control arm.

On models without 3-point hitch, remove seat and differential case top cover.

On all models, block up tractor and remove both final drive units as outlined in paragraph 102 or 104. Remove brake backing plates and brake discs and withdraw both differential output shafts.

On Series 3000 and 4000 tractors with "Syncro-Range" transmission, place transmission shift control lever in "TOW" position, turn differential assembly until one of the flats in differential housing is uppermost, then remove transmission oil pump with lines attached.

Fig. JD68—To remove the drawbar frame, differential assembly or load control shaft seals or bushings it is first necessary to remove load control shaft. Remove set screws (A) and retaining caps (B), then drive out load control shaft from RIGHT side using a brass drift.

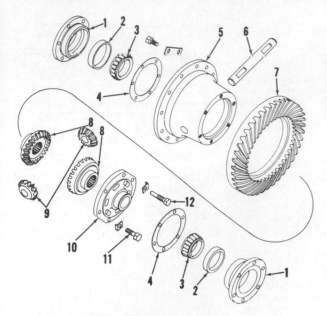

Fig. JD68A — Exploded view of bevel ring gear and differential assembly used on models not equipped with differential lock. Ring gear (7) is available only in a matched set with bevel pinion. Shaft (6) is retained in housing by the special cap screw (12) when unit is assembled.

1. Quill
2. Bearing cup
3. Bearing cone
4. Shim
5. Housing
6. Pinion shaft
7. Bevel gear
8. Axle gear
9. Differential pinion
10. Cover
11. Cap screw
12. Special cap screw

Series 3000-4000

97A. To overhaul the removed differential assembly, index housing and cover, remove retaining cap screws and lift off cover. Refer to Fig. JD-68A for typical differential assembly used on models without differential lock; or Fig. JD68B for models with differential lock. Refer to paragraph 101 for disassembly of differential lock components on models so equipped.

On models not equipped with differential lock, one cover retaining cap screw (12—Fig. JD68A) is extended to lock differential pinion shaft to housing. Models with differential lock use a separate dog-point set screw (22—Fig. JD68B).

Thrust washers are not used on differential pinions or axle gears. Both differential pinions and the pinion shaft should be renewed as a set, if any of the three are damaged. Renew axle side gears or differential housing if worn or scored in thrust areas. Overhaul differential locking mechanism as outlined in paragraph 101 on models so equipped. Refer to paragraph 98 for installation of main drive bevel gear if renewal is indicated.

When reassembling models without differential lock, make sure the special cap screw is installed in the proper hole to secure differential pinion shaft, tighten all cap screws to a torque of 35 ft.-lbs. and bend up lock plates.

Place a chain around differential housing as close to bevel gear as possible, attach a hoist and lift the differential enough to relieve the weight on carrier bearings. Remove both bearing quills using care not to lose, damage or mix the shims located under bearing quill flanges. Differential assembly may now be removed.

Overhaul the removed differential as outlined in paragraph 97 or 97A.

When installing, place an additional 0.010 shim on left bearing quill, tighten the retaining cap screws and measure differential end play using a dial indicator. Preload the carrier bearings by removing shims equal in thickness to the measured end play plus 0.002-0.005. Shims are available in thicknesses of 0.003, 0.005 and 0.010.

After the correct carrier bearing preload is obtained, attach a dial indicator, zero dial indicator button on one bevel ring gear tooth and check the backlash between bevel ring gear and pinion. Proper backlash is 0.011-0.015. Moving one 0.005 shim from one bearing quill to the other will change backlash by about 0.010.

DIFFERENTIAL OVERHAUL

Series 5000

97. To overhaul the removed differential assembly, index the differential case halves, bend down the locking tabs, remove the retaining cap screws and separate the two halves. Differential gears, pinions, thrust washers and spider can now be removed.

Examine the parts for wear, scoring or other damage. If spider or any of the differential pinions are unserviceable, all five parts should be renewed. New differential pinion thrust washers are 0.030-0.032 in thickness, renew if excessively worn or scored. Renew differential case or axle gears if worn, scored or otherwise damaged.

When reassembling, make sure index marks on case halves are aligned, tighten retaining cap screws to a torque of 85 ft.-lbs. and bend up lock plates.

Refer to paragraph 98 for installation of main drive bevel gear, if renewal is required.

NOTE: Some 5000 tractors are equipped with a hydraulically actuated differential lock. Refer to paragraphs 99 through 101 for information on differential lock.

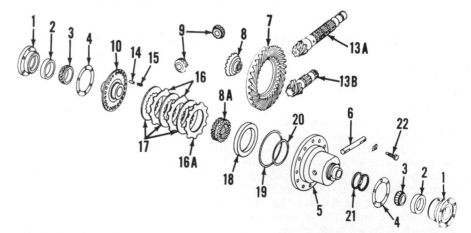

Fig. JD68B — Exploded view of main drive bevel gears, differential lock and differential assembly used on 3020 models so equipped. Other models with differential lock are similar in major details. Refer to Fig. JD68A for legend except for the following.

8A. Splined axle gear	14. Dowel	18. Piston
13A. Bevel pinion (Syncro-Range)	15. Return spring	19. Piston ring
	16. Clutch plates	20. Piston ring
13B. Bevel pinion (Power Shift)	16A. End clutch plate	21. Sealing rings
	17. Clutch discs	22. Lock screw

BEVEL RING GEAR
All Models

98. The main drive bevel ring gear and pinion are available as a matched set only. To renew the ring gear and pinion, first remove the differential assembly as outlined in paragraph 96.

Remove the cap screws retaining main drive bevel gear to differential housing and remove gear using a drift and heavy hammer. The main drive bevel gear is a press fit on housing. Heat ring gear evenly to a temperature of approximately 300° F. and position the gear. Install the retaining cap screws and lock plates. Tighten cap screws to a torque of 170 ft.-lbs. on Model 5010; or 85 ft.-lbs. on other models. Renew pinion shaft as outlined in paragraph 90D for models with "Syncro-Range" transmission, or 95D for models with power shift.

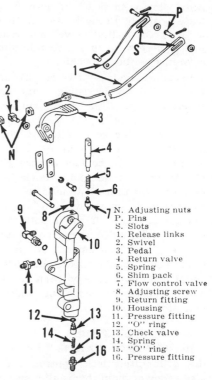

N. Adjusting nuts
P. Pins
S. Slots
1. Release links
2. Swivel
3. Pedal
4. Return valve
5. Spring
6. Shim pack
7. Flow control valve
8. Adjusting screw
9. Return fitting
10. Housing
11. Pressure fitting
12. "O" ring
13. Check valve
14. Spring
15. "O" ring
16. Pressure fitting

Fig. JD68E — Exploded view of differential lock control valve and linkage.

DIFFERENTIAL LOCK

Tractors may be optionally equipped with a hydraulically actuated differential lock which may be engaged to insure full power delivery to both rear wheels when traction is a problem. The differential lock consists essentially of a foot operated control and regulating valve and a multiple disc clutch located in differential housing, which locks the left axle gear to differential case.

OPERATION AND ADJUSTMENT

99. Refer to Fig. JD68C. When pedal (7) is depressed, pressurized fluid from the main hydraulic system is directed to clutch piston (3), locking axle gear (1) to differential case, and both differential output shafts and main drive bevel gear turn as a unit. Equal power is thus transmitted to both rear wheels despite variations in traction.

To release the differential lock, slightly depress either brake pedal (9), which releases system pressure by acting through linkage (8).

99A. **ADJUSTMENT.** Differential lock valve and linkage should be adjusted for operating pressure and pedal free play, and release linkage adjusted for length. Proceed as follows:

Operating pressure will only need to be checked if valve has been overhauled, system does not operate properly, or if incorrect pressure is suspected. To check the pressure, refer to Fig. JD68D. Install a 0-1000 psi pressure gage on valve port at the location shown. With engine operating at rated speed, depress differential lock actuating pedal. Gage pressure should be 420-480 psi; if it is not, disassemble valve as outlined in paragraph 100, and add or remove adjusting shims (6—Fig. JD68E) as required.

After valve has been disassembled, or if linkage adjustment is required, disconnect release link swivel (2) from pedal (3) and depress pedal until heavy spring pressure is encountered, but not far enough for valve linkage to snap over-center. Release the pedal and turn adjusting screw (8) in or out until it barely clears

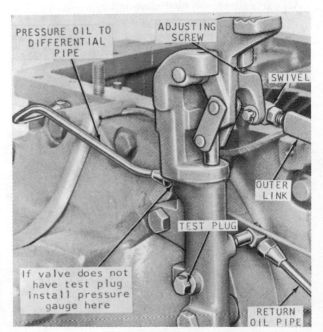

Fig. JD68C — Schematic view of hydraulically actuated differential lock available on late models. Refer to text.

1. Axle gear	6. Pressure line
2. Clutch plates	7. Pedal
3. Piston	8. Links
4. Pressure line	9. Brake pedals
5. Valve	

Fig. JD68D — Differential lock control valve and associated parts. Outer link connects to brake pedals and releases differential lock when either brake is applied.

pedal stop with pedal in normal release position. Recheck several times before reconnecting swivel, and readjust if necessary, to assure minimum clearance without touching.

Reconnect release swivel (2) and back off both adjusting nuts (N) several turns. With engine running, fully depress pedal (3) until control valve is in locked position. Lightly pull linkage (1) to rear until front ends of slots (S) contact release pins (P) in brake pedals and all slack is removed; then turn both nuts (N) into contact with swivel without moving linkage or pedal. Tighten both nuts (N) securely. When properly adjusted, the slightest movement of either brake pedal should release the differential lock, but valve will not release unintentionally because of linkage vibration.

OVERHAUL

100. **CONTROL VALVE.** To remove the differential lock control valve, bleed down main hydraulic system pressure by actuating brakes with engine not running. Disconnect release linkage and the three oil pipes at the valve. Remove the two retaining cap screws and lift off the valve, noting the location of the two spacer bushings on retaining cap screws.

Remove inlet fitting (16—Fig. JD-68E). Disconnect pedal to valve linkage, remove pedal pivot pin and lift off pedal; then withdraw return valve (4), spring (5), shim pack (6), flow control valve (7), check valve (13) and spring (14) from housing bore.

All parts are renewable individually. Examine parts for wear or scoring and springs for distortion, and renew any parts which are questionable. Keep shim pack (6) intact for use as a

Fig. JD68G — Differential lock piston can be removed after lifting out clutch plates.

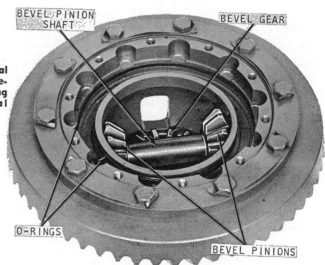

Fig. JD68H — Differential assembly with piston removed, showing sealing rings and differential gears.

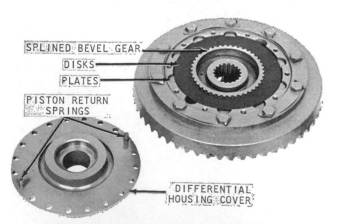

Fig. JD68F — Differential unit with housing cover removed, showing clutch plates and return springs.

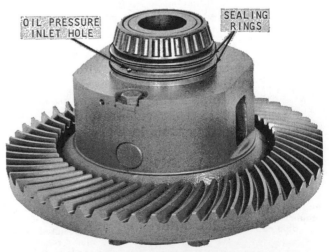

Fig. JD68J — Assembled differential unit, showing pressure passage seals and pressure passage.

starting point when readjusting operating pressure. Renew "O" rings whenever valve is disassembled.

Assemble by reversing the disassembly procedure, using Fig. JD68E as a guide. Make sure spacer bushings are properly positioned between valve and axle housing, reinstall valve and tighten retaining cap screws to a torque of 130 ft.-lbs. Check and adjust operating pressure and linkage as outlined in paragraph 99A.

101. **DIFFERENTIAL CLUTCH.** The multiple disc differential clutch can be overhauled after removing the unit as outlined in paragraph 96. Remove the cover retaining cap screws and lift off cover and the two piston return springs as shown in Fig. JD68F.

Clutch discs, plates and splined bevel gear can now be lifted from housing. Remove piston (Fig. JD68G) with air pressure or by grasping two opposing strengthening ribs with pliers. Renew sealing "O" rings (Fig. JD68H) and blow out oil passage in differential housing whenever piston is removed. Overhaul differential assembly as outlined in paragraph 97 or 97A, if required, while differential clutch is disassembled.

Examine sealing surface in bore of right bearing quill and renew quill if sealing area is damaged. Renew the cast iron sealing rings (Fig. JD68J) on differential housing if broken, scored or badly worn. Check clutch plates and discs and renew if scored,

warped, discolored by heat, or worn to a thickness of 0.100 or less.

The clutch plate to be installed next to piston has ten external lugs; the remainder of clutch plates have eight lugs. When assembling clutch, first install splined bevel gear and the plate with ten lugs, then alternately install the internally splined clutch discs and remainder of clutch plates, either side up. Align the external lugs on clutch plates, leaving two blank spaces for clutch return springs. Assemble springs over guide dowels in cover, using light, clean grease to hold them in position. Reinstall cover, tighten cover retaining cap screws to a torque of 35 ft.-lbs. and bend up lock plates.

REAR AXLE AND FINAL DRIVE

Series 3000 and 4000 tractors are available in high clearance (Hi-Crop) models equipped with drop housings which contain a final reduction bull gear and pinion. All other models have a planetary reduction final drive gear which is located at inner ends of rear axle housings.

All Models Except Hi-Crop

102. **REMOVE AND REINSTALL.** To remove the final drive unit, first drain the transmission and hydraulic fluid, suitably support rear of tractor and remove rear wheel or wheels.

Remove fenders and light wiring on models so equipped. On tractors with breakaway couplings for hydraulic remote cylinders, remove operator's platform and interfering hydraulic lines.

Support final drive assembly with a hoist, remove the attaching cap screws and swing the unit from transmission housing. When reinstalling, tighten the retaining cap screws to a torque of 170 ft.-lbs.

103. **OVERHAUL.** To disassemble the removed final drive unit, remove lock plate (20—Fig. JD69) and cap screw (19) at center of planet pinion carrier and withdraw the planet carrier assembly.

Planet pinion shafts (13) are retained in carrier by snap ring (12). To remove, expand the snap ring and, working around the carrier, tap each shaft out while snap ring is expanded.

Withdraw the parts, being careful not to lose the 31 loose bearing rollers in each planet pinion. With all rollers installed and in a horizontal position, strike tooth portion of pinion straight downward sharply on a wood block or bench. If rollers are dislodged, rollers and/or pinion are worn and should be renewed. Examine shaft for wear and renew if indicated. New dimensions are as follows:

Series 3010-4010
Planet pinion ID........2.3835-2.3843
Carrier pinion shaft
  bore ID 1.9458-1.9478

Series 3020-4020
Planet pinion ID........2.5226-2.5240
Carrier pinion shaft
  bore ID 2.0849-2.0869

Series 5000
Planet pinion ID........2.2467-2.2475
Carrier pinion shaft
  bore ID 1.7467-1.7487

After planet carrier has been removed, axle shaft (1) can be removed from inner bearing and housing by pressing on inner end of axle shaft. Remove bearing cone (5) and spacer (4) if they are damaged or worn. When assembling, heat spacer (4) and bearing (5) in oil to approximately 300 degrees F. and install on axle shaft making sure they are fully seated. Heat inner bearing cone (11) in oil to a temperature of 300 degrees F. Have planet carrier assembled and ready to install. Insert bearing on shaft and install and partially tighten

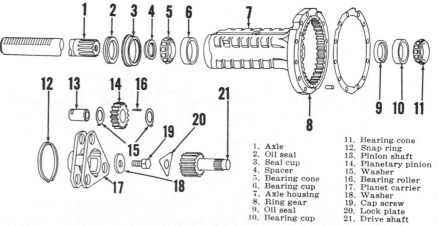

1. Axle
2. Oil seal
3. Seal cup
4. Spacer
5. Bearing cone
6. Bearing cup
7. Axle housing
8. Ring gear
9. Oil seal
10. Bearing cup
11. Bearing cone
12. Snap ring
13. Pinion shaft
14. Planetary pinion
15. Washer
16. Bearing roller
17. Planet carrier
18. Washer
19. Cap screw
20. Lock plate
21. Drive shaft

Fig. JD69—Exploded view of planetary type final drive assembly used on all Series 3000 and 4000 tractors except Hi-Crop. Series 5000 is similar.

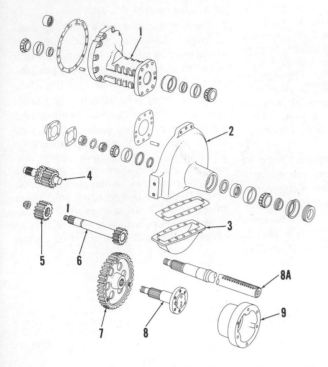

Fig. JD69A — Exploded view of "Hi-Crop" final drive assembly of the type used on Series 3000 and 4000 so equipped.

1. Shaft housing
2. Gear housing
3. Cover
4. Differential output shaft
5. Drive shaft gear
6. Final drive shaft
7. Bull gear
8. Wheel axle shaft
8A. Wheel axle shaft
9. Hub

the retaining cap screw (19). Leave a barely noticeable end play in axle bearings. Temporarily install lock plate (20) and, using a torque wrench calibrated in inch-pounds, check and record the rolling torque of the axle and planetary assembly. Remove lock plate (20) and tighten cap screw (19) until a rolling torque 20 to 70 inch-pounds greater than the previously recorded figure is obtained. Reinstall lock plate (20).

105. OVERHAUL. To disassemble the removed unit, remove the six stud nuts retaining drop housing to shaft housing, remove the two retainer plugs and thread jack screws into retainer plug holes. Tighten jack screws alternately to force the housings apart.

Remove the cotter pin and slotted nut from inner end of final drive shaft, then remove shaft using a brass drift. When reinstalling, heat bearing cones in oil to a temperature of 300 degrees F. for easy installation and tighten inner nut to provide an end play of 0.004 on shaft bearings.

To disassemble the removed drop housing, first remove the bull gear cover and wheel axle inner bearing cover, then remove inner bearing nuts with a spanner wrench. Unseat snap ring on inner side of bull gear, then press out rear axle shaft. Axle shaft is equipped with two oil seals, an inner seal which is pressed into housing, and a two-piece outer seal in housing and on shaft. When assembling, heat the bearing cones in oil to a temperature of 300 degrees F. to facilitate installation. Install axle shaft nut and tighten until drop housing wall is deflected 0.002 in area of shaft bearing. Do not forget to install and seat the bull gear retaining snap ring on axle shaft. Tighten the drop housing retaining stud nuts to a torque of 275 ft.-lbs. and fill each final drive gear housing with seven pints of SAE 90, multi-purpose gear lubricant.

BRAKES

All Hi-Crop Models

104. REMOVE AND REINSTALL. NOTE: If final drive bull gear, wheel axle shaft, bearings or seals ONLY is to be serviced. This can be accomplished by removing wheel and tire unit and drawbar or draft link, then referring to paragraph 105 for disassembly. For any other service, the entire final drive unit must be detached from transmission housing as follows:

First drain the transmission and hydraulic fluid, suitably support rear of tractor and remove rear wheel. Remove entire drawbar on tractors so equipped, or draft link on models with three-point hitch. Remove operator's platform and interfering hydraulic lines, support final drive assembly with a hoist, remove attaching cap screws and swing the unit from transmission housing. When installing, tighten the retaining cap screws to a torque of 150 ft.-lbs.

OPERATION AND ADJUSTMENT
All Models

106. The hydraulically actuated single disc brakes are located on the differential output shafts and are accessible after removing the final drive units as outlined in paragraph 102 or 104 and the output shaft and backing plate.

Power is supplied by the system hydraulic pump through foot operated control valves when engine is running, or manually by means of master cylinders when hydraulic system pump is inoperative.

NOTE: Series 5000 is equipped with a nitrogen-filled accumulator to assist in providing pressure fluid when main hydraulic system is inactive.

The only adjustment provided is an adjustable yoke on upper end of each operating valve. This adjustment is used to equalize the height of the

Fig. JD70—Adjust the operating rod for each brake to a length of 2⅞ inches as shown.

brake pedals. To adjust, loosen the locknut and turn operating rod in or out until center of pedal connecting pin is 2⅞ inches from surface of boot retainer and both pedals are equal in height.

For service of the hydraulic pump, refer to paragraph 121. To service control valve, actuating cylinders and brake discs, refer to the appropriate following paragraphs.

NOTE: Leakage or wear of the brake control valves can usually be determined by performing the leakage test outlined in paragraph 119. Other tests of the main hydraulic system which might affect brake operation are covered in paragraphs 116 through 118.

BLEEDING

All Models

107. When brake system has been disconnected or disassembled, bleed the system as follows:

Start the engine, loosen lock nut on bleed screw (B—Fig. JD71) and back out the bleed screw two full turns. Bleed return passage is internal, retighten lock nut to prevent external leakage, depress brake pedals and hold in depressed position for approximately two minutes to flush air from system. Tighten bleed screw then release the pedal. Bleed opposite brake following the same procedure.

To test the system, stop the engine and depress each brake pedal once. Solid pedal action should be obtained on next application. If pedal action is spongy or pedal travel exceeds four inches, repeat the bleeding operation.

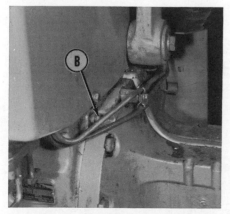

Fig. JD71—To bleed the brakes, loosen the bleed screw (B) and actuate the valve. Bleeding action is internal.

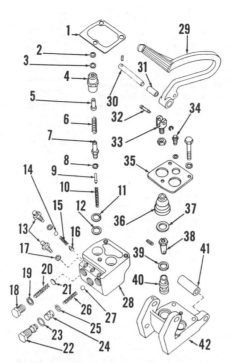

Fig. JD72—Exploded view of the hydraulic power brake operating valve. A master piston (4) applies the brakes if power is not available.

1. Gasket	22. Plug
2. Backup ring	23. "O" ring
3. "O" ring	24. Filter
4. Manual piston	25. "O" ring
5. Valve plunger	26. Spring
6. Spring	27. Check valve
7. Nipple	ball
8. "O" ring	28. Housing
9. Brake valve	29. Pedal
10. Spring	30. Shaft
11. Backup ring	31. Bushing
12. "O" ring	32. Pin
13. Connector	33. Yoke
14. Steel ball	34. Connector
15. Spring	35. Boot retainer
16. Check valve	36. Boot
stop	37. Washer
17. "O" ring	38. Operating rod
18. Plug	39. Seal
19. "O" ring	40. Guide
20. Spring	41. Spacer
21. Valve disc	42. Bracket

CONTROL VALVE

All Models

108. To remove the control valve, disconnect the pressure and discharge lines at valve housing. On Series 5000, remove right platform extension. On all models, remove the attaching cap screws and lift off the control valves and pedals as a unit.

To disassemble the removed unit, remove the two operating rod to pedal connecting pins (32—Fig. JD72) and drive the roll pin from pedal shaft (30). Tap out the shaft and remove the two pedals. Remove the cap screws attaching pedal bracket (42) to valve body (28) and lift off the bracket and operating rod assemblies. It will not be necessary to disassemble operating rods unless seal or parts renewal is indicated.

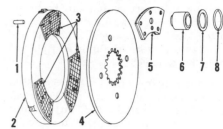

Fig. JD73—Exploded view of the wet type hydraulic brake parts which are located on differential output shafts.

1. Dowel	5. Pressure plate
2. Backing plate	6. Piston
3. Facings	7. Washer
4. Disc	8. "O" ring

Use Fig. JD72 as a guide when disassembling the control valve assembly. Manual brake pistons (4), brake valve plungers (5) and plunger return spring (6) can be lifted out. Use a deep socket to remove inlet valve nipple (7). Clean the parts thoroughly and check against the specifications which follow:

Valve spring test data

Brake valve return
  spring . . . . 0.6-0.8 lb. @ $2\frac{3}{16}$ in.
Valve plunger return
  spring . . . . 39-47 lb. @ $1\frac{11}{16}$ in.
Check valve return
  spring . . . . 7-9 oz. @ 1 in.
Pedal bushing ID . . . . . . 0.686 -0.690
Manual piston OD . . . . . . 1.109 -1.111
Manual piston bore OD . . 0.374 -0.376
    (Early models)
Brake valve plunger OD 0.3725-0.3735
    (Late models)
Brake valve plunger OD 0.5655-0.5665

When reassembling, use new "O" rings and gaskets. Use Fig. JD72 as a guide. Tighten the inlet valve nipple to a torque of 40 ft.-lbs. When control valve assembly has been installed, check for equal pedal height and bleed system as outlined in paragraph 107.

DISCS AND SHOES

All Models

109. To remove the brake discs or operating cylinders, first remove the final drive unit as outlined in paragraph 102 or 104, remove the backing plate, output shaft and brake disc. The three stationary shoes are riveted to the backing plate. The three actuating shoes are pressed on the op-

erating cylinders which can be withdrawn from transmission housing after disc is removed. Facings are available in sets of three and should only be renewed as a set.

Operating pistons are 2.2495-2.2505 in diameter and have a diametral clearance of 0.0025-0.0065 in cylinder bores. Refer to Fig. JD73 for breakdown of brake parts.

BRAKE ACCUMULATOR

Series 5000

109A. R&R AND OVERHAUL. To remove the brake accumulator, bleed fluid and pressure from hydraulic system and brakes. Remove right platform extension, disconnect hydraulic pipe from bottom side of accumulator; then, unbolt and remove accumulator.

CAUTION: Overhaul of brake accumulator should not be attempted unless charging equipment is available. Gas side of accumulator piston is charged to 475 - 525 psi with NITROGEN gas. The charging valve is required to bleed cylinder before disassembly can be safely attempted.

Before attempting to disassemble the accumulator, remove cap (1—Fig. JD74A), install charging valve and turn valve depressing lever to exhaust gas from cylinder. The Nitrogen gas is harmless and non-inflamable. After the charging gas is completely exhausted, remove gas valve body (3) from cylinder. Invert cylinder and push end cap (15) into cylinder far enough to expose snap ring (16). Remove snap ring, end cap (15) and piston (11).

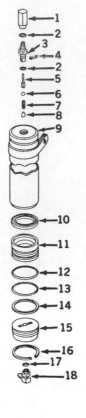

Fig. JD74A — Exploded view of brake accumulator used on Series 5000.

1. Cap
2. "O" ring
3. Gas valve body
4. Roll pin
5. Pin
6. Ball
7. Spring
8. Stop
9. Cylinder
10. Packing
11. Piston
12. "O" ring
13. "O" ring
14. Backing ring
15. Cylinder cap
16. Snap ring
17. "O" ring
18. Hydraulic fitting

Remove packing and "O" rings from piston and end cap and use new parts when unit is reassembled. Remove roll pin (4) from gas valve body (3) and disassemble the valve.

Check the cylinder, cylinder cap and piston for scoring, pitting, dents or wear. Diameters of new parts are as follows:

Cylinder ID 2.998-3.001
Cylinder cap OD 2.994-2.998
Piston OD 2.993-2.996

Assemble by reversing the disassembly procedure, using Fig. JD74A as a guide; and installing new "O" rings, backing ring and packing. Recharge the cylinder using approved charging equipment and Nitrogen ONLY, to a pressure of 500 psi. Remove charging valve and check by immersing the charged accumulator in water. When it has been determined there are no leaks, reinstall valve cap (1) and install the accumulator by reversing the removal procedure.

POWER TAKE-OFF

OPERATION

Series 3010-4010

110. The power take-off shaft is driven at either of two ASAE speeds at 1900 engine rpm by an independently controlled, flywheel mounted single disc clutch. See paragraphs 80 through 87 for service and adjustment procedures on clutch and linkage.

Speed selection is automatically controlled by the installation of the output shaft. See Fig. JD75 for a schematic view of the shifter mechanism. A spring loaded pto shifter rod is located in the center of the hollow pto countershaft and is pinned to the shifter collars as shown. The end of the shifter rod extends through the rear of the output shaft mounting

Fig. JD74—Transmission housing with final drive removed, showing brake shoes (S) and cylinder (C). The three cylinders are internally bled by the bleed screw (B).

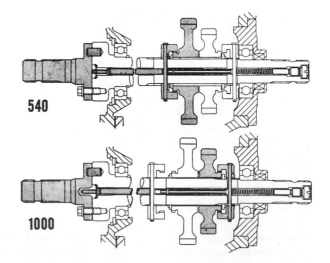

Fig. JD75 — The power take-off assembly is automatically shifted to either of the two ASAE output speeds when stub shaft is attached. Note that flat front face of 540 rpm stub shaft pushes the shifter rod forward to engage the splined collar over the large 540 rpm driven gear. Recess in 1000 rpm stub shaft allows spring to engage smaller driven gear.

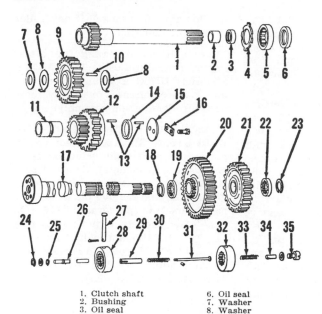

Fig. JD76 — Exploded view of power take-off gears and shafts.

11. Inner race
12. Gear
13. Bearing roller
14. Spacer
15. Plate
16. Lock
17. PTO shaft
18. Snap ring
19. Thrust washer
20. 540 rpm gear
21. 1000 rpm gear
22. Thrust washer
23. Snap ring
24. Snap ring
25. Packing
26. Shifter rod
27. Shifter pin
28. Collar
29. Shifter tube
30. Spring
31. Pin
32. Collar
33. Spring
34. Pin
35. Screw

1. Clutch shaft
2. Bushing
3. Oil seal
4. Washer
5. Bearing
6. Oil seal
7. Washer
8. Washer
9. Idler gear
10. Bearing roller

Fig. JD78 — Exploded view of PTO front bearing quill and associated parts used on Series 3020 and 4020.

1. Snap ring
2. Bearing
3. Oil seal
4. Gasket
5. Bearing quill
6. Guard

flange. When the 24 spline, 1000 rpm output shaft is installed, shifter rod extension fits in the recess of stub shaft and spring pressure engages the 1000 rpm shifter collar. When the six spline, 540 rpm output shaft is installed, the shifter rod is moved forward to engage the 540 rpm shifter collar.

Series 3020-4020

110A. Series 3020 and 4020 tractors are available with either a single-speed (1000 rpm) or dual-speed (540-1000 rpm) power take-off. ASAE speeds are obtained at 2100 engine rpm on Series 3020; or 1900 engine rpm on Series 4020.

Fig. JD77 — Power take-off gears installed on front of transmission housing.

1. Plate
2. Gear
3. Shifter pin
4. Collar
5. Bearing
6. PTO shaft
7. 540 rpm gear
8. 1000 rpm gear

The power take-off is driven by an independently controlled, **flywheel** mounted single disc clutch on "Syncro-Range" models; or a hydraulically actuated multiple disc clutch in power shift models. See paragraphs 80 through 87 for service and adjustment procedures on "Syncro-Range" clutch and linkage; and paragraphs 92A through 94D for service on power shift valve and clutches.

On all models, the front pto and transmission pto shaft operates at only 1000 rpm ASAE speed; reduction to the 540 rpm speed being accomplished by reduction gears at output end.

Series 5000

110B. The power take-off shaft is driven at 1000 rpm ASAE speed at 1900 engine rpm by an independently controlled, hydraulically actuated, multiple disc clutch mounted on front end of transmission housing. A manually operated, collar type disconnect clutch is also included, which interrupts power flow ahead of the multiple disc clutch. Refer to paragraph 86B for overhaul of the disconnect clutch; or paragraphs 112D through 112G for overhaul of the multiple disc clutch and control valve.

R&R AND OVERHAUL

Series 3010-4010

111. **PTO FRONT BEARING.** To renew the pto front bearing or oil seal, drain the transmission and hy-

draulic system fluid, remove front pto guard, then unbolt and withdraw pto front bearing quill. Protect front end of shaft while pulling bearing.

112. **PTO GEARS, SHIFTER AND COUNTERSHAFT.** To renew the pto gears, countershaft or shifter mechanism, first drain the transmission and hydraulic system fluid, then detach (split) transmission from clutch housing as outlined in paragraph 85A. If service is required on pto clutch shaft and gear, refer to paragraph 85B.

Lift off idle gear (9—Fig. JD76) being careful not to lose the loose needle bearings (10). Remove the cap screws retaining washer (15) and remove pto gear (12). Catch the loose needle bearings (13) which will drop when gear is withdrawn. Pull the front bearing on pto countershaft (17), and remove shifter spring retaining screw (35), follower and spring (33). Remove the pin retaining the 1000 rpm shifter collar (32), remove collar, snap ring (23), thrust washer (22) and the two pto drive gears. Withdraw thrust washer (19), remove snap ring (18), pin (27) and

Fig. JD78A — Rear face of clutch housing showing pto input gear and brake (B) which operates in coned surface of gear (2—Fig. JD77). Idler gear remains with transmission housing in series 4010.

the 540 rpm shift collar (28). Remove rear oil seal quill and unseat the bearing retaining snap ring, then withdraw the pto countershaft from the rear.

NOTE: If pin (27) is bent because of improper installation of output shaft, and cannot be withdrawn, remove rear oil seal quill and snap ring; then, with a heavy drift, bump the pto shaft to the rear. The pin (27) must be sheared before shaft can be removed.

Assemble by reversing the disassembly procedure. Check to make sure the shifter mechanism works properly. The spacer (14) between the two rows of loose needle bearings fits in center groove of inner race (11). When assembling, install spacer and position needle bearings on race, holding each row in place with a rubber band while gear is being installed.

Series 3020-4020

112A. **PTO FRONT BEARING.** To renew the pto front bearing or oil seal, drain the transmission and hydraulic system fluid. Remove front pto guard (6—Fig. JD78), then unbolt and withdraw quill (5). Bearing (2) is retained in quill by snap ring (1), and two punch holes are provided in quill to aid in bearing removal. Working through the holes provided, drift bearing rearward out of quill and install with a piloted arbor and a press. Install oil seal (3) with lip facing toward bearing.

112B. **PTO DRIVE GEARS AND CLUTCH SHAFT.** To renew the pto drive gears or transmission pto shaft,

first drain transmission and hydraulic system fluid; then detach (split) transmission from clutch housing as outlined in paragraph 85A (Syncro-Range Transmission) or 93A (Power Shift Models).

On "Syncro-Range" models, remove pto clutch shaft (3—Fig. JD78A) as outlined in paragraph 85B, if renewal is required. PTO countershaft idler gear (1) and bearings can be slipped from shaft after tractor is split. Hold front and rear washers in contact with gear to retain the loose needle rollers, as gear is withdrawn. Remove snap ring (Fig. JD78B) and pto drive gear from transmission pto shaft. Remove cap screws, retaining plate and idler gear from front of transmission housing, being careful to catch the loose needle rollers as gear is withdrawn.

On power shift models, the entire gear train is on front wall of transmission case as shown in Fig. JD78C. To disassemble the gears, withdraw pto clutch gear (1) and bearings (2) as a unit. Withdraw idler gear (3) and thrust washer (4). Catch the loose needle rollers as gear (3) is withdrawn. Remove and save spring (8), unseat and remove snap ring (5); then withdraw thrust washer (6), brake idler gear (7) and the contained loose needle rollers. Remove snap ring (10) and pto drive gear (9).

On all models, remove transmission pto shaft, if necessary, as outlined in paragraph 112C.

Renew any gears which are chipped, worn or otherwise damaged. Examine bearing surfaces of gear hubs and idler shafts for wear or scoring. Renew loose needle rollers in complete

sets for any one bearing, if any roller is worn or damaged.

Assemble by reversing the disassembly procedure, using Figs. JD78A, JD78B or JD78C as a guide. Use clean multipurpose grease to retain loose needle rollers during assembly.

112C. **PTO OUTPUT SHAFT AND REDUCTION GEARS.** To remove the pto output shaft (and reduction gears on dual speed models), detach transmission from clutch housing (paragraph 85A) and remove pto drive gear as outlined in paragraph 112B. On dual speed models, remove snap ring (1—Fig. JD78D) and withdraw output shaft (2). Unbolt and remove rear bearing quill (5), then lift out the loose 540 rpm drive gear (12).

Compress snap ring (16) as shown in Fig. JD78F and unseat from its groove; then tap transmission pto shaft rearward as shown in Fig. JD78G. When bearing (17—Fig. JD78D) is free from housing bore, withdraw the pto shaft. Withdraw idler shaft (22) while removing gear as shown in Fig. JD78H. NOTE: Be careful not to lose the loose needle rollers (25—Fig. JD78D).

On single speed pto models, the transmission pto shaft and rear bearing quill will be removed as a unit,

Fig. JD78C — Front face of transmission housing used on power shift models, showing pto drive train.

1. PTO clutch gear	7. Brake idler gear
2. Bearings	8. Brake return spring
3. Idler gear	9. PTO drive gear
4. Thrust washer	10. Snap ring
5. Snap ring	11. PTO shaft
6. Thrust washer	

Fig. JD78B — Front face of transmission housing on Syncro-Range 3020 tractor, showing pto drive and idler gear. Series 4020 is similar. Refer to Fig. JD78C for power shift models.

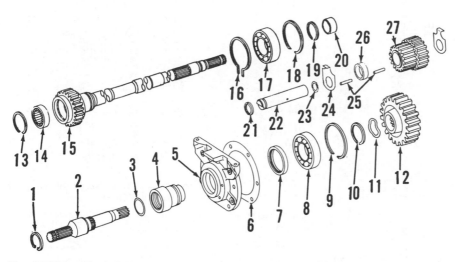

Fig. JD78D — Exploded view of pto reduction gears, output shaft and associated parts used on all late models with dual speed pto. On single speed 3020 and 4020 models, a single shaft is used in place of pto shaft (15) and stub shaft (2); and the gears are omitted.

1. Snap ring
2. Stub shaft
3. "O" ring
4. Pilot housing
5. Bearing quill
6. Gasket
7. Oil seal
8. Rear bearing
9. Snap ring
10. Snap ring
11. Spring washer
12. 540 rpm gear
13. Snap ring
14. Bearing
15. Transmission pto shaft
16. Snap ring
17. Bearing
18. Snap ring
19. Snap ring
20. Bushing
21. Snap ring
22. Idler shaft
23. "O" ring
24. Thrust washer
25. Bearing roller
26. Spacer
27. Idler gear

after removing pto drive gear and cap screws retaining quill to rear of transmission housing.

Bushing (20) has an installed ID of 1.751-1.754. Renew bushing using a piloted driver, if worn or scored. Examine gears and bearings for wear, scoring or other damage and renew as required.

When assembling, use clean grease to position bearing rollers in idler gear and thrust washers (24) in transmission housing, and assemble by reversing the disassembly procedure.

Series 5000

112D. PTO DRIVE GEARS AND CLUTCH. To remove the pto drive gears and multiple disc clutch, first detach (split) transmission from clutch housing as outlined in paragraph 86A. Remove pto drive shaft, gear and shift mechanism as in paragraph 86B.

To remove the pto clutch assembly, remove access plate from bottom of clutch housing, remove the three cap screws retaining front bearing cover (37—Fig. JD78J) and remove the

cover. Unseat and remove snap ring (34) from outer race of bearing (33) and remove clutch assembly and front bearing as a unit.

Remove snap ring (35) from front of clutch hub and remove bearing (33) using a suitable knife-edge puller. Thrust washer (32) and drive gear (31) can be lifted off after bearing is removed. Unseat the large snap ring (26) and remove backing plate (25) and multiple disc clutch plates and discs.

Clutch piston must be disassembled in a press, using a suitable straddle-mounted fixture and compressing spring until the retaining snap ring can be unseated.

NOTE: A fixture can be constructed using about 6 inches of 2-inch ID steel pipe. Make sure both ends of pipe are square with center-line. Machine a 1" × 1½" notch in one end of pipe for working space to unseat snap ring.

Using the fixture, depress spring retainer (29) and remove snap ring (30). Release spring pressure carefully after snap ring is removed, to prevent injury or damage, until all pressure is removed from spring (27). Remove the fixture, retainer (29), steel ball (28) and spring (27); then remove piston (22) using compressed air or other suitable means. Steel ball (28) fits a recess in clutch hub and retainer to prevent retainer turning on hub.

Inspect all parts for wear, scoring or other damage. Examine bushing (10) for wear or grooving at sealing ring area. Make sure oil passage is aligned if bushing must be renewed. Carrier bushing (11) has an installed

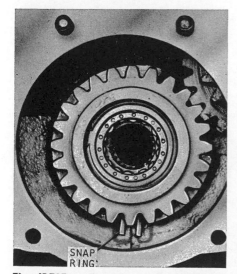
Fig. JD78F — To remove transmission pto shaft after removing front drive gear and output bearing quill, first compress and unseat snap ring shown.

Fig. JD78G — Tap transmission pto shaft rearward after unseating snap ring shown in Fig. JF78F.

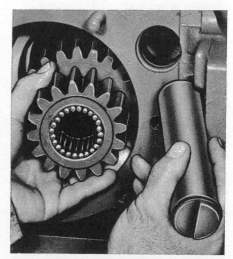
Fig. JD78H — When removing reduction idler gear and shaft, be careful not to lose the loose needle bearings.

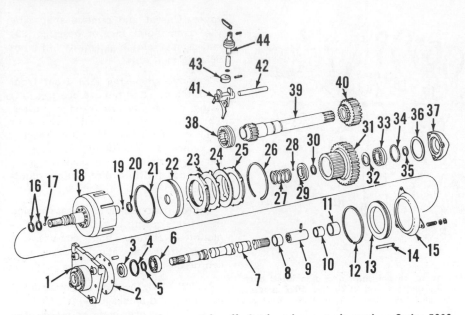

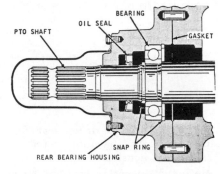

Fig. JD78K — Cross sectional view of Series 5000 pto clutch.

1. Output shaft	7. Spring retainer
2. Clutch piston	8. Snap ring
3. Clutch drum	9. Cap screw
4. Backing plate	10. Brake piston
5. Drive gear	11. Brake shoe
6. Driven gear	

Fig. JD78J—Exploded view of power take-off clutch and gear train used on Series 5000.

1. Bearing quill	12. Sealing ring	23. Clutch plate	34. Snap ring
2. Gasket	13. Brake piston	24. Clutch disc	35. Snap ring
3. Oil seal	14. Guide dowel	25. Backing plate	36. Gasket
4. Snap ring	15. Brake shoe	26. Snap ring	37. Bearing cover
5. Snap ring	16. Sealing rings	27. Clutch spring	38. Shift collar
6. Bearing	17. Passage plug	28. Lock ball	39. Clutch shaft
7. Output shaft	18. Clutch drum	29. Spring retainer	40. Drive gear
8. Bushing	19. Passage plug	30. Snap ring	41. Shift fork
9. Coupling	20. Sealing ring	31. Driven gear	42. Shift shaft
10. Bushing	21. Sealing ring	32. Thrust washer	43. Shift cam
11. Bushing	22. Clutch piston	33. Bearing	44. Shift lever

Fig. JD78L—Cross sectional view of Series 5000 pto output shaft, showing construction details.

ID of 2.003-2.005. Inspect clutch plates and discs for heat discoloration, warping or scoring. Externally lugged clutch plates are 0.090 in thickness. Internally splined discs are 0.112-0.118; renew discs if worn to a thickness of 0.080 or less.

Spring (27) should have a free length of 2⅞ inches and test 523-627 lbs. when compressed to a height of 1²¹⁄₃₂ inches. Brake shoe return springs should have a free length of 1¾ inches and test 22.5-27.5 lbs. when compressed to a height of 1⅛ inches.

Assemble by reversing the disassembly procedure, using Figs. JD78J and JD78K as a guide. Make sure locking ball (28—Fig. JD78J) is properly positioned when installing spring retainer.

112E. OUTPUT SHAFT AND REAR BEARING. Output shaft (7—Fig. JD78J), bearing (6), seal (3) and rear bearing housing (1) can be removed as a unit after draining transmission and removing the retaining cap screws. Tighten cap screws to a torque of 35 ft.-lbs. when reinstalling.

112F. PTO CONTROL VALVE. Refer to Fig. JD78M for an exploded view of control valve and associated parts.

Control valve assembly can be removed as a unit from clutch housing after disconnecting control cable (13). If spool (2) is damaged, worn or scored, detach clutch housing from transmission as outlined in paragraph 86A, and examine spool bore in transmission pump (1). If bore is damaged, renew oil pump as outlined in paragraph 89C.

To remove valve spool (2) or spring (4), drive out roll pin (3). The variable rate spring (4) controls the pto operating pressure; when reinstalling valve, adjust as outlined in paragraph 112G.

112G. PTO VALVE ADJUSTMENT. The pto control valve linkage should be adjusted to provide 60 psi clutch pressure at rated speed when clutch is engaged; and sufficient brake pressure to stop pto shaft rotation at slow idle speed with clutch disengaged. To adjust the linkage, proceed as follows:

With tractor completely assembled and at operating temperature, remove port plug (P—Fig. JD78N) and install a suitable pressure gage. With engine speed at 2150 rpm and pto clutch engaged, gage pressure should be 60 psi. If pressure is higher or lower than recommended, loosen clamp screw (C) and move cable housing backward or forward in clamp until

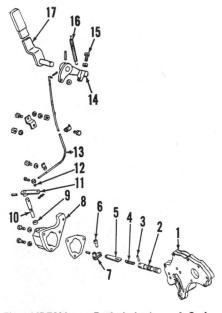

Fig. JJD78M — Exploded view of Series 5000 pto shift valve and associated parts.

1. Transmission pump	10. Shift shaft
2. Valve spool	11. Shift lever
3. Roll pin	12. Swivel
4. Spring	13. Shift cable
5. Link	14. Lever arm
6. Pin	15. Stop screw
7. Shift arm	16. Spring
8. Housing	17. Lever
9. Oil seal	

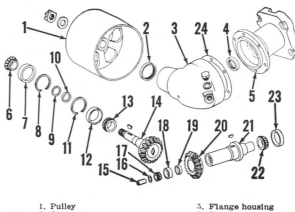

Fig. JD78P — Exploded
view of belt pulley at-
tachment.

9. Shim
10. Spacer
11. Snap ring
12. Bearing cup
13. Bearing cone
14. Pulley shaft
15. Shaft
16. Shim
17. Bearing cone
18. Bearing cup
19. Cupped plug
20. Gear
21. Input shaft
22. Bearing cone
23. Bearing cup
24. Shim

1. Pulley
2. Oil seal
3. Housing
4. Oil seal
5. Flange housing
6. Bearing cone
7. Bearing cup
8. Snap ring

Fig. JD78N — To adjust pto valve on
Series 5000, install a gage in plug hole
(P), loosen cap screw (C), and move
cable housing in clamp as shown. Refer
to text.

correct pressure is obtained; then, tighten clamp screw (C).

When gage pressure has been correctly adjusted, move hand throttle to slow idle position. Remove steering support rear access door, loosen locknut on stop screw (15—Fig. JD78M) and turn screw clockwise as far as possible. Move pto control lever to disengaged position and, while observing pto output shaft, turn screw counter-clockwise until pto shaft completely stops. Turn screw counter-clockwise one additional turn and tighten the locknut.

BELT PULLEY

All Models So Equipped

113. To disassemble the belt pulley attachment, remove the pulley, drain the gear housing, then unbolt and remove the mounting flange housing (5—Fig. JD78P). Withdraw the drive shaft (21) and gear assembly. Remove the pulley shaft oil seal (2), outer bearing cone (6) and adjusting shims (9), then remove the pulley shaft from the gear housing.

When assembling, adjust pulley shaft to 0.000-0.003 end play by means of the shims (9) located underneath outer bearing cone. Shims are available in thicknesses of 0.003, 0.005 and 0.010.

Assemble the unit, adjust the input shaft bearings to zero end play, insert the 1000 rpm pto stub shaft into pulley drive and check the gear backlash which should be 0.003-0.005 when measured at stub shaft flange. Adjust the backlash by adding or removing

shims (16) between input shaft inner bearing cone (17) and gear housing (3), then add or remove shims (24) between gear and flange housings to provide 6 to 9 inch-pounds rolling torque in the assembled pulley unit.

HYDRAULIC SYSTEM

The hydraulic lift system working fluid is supplied by the tractor main hydraulic system which also provides fluid for the power steering and power brakes. Working fluid for the hydraulic units is available at all control valves at a constant pressure of 2250 psi.

MAIN HYDRAULIC SYSTEM
All Models

114. OPERATION. The main hydraulic pump is mounted underneath the tractor radiator and coupled to front of engine crankshaft. This variable displacement, radial piston pump provides only the fluid necessary to maintain system pressure. When there are no demands on the system, pistons are held away from the pump camshaft by fluid pressure and no flow is present. When pressure is lowered in the supply system by moving a control valve or by leakage, the stroke control valve in the pump meters fluid from the camshaft reservoir, permitting the pistons to operate and supply the flow necessary to maintain system pressure. A maxi-

mum of 18 gallons per minute is available at full stroke.

The transmission pump provides pressure lubrication for the transmission gears and shafts; on Series 5000, supplies operating fluid for the power take-off; and on Power Shift models, operating fluid for transmission and power take-off. On all models, excess fluid from transmission pump passes through the full flow system filter to the inlet side of the main hydraulic system pump. If no fluid is demanded by the main pump, the fluid passes into the oil cooler then back to reservoir in transmission housing. Transmission pump capacities at 1900 engine rpm are as follows:

Series 5000 ....10 gpm
Other Syncro-Range Models... 6 gpm
All Power Shift Models.......13 gpm

The oil cooler is mounted in front of the tractor radiator and contains cooling fins to control fluid temperature. It also provides 1½ gallons of reserve hydraulic fluid to the main pump for peak load demands.

115. RESERVOIR AND FILTER. The hydraulic system reservoir is the transmission housing and the same fluid provides lubrication for the transmission gears and differential and final drive units. The manufacturer recommends that only John Deere Type 303 Special Purpose Oil be used in the system. Reservoir capacity is 11 gallons. To check the fluid level, stop the tractor on level ground and check to make sure that fluid level is in "SAFE" range on dipstick (D—Fig. JD79).

The oil filter element (Power Shift models have two filters) is located on left side of transmission housing as shown at (F—Fig. JD80). The filter cartridge may be renewed without draining the fluid reservoir by removing the filter cover and extracting the element.

The transmission housing contains a check valve at the point where the pressure line from transmission pump enters housing. The purpose of this valve is to prevent the fluid from draining back into reservoir from oil cooler when transmission pump is not operating. To renew the check valve or spring, first drain the fluid and remove the rockshaft housing (or transmission rear cover) as outlined in paragraph 125. Remove the transmission pump outlet (left) line, then thread the outlet pipe bushing in housing with a thread tap. This bushing also serves as the check valve seat. Install a cap screw in tapped hole to serve as a puller and remove the bushing. Check valve and spring may then be withdrawn. Thoroughly clean the metal chips from housing and bore, and use a new bushing when reinstalling the valve.

A surge pressure relief valve (Early Models) and a filter bypass valve are located in the external housing at rear of filter cover. To remove the filter bypass valve, remove plug (P—Fig. JD80) and withdraw the valve assembly. Service the surge pressure valve by removing plug (R).

116. SYSTEM TESTS. Efficient operation of the tractor hydraulic units requires that each component of the main supply system functions properly. A logical procedure for testing the system is therefore required. The indicated system tests include transmission pump flow test, system pressure tests and leakage tests as outlined in the following three paragraphs. Unless the indicated repairs of hydraulic units is obvious because of breakage, these tests should be performed before proceeding with repairs on the individual hydraulic units.

117. TRANSMISSION PUMP FLOW AND SURGE PRESSURE. A quick test of transmission pump operation can be performed by removing the fluid filter (F—Fig. JD80) and turning the engine over with the starter. If pump operation is satisfactory, a generous flow of fluid will be pumped into the filter housing.

An alternate method which may be used is as follows: Disconnect the main hydraulic pump supply line at filter relief valve housing (on Syncro-Range models; or regulator valve housing on Power Shift models). Plug the line to prevent loss of fluid from oil cooler. Loosen locknut on connecter elbow, turn open end of elbow downward and attach a short pipe which contains a 0-250 psi pressure gage and a gate valve. The pressure gage must be located between the gate valve and tractor system. Disengage the transmission clutch, start the engine and adjust engine speed to 1900 rpm. Place a clean container beneath pipe opening, engage clutch and check fluid flow for a period of 15 seconds.

The amount of flow should be approximately 1½ gallons on Series 5000, 1 gallon on other Syncro-Range models; or 3 gallons on power shift models.

Use the gate valve to partially restrict the fluid flow until gage pressure registers 50 psi, then again check the flow for 15 seconds. Flow should not drop appreciably at 50 psi. If flow drops noticeably, remove and overhaul the transmission pump.

NOTE: On Series 5000, pto control valve must be correctly adjusted as outlined in paragraph 112G, and pto clutch engaged, when making the test.

Series 3010 tractors below Serial Number IT28972 and Series 4010 tractors below Serial Number 2T33359 are equipped with a Surge Pressure Relief Valve located between the check valve and filter to provide protection for the oil cooler. Test the surge pressure relief valve as follows:

With engine still running at 1900 rpm, slowly close the gate valve while watching the pressure. When pressure reaches 90 to 130 psi, the surge pressure relief valve should open and cut off the external flow. If pressure is not within the specified limits, or if flow is excessive, remove and service the surge pressure relief valve as outlined in paragraph 115.

Fig. JD80—Transmission fluid filter is located on left side of transmission housing as shown.

F. Filter
P. Bypass valve
R. Surge pressure valve
S. Supply line

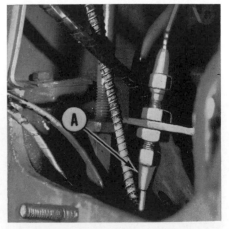

Fig. JD81—To check the main hydraulic system pump pressure, remove the lower instrument panel access door, disconnect brake supply line (A) and attach master gage.

Fig. JD79 — Operators platform showing transmission filler cap (F), dipstick (D) and rockshaft selector lever (S).

CAUTION: DO NOT attempt the surge pressure test on Series 3010 tractors above Serial Number IT28971; Series 4010 tractors above Serial Number 2T33350; nor any 3020 or 4020 tractor.

118. SYSTEM PRESSURE TEST. To check the system pressure, disconnect the brake pressure line on bracket as shown in Fig. JD81 and install a 3000 psi pressure gage in line. Make sure that all hydraulic control levers are in neutral, start tractor engine and operate at 1900 rpm. The pressure gage needle should immediately rise to 2250 psi then remain stationary.

Reduce engine speed to 1500 rpm and operate the remote cylinder operating levers and rockshaft lever at the same time. The gage needle should drop momentarily then immediately rise to 2000 psi or above and maintain this pressure while action continues.

Adjust the standby pressure to the specified 2250 psi as follows: Remove the left grille screen, loosen the locknut on stroke control valve on top of main hydraulic pump, and turn the

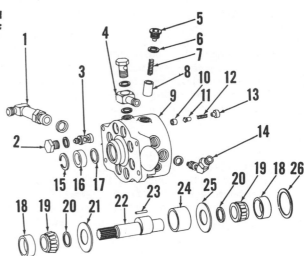

Fig. JD84 — Exploded view of main hydraulic system pump.

1. Elbow
2. Plug
3. Inlet valve
4. Connector
5. Plug
6. "O" ring
7. Piston spring
8. Piston
9. Housing
10. Valve seat
11. Discharge valve
12. Spring
13. Guide
14. Elbow
15. Snap ring
16. Oil seal
17. Packing
18. Bearing cup
19. Bearing cone
20. Spacer
21. Washer
22. Camshaft
23. Bearing roller
24. Race
25. Washer
26. Shims

Fig. JD82—Main system pump operating pressure is adjusted by means of adjusting screw (A) located on top of hydraulic pump stroke valve housing.

Fig. JD83—To remove the main hydraulic pump it is first necessary to remove lower frame plate (P).

screw in or out until specified pressure is obtained. See Fig. JD82.

If system pressure is below the specified 2000 psi pressure while testing, check and adjust the metering valves on control valve housings as outlined in paragraphs 127 and 137.

118A. PRIORITY VALVE TEST. Leave gage attached as outlined in paragraph 118. If tractor is equipped for remote cylinder operation, connect a hose to one breakaway coupling and return free end of hose to transmission filler opening. If tractor is not equipped with remote valve, disconnect rockshaft cylinder hydraulic line at either end and attach a jumper hose leading back to reservoir.

Start engine and operate at 600 RPM (Slow Idle), move control lever to pressurize the disconnected hydraulic line and note pressure gage reading. Gage will register priority valve setting which should be 1800-1900 psi. If pressure is not within recommended range, remove and adjust the valve as outlined in paragraph 9B.

119. LEAKAGE TEST. To check for leakage at any of the system valves, move all valves to neutral and run the engine for a few minutes at a speed of 1900 rpm. Check all of the hydraulic unit return pipes individually for heating. If the temperature of the return pipe is appreciably higher than the rest of the lines the valve is probably leaking. Disconnect the return line and measure the flow from the line for a period of one minute. Leakage should not exceed ½ pint; if it does, overhaul the system valves as outlined in the appropriate sections of this manual.

120. LINES AND FITTINGS. Flared, seamless steel tubing is used for all hydraulic system components. Fittings have SAE Straight Tubing threads with "O" ring seals. Do not attempt to substitute pipe thread fittings for components or test equipment.

121. MAIN HYDRAULIC PUMP. When external leaks, or failure to build or maintain operating pressure indicates a faulty pump, the main hydraulic pump must be removed for service as follows: To remove the pump on tricycle models, first remove lower frame plate (P—Fig. JD83) from underneath tractor frame. On axle types remove front axle assembly and pivot bracket as outlined in paragraph 5, then remove lower frame plate.

Disconnect and remove the drive coupler from between crankshaft pulley and pump drive connecter. Disconnect the main pump supply line at rear end and allow fluid from oil cooler to drain into a clean container. Disconnect main supply line, oil cooler line and pressure line from pump and remove fuel line clamp from pump support. Remove support retaining bolts and lower the pump and support from its position in the tractor frame. Pump can now be detached from support.

Install the pump by reversing the disassembly procedure and tighten the retaining cap screws as indicated below:

Pump to support screws....85 ft.-lbs.
Support to tractor frame....85 ft.-lbs.
Drive coupler cap screws...32 ft.-lbs.
Coupler to pump clamp
    bolts ....20 ft.-lbs.

122. OVERHAUL. Before disassembling the removed pump, attach a dial indicator and measure the pump shaft end play which should be 0.002-0.006. Record the measurement for use in adjusting pump bearings when reassembling pump. Adjusting shims (26—Fig. JD84) are available in thicknesses of 0.006 and 0.010.

To disassemble the pump, remove the four cap screws retaining the stroke control valve housing to front of pump and remove the housing. Do not lose the bearing shims (26) located under the shaft bearing cup.

Although the piston valves and pistons are available as individual parts, it is good shop practice once they have been installed and used, to install them in their original locations. Use a compartmented pan or other means to keep them identified when pump is disassembled. Remove the discharge valve guides (13), springs (12) and valves (11) from front of pump, then remove the piston plugs (5), springs (7) and pistons (8).

Inlet valves (3) are pressed into the housing. To remove the inlet valves, remove plugs (2) and use a small pin punch working through discharge valve seat. Discharge valve seats can be removed with a punch by working from inlet valve side of housing after inlet valve assembly is removed. Examine the seats and do not remove unless renewal is indicated. The pump camshaft can be removed as an assembly consisting of shaft and bearings by tapping on spline end with a soft faced hammer. Remove snap ring (15) and pry out the shaft seal (16). When installing, drive the seal into housing only far enough to install snap ring to keep from restricting or plugging outlet hole in housing at this point.

Inspect the seating surfaces of inlet valve assembly to make sure they are in good condition. Measure the lift of each inlet valve in turn, and check for looseness of stem in guide. There should be no apparent play of valve stem and total lift should be between 0.060 and 0.082. If looseness is apparent, lift is greater than 0.082 or spring is broken or damaged, renew the valve assembly. Check the piston springs with a spring tester at a compressed length of 1⅜-inch. Springs must be matched to 0.2 lb. Check the remainder of the pump parts against the values which follow:

Housing Bore Dimensions

Drive Shaft Bore
   at "O" Ring ....1.305 -1.305
Piston Bore ID........0.8747-0.8753
Inlet Valve Bore ID...0.9995-1.0005
Discharge Valve
   Bore ID 0.6245-0.6255
Pump Piston OD......0.8740-0.8744
Drive Shaft Cam OD..1.9632-1.9638
Cam Race ID........2.3400-2.3406

Assemble by reversing the disassembly procedure. Immerse all of the parts, including "O" rings and seals, in clean hydraulic fluid when assembling. Tighten the piston plugs (5) to a torque of 100 ft.-lbs.

Check the stroke control housing as outlined in paragraph 123. Install the housing and tighten the retaining cap screws to a torque of 85 ft.-lbs. Check the pump shaft for the specified 0.002-0.006 end play, and adjust if necessary by adding or removing shims (26). After pump is installed, adjust pump pressures as outlined in paragraph 118.

123. STROKE CONTROL VALVE HOUSING. The valves located in the stroke control valve housing, control the pump output as follows:

The closed hydraulic system has no discharge line except through the operating valves. Peak pressure is thus maintained for instant use. Pumping action is halted when line pressure reaches a given point by pressurizing the pump housing underneath the pistons, in the camshaft reservoir. For a brief description of the control mechanism refer to Fig. JD85.

The drilled passage (4) in stroke control housing is ported to the inlet side of the pump pistons. Passage (5) is ported to the pressurized fluid supply line. The valve ports (6) and the spring area of the stroke control valve (3) are connected to the camshaft reservoir (crankcase) by drilled passages not shown. Valve ports (6) are also connected to inlet passage (4) when outlet valve (2) is open as shown. As pressure increases in the pump, the crankcase outlet valve (2) is moved upward against spring pressure until outlet ports (6) are closed by the valve lands. This movement takes place at approximately 2000 psi system pressure. Pumping action continues until system pressure reaches the standby pressure of approximately 2250 psi at which time the stroke control valve (3) moves against spring pressure to open the poppet valve (7). When the stroke

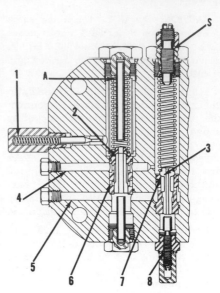

Fig. JD85—Cross sectional view of stroke control valve housing. See text for operation.

1. Relief valve
2. Crankcase outlet valve
3. Stroke control valve
4. Inlet passage
5. Pressure passage
6. Valve port
7. Valve seat
8. Lockout screw
A. Adjusting shim
S. Adjusting screw

control valve opens, the crankcase is pressurized and the pump pistons are held away from the camshaft by hydraulic pressure, thus stopping pump action. When system pressure drops below 2000 psi, ports (6) again open, relieving crankcase pressure, and pumping action is resumed.

The crankcase outlet valve pressure is controlled by means of shims (A) at upper end of valve spring. The stroke control valve pressure is controlled by means of the adjusting screw (S).

The safety valve (1) is tapped into the pressure passage and is vented into the camshaft reservoir (crankcase). It is set to open at 2650-2950 psi and only operates if the stroke control valve should stick or if filter (11 — Fig. JD86) should become plugged. Safety valve is not used on late models. Shut-off screw (23) can be screwed into pump housing to manually open the stroke control valve, locking the pump out of stroke.

To provide the correct loading of crankcase outlet valve spring (10), the manufacturer has designed a special tool (No. JDH-19). To use the tool, remove the outlet valve lower plug (16) and install the spring-loaded special tool. Add or remove shims (8) until the scribe line on tool plunger is flush with outer face of tool. Shims are available in thicknesses of 0.032 and 0.010. If the special tool is not available, use the same

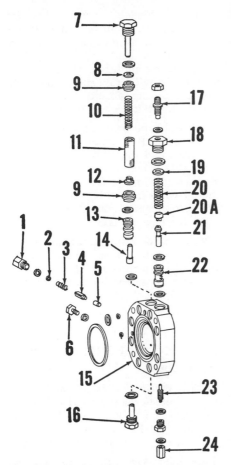

Fig. JD86—Exploded view of stroke valve housing showing component parts. Relief valve (1 through 5) is omitted on later models.

1. Plug	14. Outlet valve
2. Shim	15. Housing
3. Spring	16. Plug
4. Relief valve	17. Adjusting screw
5. Seat	18. Nut
6. Plug	19. Washer
7. Plug	20. Spring
8. Shim	20A. Guide
9. Retainer	21. Valve
10. Spring	22. Sleeve
11. Filter	23. Lockout screw
12. Guide	24. Nut
13. Valve sleeve	

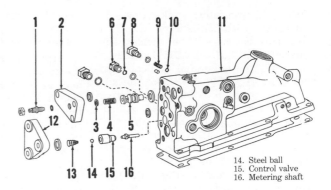

Fig. JD87 — Exploded view of control valve housing showing valves. See Fig. JD88 for control mechanism.

1. Adjusting screw
2. Flow control cover
3. Washer
4. Spring
5. Flow control valve
6. Plug
7. Shim
8. Plug
9. Spring
10. Ball
11. Housing
12. Cover
13. Valve spring
14. Steel ball
15. Control valve
16. Metering shaft

ROCKSHAFT HOUSING & COMPONENTS

All Models

125. REMOVE AND REINSTALL. To remove the rockshaft housing, first remove the rockshaft housing covers, driver's seat and platform. Disconnect the three-point lift links on tractors so equipped. Disconnect and remove all interfering wiring and hydraulic lines and disconnect the rockshaft control valve cable, then remove the attaching bolts and lift housing from tractor with a hoist.

When installing, make sure that the cam follower roller in valve housing is positioned to the rear of the follower arm in transmission housing, and lower the unit carefully so as not to damage or bend the draft control mechanism. Tighten the attaching bolts securely.

126. CONTROL VALVE HOUSING. To remove the control valve housing, remove the right rockshaft cover, disconnect the control cable and the hydraulic fluid lines. On models so equipped, remove the right breakaway couplings. Remove the cap screws retaining the control valve housing to rockshaft housing and lift off the complete control valve unit.

To disassemble, remove the flow control valve cover (2—Fig. JD87) and withdraw spring (4). Remove control valve cover (12) and withdraw springs (13), balls (14), valves (15) and metering shafts (16). Disassemble the operating mechanism by unscrewing operating shaft quill (15 —Fig. JD88), then removing operating link spring (11) and link (10). Differential gears (6 and 14) are pinned to their shafts. End play of operating hinges (3 and 4) is controlled by means of shims (7—Fig. JD87). Specified end play is 0.002-0.007 and shims are available in thicknesses of 0.005, 0.010 and 0.030.

127. CONTROL VALVE ADJUSTMENT. When assembling the control valve housing, it is necessary to adjust the operating hinges to a positive but minimum clearance by means of the adjusting screws (1 and 2—Fig.

quantity and thickness of shims (8) as were removed, then check system pressure after assembly, as outlined in paragraph 118.

If system pressure while operating is appreciably above or below the recommended 2000 psi, stop the tractor, remove plug (7) and add or remove one 0.010 shim (8) as indicated, then recheck after plug is installed.

If stroke control valve assembly is disassembled, use Fig. JD86 as a guide when reassembling. Pay special notice to the position of the spring guide (20A) which must be installed as shown. Make sure that all passages are clean and use new seals when assembling. Lubricate the parts thoroughly in hydraulic fluid before assembly.

Fig. JD88 — Control valve operating mechanism contained in control valve housing. See also Fig. JD87.

1. Adjusting screw
2. Adjusting screw
3. Operating hinge
4. Operating hinge
5. Operating shaft
6. Operating gear
7. Spring
8. Pinion
9. Control arm
10. Link
11. Spring
12. Load arm
13. Selector lever
14. Differential gear
15. Quill
16. Control lever

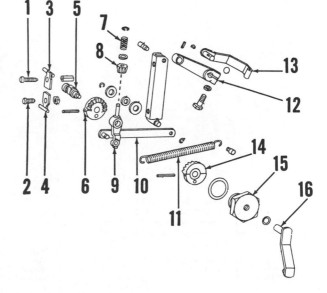

JD88). To accurately make this adjustment with a minimum of effort requires the use of the special adjusting tools JDH-8, JDH-9, JDH-10 and JDH-11 and a dial indicator as shown in Fig. JD90. To make the adjustment, proceed as follows: Remove the two covers from rear of control valve housing and remove the two springs (13—Fig. JD87) from rear of valves. Make sure that the valve balls (14) remain in housing. Back out the two screws (6 and 7—Fig. JD90) in rear of valve adjusting cover (5) and attach cover to housing as shown. Back out the valve adjusting screws (1 and 2—Fig. JD89) to make sure clearance exists, then tighten the check valve balls firmly on their seats using the screws (6 and 7—Fig. JD90). Attach the indicator (2) as shown, using the special screw (1) to retain it to operating shaft gear (6—Fig. JD88). Install the stop (3—Fig. JD90) and use the tension wire (4) to hold indicator up away from stop. Carefully turn the discharge valve adjusting screw (1—Fig. JD89) into hinge until indicator (2—Fig. JD90) just touches stop (3). Lock the discharge valve adjusting screw in position with the locknut. Zero the dial indicator on outer end of indicator as shown, then remove stop (3). Adjust the inlet valve adjusting screw (2—Fig. JD89) until total movement of indicator is 0.007-0.010 when measured with dial indicator. Be sure to use only the tension wire (4—Fig. JD90) when measuring valve movement. Because of the length of the

indicator arm, the valve clearance is barely noticeable at valve hinges and measurement is almost impossible.

After valves are adjusted, remove the tools and reinstall housing covers and valve springs. Install and time operating link arm, valve operating lever and sector gear. The linkage is correctly timed when pin on operating link arm, and the cast pointers on housing and operating lever are aligned as shown at (A—Fig. JD89). Tighten lever quill (15—Fig. JD88) securely, install cam follower link and spring, then install valve housing.

After assembly is completed, move selector lever on rockshaft housing to the lower (D) position and attach control cable (C—Fig. JD91). Start engine and check to see that rockshaft moves to full raised and full lowered position when dash control lever is moved the full length of quadrant.

The adjusting screw (A) controls the lowering speed of the rockshaft. Adjusting screw (B) controls the raising speed. These speeds, when adjusted, are independent of engine speed. Adjust the rate-of-lift valve (B) until 2-2½ seconds is required to move the rockshaft from full lowered to full raised position.

129. ROCKSHAFT HOUSING OVERHAUL. To overhaul the removed rockshaft housing assembly, first remove the control valve housing, then on all models except 3000 Utility and Series 5000, remove the piston rear cover. Piston can now be forced from cylinder by rotating the rockshaft.

On 3000 Utility and Series 5000, disconnect piston line, and remove cylinder and piston as a unit.

On all models, remove the set screw retaining cam to rockshaft, remove the retaining cap screws in ends of rockshaft and withdraw shaft from housing and arms. On 3000 Utility models, withdraw shaft from right side of housing. On all other models, withdraw shaft from left side.

Oil seals and shaft bushings can be renewed at this time. Bushings are pre-sized and should be installed with a suitable driver using care that they are not distorted. Be sure to align the oil holes in bushings and housing. Shaft to bushing clearance is 0.002-0.008 for new parts.

When assembling, tighten the two upper piston cover retaining cap screws to a torque of 115 ft.-lbs. and the four lower screws to a torque of 85 ft.-lbs.

LOAD CONTROL ARM AND SHAFT
Series 5000

129A. OPERATION. When the hydraulic selector lever is moved to the "L" (Load) position, operating depth is controlled by the draft of the implement and position of control lever.

Refer to Fig. JD91A for an exploded view of load control sensing linkage and associated parts. The lower links attach to draft link support (1) which is mounted beneath transmission housing and suspended from the two spring-steel control shafts (7 and 11). Positive or negative draft causes a controlled flexing of the shafts which is transmitted by the front shaft (11) to the pivoted arm (16) and through push rod (19) to the control valve.

Fig. JD89—After valve is adjusted, make sure that the three indicated points (A) on control arm, housing and control lever are aligned when assembling differential gears.

1. Discharge valve adjusting screw
2. Inlet valve adjusting screw
3. Operating shaft

Fig. JD90—Special tools are required to properly adjust the control valve. See text for details.

1. Tool JDH-9	5. Plate JDH-10
2. Tool JDH-11	6. Screw
3. Stop JDH-8	7. Screw
4. Tension wire	8. Dial indicator

Fig. JD91—Rear view of rockshaft housing showing points of adjustment.

A. Lowering speed adjustment
B. Raising speed adjustment
C. Control cable

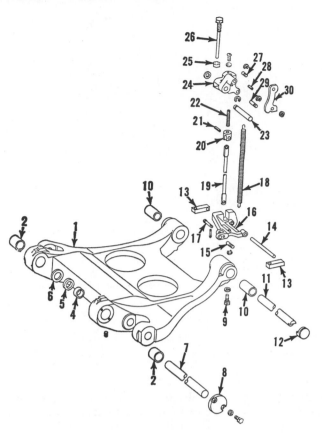

Fig. JD91A — Exploded view of load control mechanism used on Series 5000.

1. Support
2. Bushing
4. Bushing
5. Oil seal
6. Washer
7. Shaft
8. Retainer
9. Cap screw
10. Bushing
11. Shaft
12. Retainer
13. Support
14. Pivot shaft
15. Pin
16. Arm
17. Pin
18. Spring
19. Push rod
20. Rod end
21. Roll pin
22. Adjusting screw
23. Pivot shaft
24. Follower arm
25. Locknut
26. Stop screw
27. Pin
28. Roll pin
29. Pin
30. Cam follower

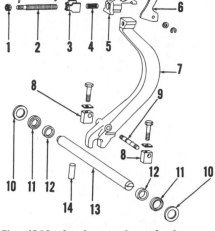

Fig. JD92—Load control mechanism used on all models except 3000 utility. Shaft (13) flexes under load to move the control valve through arm (7).

1. Nut
2. Adjusting screw
3. Lock plate
4. Spring
5. Extension
6. Follower
7. Arm
8. Bracket
9. Shaft
10. Washer
11. Seal
12. Bushing
13. Shaft
14. Stop pin

129B. REMOVE AND REINSTALL. To remove the support (1—Fig. JD-91A) or load control shafts (7 and 11), first drain transmission and remove drawbar and lower links.

Support the frame (1) on a rolling floor jack or other suitable means. Remove rear shaft retainers (8), locking cap screws (9) and front shaft retainers (12); then drive the shafts (7 and 11) either way out of draft link support frame and transmission housing.

Remove shaft washers (6) and seals (5) from the two shaft bores on each side of transmission housing and renew the seals, whenever shaft is removed.

Inspect shaft bushings (2, 4 and 10) in transmission housing and draft link support frame. If bushings must be renewed, insert a long rod through one bushing and drive the opposite bushing from housing or support. Inside diameters of shaft bushings are tapered. Install bushings in transmission housing from outside, small ID end first. Install bushings in support frame from outside, large ID end first. When properly installed, large ID ends of mating bushings in housing and support frame will be together.

Install seals (5) in transmission housing until they bottom, and washers (6) with outside of washer flush with outside of transmission housing.

Renew front shaft (11) if badly worn at contact point of control arm (16), and either shaft if worn at bushing contact area or otherwise damaged. Cutout sections on ends of front shaft should be down when shaft is installed.

Sensing linkage can be removed after draining transmission and removing seat, rockshaft housing and operator's platform. Remove locking cap screws (9) and retainers (12), and drive out front shaft (11) only far enough to clear control arm (16). Remove the cap screws securing retainers (13) to transmission housing and lift the load control arm (16) and push rod from housing as a unit. Spring (18) should require a pull of 18-22 lbs. to extend spring to a length of 18⅝ inches.

Reinstall by reversing the removal procedure, using Fig. JD91A as a guide. Use care when installing load control shaft (11) to prevent damage to oil seals (5) in housing bores.

129C. ADJUSTMENT. First check to be sure control cable is properly adjusted to allow rockshaft to move from full raised to full lowered position when control lever is moved the full length of quadrant.

Set the load selector lever in "L" (Load) position. Remove the hex head plug from top of rockshaft housing at right of seat support and, working through plug hole, turn adjusting screw (22—Fig. JD91A) until rockshaft will just start to raise when leading edge of control lever is aligned with "1½" mark on quadrant. Rockshaft should lower when leading edge of control lever reaches "2½" mark.

Move control lever toward "1½" mark as far as possible without causing lift arms to raise, loosen locknut (25) (on top of rockshaft housing rearward from plug hole) and turn stop screw (26) clockwise until rockshaft starts to raise; then counterclockwise ½ turn. Tighten locknut (25) to secure the adjustment.

All Other Models

130. OPERATION. When the hydraulic selector lever is moved to the "L" (Load) position, the operating depth of the three-point hitch is controlled by the draft of the attached implement acting in conjunction with the position of the control lever.

The amount of draft is transmitted by the lower links to the drawbar frame, then to the control valve by the load control arm and shaft shown in Fig. JD92 or JD93. The spring steel

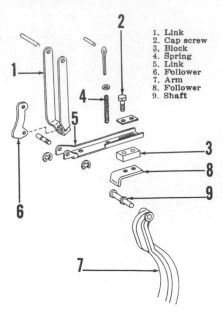

1. Link
2. Cap screw
3. Block
4. Spring
5. Link
6. Follower
7. Arm
8. Follower
9. Shaft

Fig. JD93—Load control arm and extension used on 3000 utility models. Remainder of mechanism is similar to Fig. JD92.

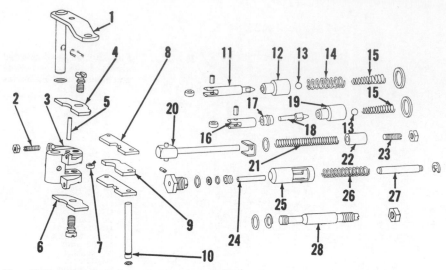

Fig. JD95—Exploded view of control mechanism and valves used in selective (remote) control valve.

1. Arm	8. Detent	15. Spring	22. Plug
2. Adjusting screw	9. Cam	16. Follower	23. Adjusting screw
3. Rocker	10. Shaft	17. Guide	24. Pin
4. Cam	11. Follower	18. Metering shaft	25. Flow control valve
5. Pin	12. Valve	19. Valve	26. Spring
6. Cam	13. Valve ball	20. Guide	27. Guide
7. Roller	14. Spring	21. Spring	28. Metering valve

shaft (13) is anchored in each side of the transmission housing and the drawbar frame is affixed to the outer ends. Positive or negative draft causes the load control shaft to deflect a predetermined amount according to the load encountered. The center arc of the flexing shaft (13) moves the straddle-mounted lower end of load control arm (7) around pivot shaft (9), thus moving the control mechanism to compensate for changes in draft.

The rockshaft follower arm (6) is attached to the load control arm, enabling one adjustment to synchronize the valves for any type of hydraulic control. Adjustment is made by means of adjusting screw (2) as outlined in paragraph 132.

Fig. JD94—To adjust the load control arm on all models except 3000 Utility and Series 5000, remove plug (P) and use a socket and extension to turn adjusting screw (A). Rockshaft housing is removed for illustration only.

131. **REMOVE AND REINSTALL.** Removal of the load control arm on tractors equipped with power take-off, can only be accomplished in conjunction with removing the differential assembly and the procedure is outlined in paragraph 96.

On tractors not equipped with PTO, removal of the load control arm can be accomplished as follows:

Drain transmission and hydraulic fluid and remove rockshaft control valve housing as outlined in paragraph 126. Remove cam follower (6—Fig. JD 92 or 93) from upper end of load control arm. Remove transmission rear cover, the two cap screws retaining control arm shaft brackets (8), then withdraw the control arm assembly out rear of housing. Install by reversing the above procedure.

If the load control shaft only, is to be renewed, proceed as follows: Drain the transmission and hydraulic fluid and remove the set screws which hold the load control shaft retainers at each side of the drawbar frame. With a brass drift, and working from right side of tractor, bump the shaft to the left and out of the tractor. The drawbar frame will be free to drop at the rear end when the shaft is removed.

Bushings in transmission housing and drawbar frame, and the housing oil seals may be renewed at this time. The inside diameter of the bushings is tapered to provide a small bearing area for the flexing shaft. Install the bushings in transmission housing with the small ID to inside of housing, and

the bushings in drawbar frame with the small ID to the outside away from transmission housing. Lubricate the shaft and install carefully to prevent damage to the oil seals. Shaft must be installed with the cut-out portions in ends of shaft to the bottom.

Check the load control shaft, shaft bushings and load control arm against the dimensions which follow:
Load control shaft OD ...0.998-1.000
Load control shaft bushing,
  small ID 1.003-1.005
Load control arm yoke,
  width 1.005-1.015

132. **ADJUSTMENT.** To adjust the control mechanism on all models except 3010 Utility, remove the plug (P—Fig. JD94). To adjust the load

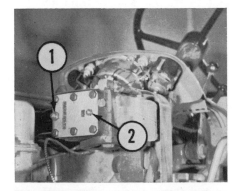

Fig. JD96—Single remote valve used on some models. When two remote valves are used, the second identical valve attaches below the one shown.

1. Metering valve adjusting screw
2. Detent adjusting screw

control arm on 3000 Utility models, remove the driver's seat and small inspection plate.

With the engine running and selector lever in "D" (Depth) position, move the three-point hitch control lever up the quadrant until leading edge of lever is aligned with the "O" mark on quadrant scale. Adjust rockshaft control cable, if necessary until rockshaft moves to the full raised position. Move quadrant lever to rear position and check to make sure that rockshaft moves to the full lowered position.

Move the selector lever to "L" (Load) position and quadrant lever until leading edge is aligned with the "1½" mark on quadrant scale. Working through the plug hole (all except 3000 Utility) or inspection plate (Utility models), adjust the internal linkage to the point where lift arms just begin to raise. Links should start to lower when quadrant lever is moved to "2" mark on quadrant scale.

Move selector lever to "L-D" (Load and Depth) position and make sure that rockshaft moves through the full range of travel.

SELECTIVE (REMOTE) CONTROL VALVES
All Models

133. **OPERATION.** Tractors are optionally equipped with one or two selective (remote) control valves for the independent operation of remote cylinders. The valves are mounted on left front of firewall and con-

trolled by levers on the instrument panel.

As with all other units of the hydraulic system, pressure is always present at the valves but no flow exists until the valve is moved. Refer to Fig. JD95 for an exploded view of the valve mechanism. Each valve assembly is equipped with two ball-check pressure and discharge valves which are unseated by plungers when valve is actuated. Flow control valve (25) limits the incoming fluid flow for feathering action when valve lever is moved only slightly, and also seats in housing for a double pressure check when valve is in neutral position. Maximum valve flow is limited by the adjustable metering valve (28). When valve lever is moved to full raising or lowering, position is maintained by notches in locking detents (8). When the attached remote cylinder reaches the end of it's stroke, the valve is centered by the flow control valve acting through plunger (24) on centering cam (9). The pressure of detent spring (21) is adjustable from front of valve housing by means of adjusting screw (23).

134. **REMOVE AND REINSTALL.** To remove the selective control valve or valves, first remove the cowl and hood, disconnect the remote cylinder lines and fluid pressure and return lines. Cap all connections to prevent

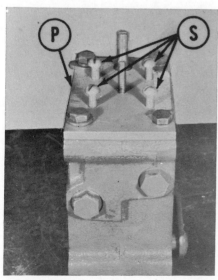

Fig. JD97—To adjust the selective (remote) valve, attach the special adjusting plate JDH-15 as shown at (P) then tighten the four screws (S) down to firmly hold the operating valve balls on their seats.

Fig. JD98—With adjusting plate installed as shown in Fig. JD97, tighten the adjusting screws (P and D) down to remove all end play, then back out the two pressure valve adjusting screws (P) ½ turn and the discharge valve adjusting screws (D) ⅜ turn. Tighten the locknuts when valve is adjusted.

dirt entry. Disconnect the operating rods at the valves then unbolt and remove the valve assemblies.

135. **OVERHAUL.** Service, in most instances, consists of adjustment as outlined in paragraph 136, or in renewing the damaged or worn parts. Before removing the valve, conduct a leakage test as outlined in paragraph 119. To disassemble the valve remove the end caps, then withdraw the springs and operating valves. Thoroughly clean the parts and examine the check balls (13—Fig. JD95) and seats (12 and 19) for wear or other damage. Make sure the flow control valve (25) slides freely in it's bore and examine the valve and seat. To disassemble the operating mechanism, drive out the roll pin which retains operating shaft (1) to rocker (3) and withdraw the parts. Remove the two special screws which retain operating cams (4 and 6) to housing, then withdraw detent and centering cam shaft and cams.

Use all new "O" rings and seals when assembling, and renew any other parts which are questionable, then adjust the valve as outlined in the following paragraph.

136. **ADJUSTMENT.** Two valve adjustments are made with the valve assembled, and in operating condition. These adjustments are the metering valve adjustment and detent spring adjustment. To synchronize the operating valves, the assembly must be removed. To make the adjustments, proceed as follows:

137. **METERING VALVE.** To adjust the metering valve, install a 3 x 8 inch double acting cylinder, operate the valve a few times to bleed air from the system, then fully retract the cylinder. With the engine running at 1900 rpm, move the control lever fully into the extend position and check the time required for the cylinder to fully extend. The time should be 1½ to 2 seconds. To adjust, loosen the locknut on the metering valve (1—Fig. JD96) and turn the valve in or out until the adjustment is correct.

138. **DETENT SPRING.** The detent spring and plunger should hold the control lever in the operating position until the attached cylinder reaches the end of it's stroke, then release and allow the lever to return to neutral. To adjust, attach a double acting cylinder to the valve and check for

improper performance. If detent fails to hold control valve in operating position, loosen the locknut on adjusting screw (2—Fig. JD96) and thread the screw into the cover approximately ½ turn and recheck. If the condition cannot be corrected, disassemble and overhaul the valve as outlined in paragraph 135.

If the detent fails to release when the cylinder reaches the end of the stroke, loosen the adjusting screw slightly and recheck. If the condition cannot be corrected, check the flow control valve and spring, or disassemble and overhaul the valve as outlined in paragraph 135.

139. VALVE SYNCHRONIZATION. To synchronize the operating valves, remove the valve unit as outlined in paragraph 134, and remove both end covers. To adjust, the four operating valve balls and the flow control valve must be firmly held on their seats. The manufacturer provides an adjusting cover as shown in Fig. JD97 to seat the valves. Install the cover and tighten the five adjusting screws finger-tight against the check balls and flow control valve. Loosen the jam nuts on the four control valve adjusting screws (2—Fig. JD95) and the two special screws retaining operat-

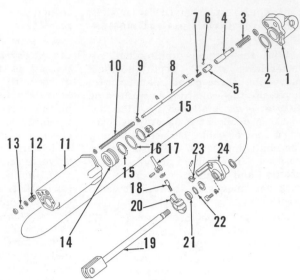

Fig. JD99 — Exploded view of the high pressure remote cylinder used on all models so equipped.

1. Cap
2. Gasket
3. Spring
4. Stop valve
5. Bleed valve
6. Ball
7. Spring
8. Stop rod
9. Washer
10. Spring
11. Cylinder
12. Spring
13. "V" packing
14. Piston
15. Backup ring
16. "O" ring
17. Lever
18. Stop screw
19. Piston rod
20. Stop
21. Oil seal
22. Backup ring
23. Arm
24. Guide

ing cams (4 and 6). Turn the adjusting screws (2) into rocker until they are seated. Back off the pressure valve adjusting screws (Upper left and lower right) ½ turn and tighten the jam nuts. Back off the discharge valve adjusting screws (Upper right and lower left) ⅜ turn and tighten the jam nuts. Hold the operating cams (4 and 6) back against the adjusting screws and tighten the retaining screws. Reassemble and install the valve and adjust metering valve and detent spring as outlined in paragraphs 137 and 138.

REMOTE CYLINDER
All Models

140. Refer to Fig. JD99 for an exploded view of the double acting, hydraulic stop, remote cylinder. To disassemble, remove end cap (1), stop rod spring (3) and valves (4 and 5), using care not to lose ball (6). Fully retract the cylinder and remove nut from piston end of piston rod (19). To remove the stop rod and springs, drive the groove pin from stop rod arm (23). The procedure for further disassembly is obvious if required for repair or renewal of parts.

NOTES

JOHN DEERE

Series ■ 3020 (SN. 123000 & up) ■ 4000
■ 4020 (SN. 201000 & up) ■ 4320
■ 4520 ■ 4620

Previously contained in I & T Shop Service Manual No. JD-53

SHOP MANUAL

JOHN DEERE

SERIES

3020 (Serial   4000   4320
No. 123000 and After)   4020 (Serial No.   4520
                        201000 and After)   4620

Tractor serial number located on rear of transmission case. Engine serial number located on front right side of engine block.

INDEX (By Starting Paragraph)

CONDENSED SERVICE DATA

GENERAL	3020 LP Gas	3020 Gasoline	3020 Diesel	4000 Gasoline	4000 Diesel	4020 LP Gas
Engine Make		OWN			OWN	
Engine Model	4241L	4241G	4270D	6362G	6404D	6362L
Number of Cylinders		4			6	
Bore-Inches	4.25	4.25	4.25	4.25	4.25	4.25
Stroke-Inches	4.25	4.25	4.75	4.25	4.75	4.25
Displacement-Cu. In.	241	241	270	362	404	362
Induction*	N-A	N-A	N-A	N-A	N-A	N·A
Cylinder Sleeves		WET			WET	

TUNE-UP						
Firing Order		1-3-4-2			1-5-3-6-2-4	
Valve Tappet Gap–						
Exhaust-Inch	0.028	0.028	0.018	0.028	0.018	0.028
Inlet-Inch	0.015	0.015	0.018	0.015	0.018	0.015
Ignition or Injection Timing–						
Static	....	....	TDC	....	TDC	....
Running	25° at 2100	20° at 2200	....	20° at 2000	....	25° at 2000
Spark Plug Electrode						
Gap-Inch	0.015	0.025	....	0.025	....	0.015
Breaker Point Gap-						
Inch	0.022	0.022	....	**	....	**
Governed Speeds-Engine RPM–						
Low Idle	800	800	800	800	800	800
High Idle	2690	2690	2660	2440	2400	2440
Loaded	2500	2500	2500	2200	2200	2200

SIZES-CAPACITIES-CLEARANCES						
Cooling system (Quarts)+	19	19	19	24	24	24
Crankcase Oil (Quarts)– Including Filters	8	8	8	8	12	8
High Clearance Final Drive Housing (Quarts)–	1¾	1¾	1¾	1¾	1¾	1¾
Transmission & Hydraulic System (Gallons)++						
Syncro-Range	8	8	8	10	10	10
Power Shift	11	11	11	14	14	14
Fuel Tank (Gallons)	33.6	29	29	34	34	45
Crankshaft Sizes and Clearances		See Paragraph 54			See Paragraph 54	
Piston, Rings & Sleeves		See Paragraph 51			See Paragraph 51	
Battery:						
Volts		12			12	
Ground Polarity		Negative			Negative	

TIGHTENING-TORQUES-Ft.-Lbs.						
Cylinder Head-Final	130	130	130	130	130	130
Main Bearing Caps	150	150	150	150	150	150
Connecting Rod Caps	85-95	85-95	100-110	85-95	100-110	85-95

*N-A = Naturally Aspirated; T = Turbocharged; T-I = Turbocharged and Intercooled.
**0.016 inch prior to engine S.N. 280001; 0.021 inch engine S.N. 280001 and after.
+Add approximately 2 quarts if equipped with heater.
++Approximate capacity only. When draining, 3 to 6 gallons may remain in case. Add 4½-5 gallons if equipped with Power Front Wheel Drive.

CONDENSED SERVICE DATA

	4020 Gasoline	4020 Diesel	4320 Diesel	Early 4520 Diesel	Late 4520 Diesel	4620 Diesel
GENERAL						
Engine Make			OWN			
Engine Model	6362G	6404D	6404T	6404T	6404A	6404A
Number of Cylinders			6			
Bore-Inches	4.25	4.25	4.25	4.25	4.25	4.25
Stroke-Inches	4.25	4.75	4.75	4.75	4.75	4.75
Displacement-Cu. In.	362	404	404	404	404	404
Induction*	N-A	N-A	T	T	T-I	T-I
Cylinder Sleeves			WET			
TUNE-UP						
Firing Order			1-5-3-6-2-4			
Valve Tappet Gap—						
Exhaust-Inch	0.028	0.018	0.028	***	***	0.028
Inlet-Inch	0.015	0.018	0.018	0.018	0.018	0.018
Ignition or Injection Timing—						
Static	....	TDC	TDC	TDC	TDC	TDC
Running	20° at 2000	....	....	....	....	....
Spark Plug Electrode						
Gap-Inch	0.025	....	....	....	....	....
Breaker Point Gap—						
Inch	**	....	....	....	....	....
Governed Speeds-Engine RPM—						
Low Idle	800	800	800	800	800	800
High Idle	2440	2400	2400	2400	2400	2400
Loaded	2200	2200	2200	2200	2200	2200
SIZES-CAPACITIES-CLEARANCES						
Cooling System						
(Quarts)+	24	24	28	28	28	28
Crankcase Oil (Quarts)-						
Including Filters	8	12	16	16	16	16
High Clearance Final						
Drive Housing (Quarts)	1¾	1¾	....	....	....	....
Transmission & Hydraulic						
System (Gallons)+ +						
Syncro-Range	10	10	14	18	18	18
Power Shift	14	14	....	16	16	16
Fuel Tank (Gallons)	34	34	44	50	50	50
Crankshaft Sizes and Clearances . . .			See Paragraph 54			
Piston, Rings & Sleeves			See Paragraph 51			
Battery:						
Volts			12			
Ground Polarity			Negative			
TIGHTENING-TORQUES-Ft.-Lbs.						
Cylinder Head-Final	130	130	130	130	130	130
Main Bearing Caps	150	150	150	150	150	150
Connecting Rod Caps	85-95	100-110	100-110	100-110	100-110	100-110

  *N-A = Naturally Aspirated; T = Turbocharged; T-I = Turbocharged and Intercooled.

  **0.016 inch prior to engine S.N. 280001; 0.021 inch engine S.N. 280001 and after.

  ***0.022 inch prior to engine S.N. 303519; 0.028 inch engine S.N. 303519 and after.

  + Add approximately 2 quarts if equipped with heater.

  + +Approximate capacity only. When draining, 3 to 6 gallons may remain in case. Add 4½-5 gallons if equipped with Power Front Wheel Drive.

FRONT SYSTEM

Three different tricycle front end units have been used: Single front wheel, Dual front wheels and "Roll-O-Matic" dual wheel tricycle units. Tricycle front end units require a special front support and steering motor assembly. Some units are convertible to adjustable axle models.

Adjustable front axle on tractors without front drive may be Low Profile, High Crop or standard height in narrow, regular or wide widths. Adjustable front axle on tractors with Power Front Wheel Drive is regular width only.

A Low Profile front axle with fixed tread width is available on some models.

TRICYCLE FRONT END UNITS

Models So Equipped

1. **REMOVE AND REINSTALL.** The spindle extension (pedestal) attaches directly to steering motor spindle by four cap screws. To remove unit, support front of tractor with a hoist using special John Deere lifting plate (JDG-3) or other suitable support attached to tractor front support. Remove retaining cap screws and roll assembly forward away from tractor. Tighten retaining cap screws to a torque of 300 ft.-lbs. when unit is reinstalled.

2. **SINGLE WHEEL TRICYCLE.** The fork mounted single wheel is supported on taper roller bearings as shown in Fig. 1. Bearings should be adjusted to provide a slight rotational drag by means of adjusting nut (4). The one-piece wheel and rim assembly (5) accommodates a 7.50x16 tire, the two-piece rim (6) a 11.00x12 tire. Wheel fork (1) is not interchangeable for the two tire sizes.

3. **DUAL WHEEL TRICYCLE.** An exploded view of dual wheel tricycle pedestal and hub is shown in Fig. 2. Horizontal axles are not renewable. Service consists of renewing complete pedestal assembly.

4. **ROLL-O-MATIC UNIT.** The "Roll-O-Matic" front wheel pedestal and associated parts are shown exploded in Fig. 3. The unit can be overhauled without removing assembly from tractor.

Support front of tractor and remove wheel and hub units. Remove knuckle caps (6) and thrust washers (5), then pull knuckle and gear units (4) from housing.

The "Roll-O-Matic" unit is equipped with a lock (2) and lock support (3) which may be installed for rigidity when desired. Renew thrust washers (5) if worn. Bushings are presized and contain a spiral oil groove which extends to one edge of bushing. When installing new bushings, use a piloted arbor and press bushings into knuckle arm so OPEN end

of spiral grooves are together as shown in Fig. (4). Bushing at spindle end of "Roll-O-Matic" unit should be pressed into arm so outer edge (B) is 1/32-inch below edge of bore. There should be a gap (C) of 1/32 to 1/16-inch between bushings and distance (A) from edge of inner bushing to edge of bore should measure 3/16-inch. Soak felt washers in engine oil prior to installation. Install one of the knuckles so wheel spindle extends behind vertical steering spindle. Pack "Roll-O-Matic" unit with multi-purpose type grease and install other knuckle so timing marks on gears are in register as shown at (M—Fig. 5). Tighten thrust washer attaching screws to a torque of 55 ft.-lbs., and bend up corners of lock plates.

ADJUSTABLE AND FIXED TREAD AXLES

Models So Equipped

5. **HOUSING AND PIVOT BRACKET.** The front axle attaches directly to front support or to a pivot

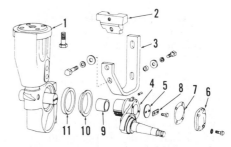

Fig. 3 — "Roll-O-Matic" spindle extension showing component parts. Lock (2) and support (3) may be installed to increase rigidity.

1. Pedestal extension	
2. Lock	7. Gasket
3. Support	8. Lock plate
4. Knuckle	9. Bushing
5. Thrust washer	10. Felt retainer
6. Cap	11. Felt washer

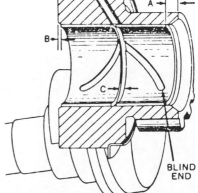

Fig. 4 — Cross-sectional view of knuckle showing details of bushing installation.

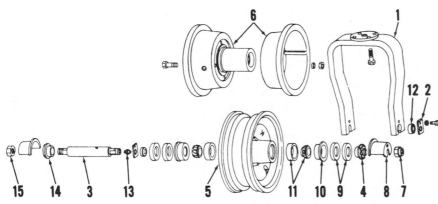

Fig. 1 — Exploded view of front wheel fork and axle assembly used on single wheel tricycle models. One-piece wheel (5) is for 7.50x16 inch tire and two-piece wheel (6) for 11.00x12 inch tire.

1. Fork	5. One-piece wheel	9. Felt washers
2. Lock plate	6. Two-piece wheel	10. Felt retainer
3. Axle	7. Nut	11. Bearing assy.
4. Bearing adjusting nut	8. Dust shield	12. Washer
		13. Grease fitting
		14. Spacer
		15. Nut

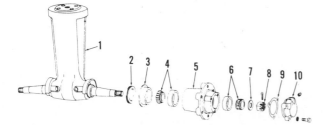

Fig. 2 — Dual wheel tricycle pedestal showing one wheel hub and associated parts.

1. Pedestal
2. Oil seal
3. Oil seal cup
4. Bearing assy.
5. Hub
6. Bearing assy.
7. Washer
8. Nut
9. Gasket
10. Cap

bracket. Refer to Fig. 6 or 7. On all models, install rear pivot pin bushing flush with bottom of chamfer and front pivot pin bushing flush with rear edge of pin hole.

6. SPINDLES AND BUSHINGS. Steering arm is splined to spindle and retained by a bolt. Refer to Fig. 6 or 7. Spindle bushings are presized. Shim washers are provided to adjust end play of spindle.

Refer to paragraph 8 for service to front axle used with power front wheel drive.

7. TIE RODS AND TOE-IN. The outer tie rod ends on axles with adjustable tread are adjustable with several holes provided. On all models, the inner ends are threaded. Make sure tie rods are equal length so tractor will turn the same in either direction. Adjust toe-in to ⅛ to ⅜-inch.

POWER FRONT WHEEL DRIVE AXLE

Models So Equipped

8. AXLE AND PIVOT BRACKET. The front axle attaches to a pivot bracket. Service will normally consist of replacing pivot pins and bushings.

9. SPINDLE (MOTOR HOUSING) AND BUSHINGS. Refer to Fig. 8 for exploded view showing parts location and arrangement. Bushings (3) are presized. When reassembling, be sure to install quad rings (6) and back-up washers (7) in order shown.

Refer to paragraph 22 for service to drive portion of power front wheel drive axle.

10. TIE RODS AND TOE-IN. Check toe-in, and adjust if necessary. Adjust tie rod ends equally to provide ⅛ to ⅜-inch toe-in. Tighten tie rod clamp bolts to 35 ft.-lbs. torque.

STEERING SYSTEM

All models are equipped with a full power steering system. No mechanical linkage exists between steering wheel and front unit; however, steering can be manually accomplished by hydraulic pressure when tractor hydraulic unit is inactive. Power is supplied by the same hydraulic pump which powers lift and brake systems. A pressure control (priority) valve is located in outlet line from main hydraulic pump. Valve gives steering system first priority on hydraulic flow.

OPERATION

All Models

11. The power steering system consists of tractor hydraulic supply system described in paragraph 193, plus steering control unit and steering cylinders or motor described in this section (See Fig. 9).

The control unit contains a double acting piston (2) of approximately equal

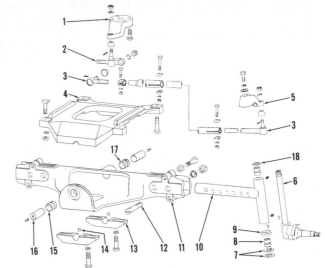

Fig. 6 — Exploded view of adjustable front axle used on some models.

1. Steering motor arm
2. Inner tie rod connector
3. Tie rod ends
4. Axle pivot bracket
5. Steering arm
6. Spindle
7. Thrust washers
8. Bushing
9. Dust shield
10. Knee
11. Axle housing
12. Lockbolt
13. Pivot pin clamp
14. Dowel
15. Rear bushing
16. Pivot pin
17. Front bushing
18. Shim washer

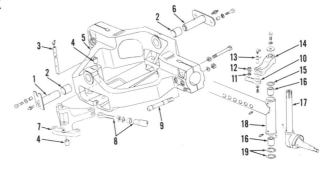

Fig. 7 — Exploded view of adjustable front axle used on models with steering cylinders.

1. Rear pivot pin
2. Bushing
3. Bellcrank pivot pin
4. Bushings
5. Axle housing
6. Front pivot pin
7. Bellcrank
8. Tie rod
9. Lockbolt
10. Tie rod end
11. Washer
12. Bushing
13. Pin
14. Steering arm
15. Shim washer
16. Bushings
17. Spindle
18. Knee
19. Thrust washers

Fig. 5 — Make sure timing marks (M) are aligned when installing knuckles in "Roll-O-Matic" unit.

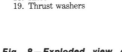
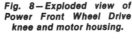

Fig. 8 — Exploded view of Power Front Wheel Drive knee and motor housing.

1. Knee
2. Cap
3. Bushing (2 used)
4. Thrust washers
5. Pivot pin (2 used)
6. Quad rings
7. Back-up rings
8. Tube
9. Cap
10. Motor housing
11. Steering arm

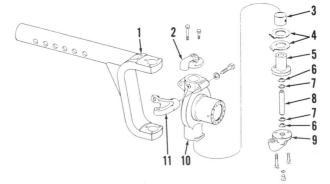

displacement to the two operating pistons (14). In addition, it contains two pressure valves, two return valves and a check valve which actuate the power assist.

Models 4520 and 4620 use two, single acting displacement type steering cylinders; all other models use a steering motor, consisting of a main housing, steering spindle (15) and two rack-type double piston assemblies (14).

When control unit is in neutral position, there is no fluid flow, but oil at pump pressure is available at pressure line (10). When steering wheel is turned for a right-hand turn, first movement of steering shaft (1) and operating collar (4) causes lower operating valve lever (7) to unseat inlet valve (8) and return valve (11). Fluid at pump pressure then enters area below steering valve piston (2) forcing trapped fluid above piston through right-hand steering motor line (16) into right-rear and left-front steering cylinders. At the same time, unseating of return valve (11) allows fluid in opposite ends of steering motor cylinders to return to sump.

When steering wheel is turned for a left-hand turn, first movement of steering shaft (1) and operating collar (4) causes upper operating valve lever (6) to unseat inlet valve (5) and return valve (13). Fluid at pump pressure then passes through inlet valve (5) and left-hand steering motor line (17) into left-rear and right-front steering cylinders. Fluid at opposite ends of cylinders is forced back through line (16) to area above steering valve piston (2). Fluid below the piston returns to sump through unseated return valve (13).

In manual steering, no pressure at inlet, spring pressure seats inlet check valve ball (9) and return check valve (12), trapping fluid between control unit and steering motor. The piston (2) is manually moved by steering shaft screw (3) and manual steering is accomplished by exchange of trapped fluid between piston and steering motor.

TROUBLESHOOTING
All Models

12. Check fluid level before proceeding.

NO POWER STEERING. If Fluid reservoir is full, operate another function, such as rockshaft or a remote control cylinder to determine if problem is lack of system pressure. If other function also fails to operate, refer to paragraph 193 HYDRAULIC SECTION for possible causes.

NO OR POOR POWER STEERING TO LEFT.
Leaking right pressure or right return valve
Steering motor piston seal failure

NO OR POOR POWER STEERING TO RIGHT
Upper or lower steering valve piston rod seal failure
Left return or left pressure valve leaking
Steering check valve piston seal failure
Steering motor piston seal failure

NO OR POOR MANUAL STEERING TO LEFT
Steering valve cylinder piston seal failure
Synchronizing valve failure
Lower steering valve piston rod seal failure
Inlet check valve seat or seal failure
Steering check valve ball seat failure
Steering check valve piston stuck
Steering motor piston seal failure

NO OR POOR MANUAL STEERING TO RIGHT
Upper steering valve piston rod seal failure
Steering valve cylinder piston seal failure
Synchronizing valve leaking
Steering check valve piston seal failure
Steering motor piston seal failure

STEERING WANDERS TO LEFT OR RIGHT
Upper steering piston rod seal
Steering valve cylinder piston seal
Steering valve operating shaft collar assembly loose
Left pressure and return valve leaking
Steering check valve piston seal failure
Improper valve adjustment
Steering motor piston seal failure

FREQUENT SYNCHRONIZATION
Upper steering valve piston rod seal failure

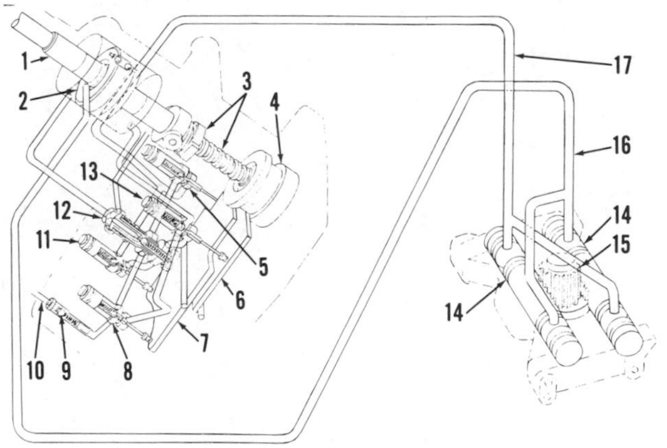

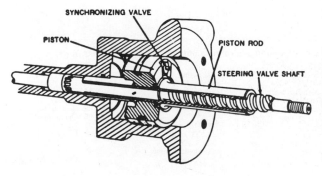

Fig. 9—Schematic view of power steering system showing operating parts. Some models use two, single acting steering cylinders in place of steering motor.

1. Steering shaft
2. Piston
3. Actuating screw
4. Actuating collar
5. Inlet valve
6. Operating lever
7. Operating lever
8. Inlet valve
9. Check valve
10. Pressure inlet line
11. Return valve
12. Check valve
13. Inlet valve
14. Operating pistons
15. Steering spindle
16. Steering line
17. Steering line

Fig. 10—Cross-sectional view of steering valve operating piston, cylinder and steering shaft. Piston is moved up or down in cylinder by helical thread on steering shaft. The synchronizing valve which corrects for internal leaks is shown in Fig. 11.

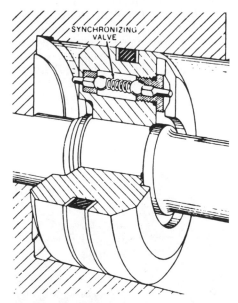

Fig. 11—Steering valve piston must be synchronized with steering motor for full turning action. Synchronization is automatically accomplished. When control valve piston reaches end of its stroke, the extended rod unseats ball check valve allowing pressurized fluid to flow through piston until motor and valve are synchronized.

Steering check valve piston seal
failure
Steering motor piston seal failure
Synchronizing valve failure
EXCESSIVE STEERING WHEEL
FREE PLAY
Steering valve shaft nut and rod loose
Collar loose
Air in steering system

BLEEDING
All Models

13. To bleed air from system, engine must be shut off at least one hour prior to bleeding. Remove cowling and attach a transparent hose to bleed screw (Fig. 12), running free end back to reservoir. Before starting engine, open bleed screw and turn steering wheel to full right. Start engine and run at slow idle speed. When fluid is free of air, close bleed screw. Turn steering wheel full left and open bleed screw, then turn steering wheel very slowly, so front wheels do not turn, to full right. Close bleed valve and allow front wheels to turn full right.

Repeat procedure, if necessary, until air-free fluid is being returned to reservoir.

PRESSURE CONTROL (PRIORITY) VALVE
All Models

14. **OPERATION.** The pressure control (priority) valve is mounted under engine cowl on right side. The valve cuts off hydraulic flow to hydraulic lift system whenever system pressure drops below 1650-1700 psi, thus giving priority to steering and brakes.

15. **R&R AND OVERHAUL.** To remove valve, first remove cowl and hood. Relieve hydraulic pressure, then disconnect pressure and bleed lines. Unbolt and remove valve assembly. Refer to Fig. 13 for exploded view of valve assembly. Remove inlet connector (9), valve (8), orifice (7), shim (6) and spring (2) from body (1).

Inspect all parts for wear or other damage. Spring should have approxi-

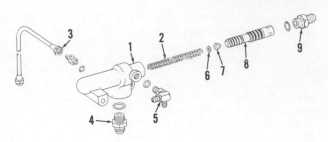

Fig. 13—Exploded view of steering priority valve. Shims (6) adjust operating pressure.

1. Body
2. Spring
3. Bleed line
4. Outlet to hydraulic systems
5. Outlet to steering & brakes
6. Shim
7. Orifice
8. Valve
9. Inlet from pump

mate free length of 4⅝ inches and test 45-55 lbs. when compressed to height of 3½ inches. Shims (6) are used to adjust spring tension and operating pressure.

STEERING CONTROL UNIT
All Models

16. **REMOVE AND REINSTALL.** To remove steering control unit, first relieve system hydraulic pressure. Remove steering wheel with suitable puller. Remove hood and cowl, then dis-

connect electrical wiring and remove dash. Disconnect steering valve hydraulic lines, being sure to cap all fittings, then unbolt and remove steering control unit.

When reinstalling, tighten steering wheel nut to 50 ft.-lbs. torque. Bleed steering system as outlined in paragraph 13.

17. **OVERHAUL.** To disassemble removed steering control unit, first remove lower cover (40–Fig. 14).

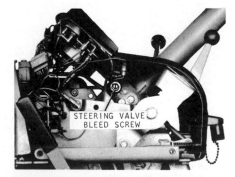

Fig. 12 — To bleed steering system, remove cowl and attach bleed line to bleed screw. Refer to paragraph 13 for procedure.

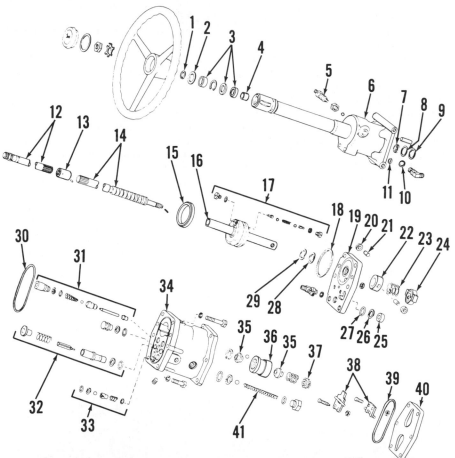

Fig. 14 — Exploded view of steering valve assembly showing component parts.

1. Snap ring	13. Coupling	22. Collar	32. Check valve
2. Washer	14. Steering valve	23. Spring	33. Inlet check valve
3. Seal assy.	operating shaft	24. Nut	34. Steering valve
4. Bushing	15. Seal	25. Check valve piston	housing
5. Bleed screw	16. Piston	26. Back-up washer	35. Bearing race
6. Steering cylinder	17. Synchronizing valve	27. "O" ring	36. Operating collar
7. Thrust bearing	assy.	28. Back-up washer	37. Nut
8. Back-up washer	18. Seal ring	29. "O" ring	38. Operating levers
9. "O" ring	19. Steering valve cover	30. Seal ring	39. Seal ring
10. "O" ring	20. Roller	31. Control valve	40. Cover
11. Seal	21. Pin	(4 used)	41. Check valve spring
12. Steering shaft			

Check valve adjustment prior to complete disassembly, especially if previous tests have indicated incorrect valve adjustment. (Refer to paragraph 18 for adjustment procedures.)

Remove cotter pin and nut (37–Fig. 14) from lower end of steering shaft, then unbolt and withdraw valve housing (34) and operating collar (36). Do not remove collar before removing housing to prevent damaging valve levers. Remove check valve assembly (32) as housing is removed (see Fig. 15).

Remove nut (24–Fig. 14), spring (23) and collar (22). The two rollers and pins in collar must be removed before collar can be withdrawn. Temporarily install steering wheel and turn wheel counterclockwise to force cylinder cover (19) from cylinder housing (6), then withdraw cover, piston and lower steering shaft from cylinder housing as a unit. Upper steering shaft is retained in steering column by a snap ring. Remove shaft if service is required on oil seal, bushing, shaft or housing. Withdraw springs from operating valves in valve housing, remove valve balls and inspect balls and seat for line contact.

NOTE: If inlet check valve spring or guide damage is noted, there has been excessive flow through the assembly. Check for foreign particles in a control valve orifice or for particles holding control valve or control valve ball off its seat. Check for seat damage due to foreign material.

Check synchronizing valve assembly for sticking. This can prevent proper synchronizing between valve and motor. Renew any parts that are damaged or worn.

Clean all parts by washing in clean solvent and immerse all parts including "O" rings and back-up washers in clean hydraulic fluid before assembly.

When reassembling, make sure synchronizing valves in piston are at one o'clock position and rod holes are horizontal as shown in Fig. 16. Tighten spring loaded nut (Fig. 17) on operating shaft until a gap of approximately 5/16-inch exists between nut and collar. This tension provides friction which gives a feeling of stability to the steering effort. Lever plugs control end play of pivot shafts on operating levers (38–Fig. 14). Adjust plugs, if necessary, until levers turn freely but end play is limited to a maximum of 0.003 inch. The ball races (35) for operating collar (36) contain 13 loose bearing balls. Tighten nut (37) to a torque of 5 ft.-lbs., loosen to nearest castellation (if necessary) and install cotter pin. Operating collar (36) must turn by hand on shaft. Tighten valve housing

cap screws to a torque of 110 ft.-lbs., and cylinder cover stud nuts to 55 ft.-lbs. Adjust valve levers as outlined in paragraph 18.

18. **ADJUSTMENT.** Install special neutral stops (JDH-3C) as shown in Fig. 18. Use side of tools marked "3000-4000-5000." The tools will hold lower edge of operating collar (E—Fig. 19) 0.030 inch from machined face of housing (H), which is neutral position. Turn steering wheel clockwise until operating collar contacts special tools, then attach a weight to steering wheel as shown in Fig. 18 to hold operating collar securely in this neutral position for steering valve adjustment.

Loosen locknuts on adjusting screws (A, B, C and D—Fig. 19) and make sure both operating levers have at least 0.003 inch free movement. Position a dial indicator on housing as shown in Fig. 18 so operating lever clearance can be checked at each adjusting screw. Adjust each screw to specified clearance using the following adjusting sequence: Adjust upper left return valve screw (A—Fig. 19) to provide 0.003 inch lever movement, then adjust upper right pressure valve screw (B) to obtain 0.001 inch movement. Adjust lower right return valve screw (C) to provide 0.003 inch lever movement, then adjust lower left pressure valve screw to obtain 0.001 inch movement. This adjustment allows pressure valves to open slightly before return valves.

Note that valve adjusting screw locations are as viewed from front side of housing with operating coller positioned at top as shown in Fig. 18. Also, adjustment can be more accurately made if operating lever shaft is pulled outward with light pressure to eliminate shaft free play as lever movement is being measured.

Fig. 15 — Remove steering check valve stop and spring as unit is separated.

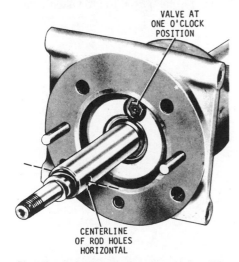

Fig. 16 — Assemble piston with synchronizing valve at one o'clock and rod holes horizontal as shown.

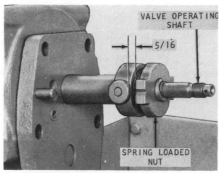

Fig. 17 — Tighten spring loaded nut on operating shaft until 5/16-inch gap exists between nut and collar.

Fig. 18 — Install positioning clamps for valve adjustment.

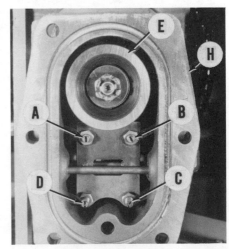

Fig. 19 — Front view of valve with cover removed. Correct positioning of operating coller (E) with relation to housing face (H) is accomplished by the two blocks JDH-3C shown in Fig. 18.

STEERING MOTOR

Models 3020-4000-4020-4320

19. **REMOVE AND REINSTALL.** To remove steering motor, remove grille screens, fuel tank and right side frame. Support front of tractor with suitable hoist. On tricycle models, remove wheels and pedestal assembly. On axle models, remove front axle and support as a unit.

Support motor with a floor jack. Relieve hydraulic system pressure, then disconnect and plug hydraulic lines. Unbolt steering motor from frame and remove from tractor.

To install steering motor, reverse removal procedure and bleed system as outlined in paragraph 13. Tightening torques are as follows:
Motor To Side Frame . . . . . . 215 ft.-lbs.
Motor To Front Frame . . . . . 150 ft.-lbs.
Steering Spindle Flange
   Bolts . . . . 300 ft.-lbs.

20. **OVERHAUL.** To disassemble re-

moved steering motor, turn unit upside down on bench and remove cap screws securing spindle retainer (12 – Fig. 20) to motor housing. Tap spindle retainer from doweled position and withdraw spindle (13), retainer and bearing (10) as a unit. Spindle bearing (10) is retained to spindle by snap ring (9) and can be removed if service is indicated on any of the components. Upper spindle bearing (8) can be removed using a suitable slide hammer puller.

NOTE: Install lower spindle bearing (10) as follows: On New Departure bearing, part number and open side of bearing goes towards oil seal (11). On Fafnir bearing, side marked "thrust" goes away from seal.

To disassemble piston assemblies, remove cap screws and washers (1), then remove snap ring (2). Cap screws may be installed in piston plug (3) to assist in pulling plug. Push pistons out of housing using brass drift or other suitable tool.

When reassembling, install "O" rings

(14) and backing washers (15) on pisto with washers toward center. Install "O ring (5) and backing washer (4) on pisto plug with washer towards outer end o bore. Install two end plugs diagonall opposite each other, and secure witł snap rings, washers and cap screws. In stall piston assemblies, teeth towarc center of motor housing, until they bot tom against previously installed enc plugs. Reassemble spindle retainer spindle and bearing. Then carefully in sert spindle in housing so V-mark or spindle flange is aligned with appro priate limit-mark on retainer as show in Fig. 21. When properly assemblec and timed, spindle should rotate 12C degrees between scribed limit marks and V-mark on spindle should point directly to the rear in mid-position. Complete assembly by reversing disassembly procedure.

STEERING CYLINDER

Models 4520-4620

21. **R&R AND OVERHAUL.** To remove either steering cylinder, disconnect hydraulic hose at cylinder. Remove pivot pins from cylinder ends and withdraw cylinder from tractor.

To disassemble cylinder, refer to Fig. 22 for an exploded view of cylinder components. Place cylinder body in a vise and unscrew end cap (10) from cylinder tube (2), then withdraw piston and rod assembly from tube. Remove cap screw and separate piston (5), bushing (6) and end cap (10) from piston rod (13). Remove "O" rings, back-up washers and oil seal.

When reassembling, dip all parts in clean hydraulic oil. Refer to Fig. 22 for location and arrangement of "O" rings, back-up washers and oil seal. Tighten piston to rod retaining cap screw to 200 ft.-lbs. torque, piston rod to yoke to 200

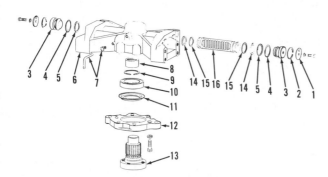

Fig. 20 – Exploded view of steering motor showing component parts.

1. Washer
2. Snap ring
3. Plug
4. Back-up washer
5. "O" ring
6. Motor housing
7. Vent tube
8. Upper bearing
9. Snap ring
10. Lower bearing
11. Oil seal
12. Retainer
13. Spindle
14. "O" ring
15. Back-up washer
16. Piston

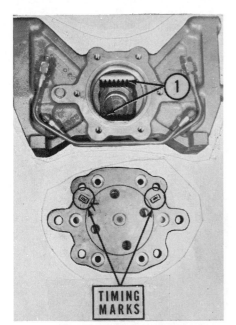

Fig. 21 – When assembling steering motor, make sure pistons are at opposite ends of cylinders as shown at (1) and appropriate timing marks are aligned on spindle and retainer.

Fig. 22 – Exploded view of steering cylinder used on Models 4520 and 4620. Two cylinders are used.

1. Pivot pin
2. Cylinder tube
3. Back-up washer (2 used)
4. "O" ring
5. Piston
6. Bushing
7. "O" ring
8. Back-up washer
9. "O" ring
10. End cap
11. Oil seal
12. "O" ring
13. Piston rod
14. Yoke
15. Pivot pin

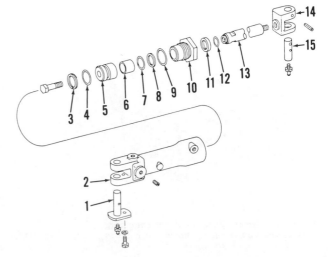

ft.-lbs. and cylinder end cap to 200 ft.-lbs.

Reinstall cylinder by reversing removal procedure. Bleed air from system as outlined in paragraph 13.

POWER FRONT WHEEL DRIVE

Models So Equipped

22. **OPERATION.** With Power Front Wheel Drive, the front wheels are driven by hydraulic fluid supplied by tractor main hydraulic system. The hydraulic power unit consists of an axial piston hydraulic motor in each front wheel hub which turns the wheel through a planetary gear reduction unit.

Power is supplied by the constant pressure, variable volume Main Hydraulic System Pump, thus providing a specified amount of torque to front wheels without regard to tractor ground speed. This design permits a "Power Assist" traction boost which need not be precisely synchronized with tractor drive gears, and which eliminates possibility of damage or overload due to wheel slippage, altered tire size or tread wear.

The control valve is electrically operated by switches contained in Shift Lever Quadrant, and automatically coupled to the directional control solenoid. On Syncro-Range and Power Shift tractors, power front drive unit is automatically disengaged above 7th gear in "Low Torque" range, above 5th gear in "High Torque" range; or whenever Inching Pedal (Power Shift Models) or Clutch Pedal (Syncro-Range Models) is depressed.

The panel mounted operating (selector) switch has three positions: Off, High Torque and Low Torque.

With operating switch in Off position, the wheels do not drive, but are allowed to rotate freely.

When operating switch is in High Torque position, oil pressure from main hydraulic pump is directed to drive motors in both front wheel motors and to front wheel drive brake.

With operating switch in Low Torque position, control valve directs pressurized oil to only one front wheel drive motor. Oil returning from this front wheel drive motor flows back to the control valve and is directed to the other front wheel drive motor. The drive motors in Low Torque drive are connected hydraulically in series and since both drive wheels displace the same volume, both wheels will rotate a similar amount with a minimum amount of slippage.

23. **CONTROL VALVE.** Refer to Fig. 23 for a cross-sectional view of front wheel control valve. The unit contains three solenoid-operated pilot valves; F

(Forward), R (Reverse) and T (Torque), and three hydraulically actuated operating valves; (Pressure Control Valve) which gives priority to other hydraulic functions if necessary, (Direction Valve) and (Torque Valve). A pressure relief valve is mounted near pressure inlet.

The three pilot valves are spring loaded in the closed position and direction valve is spring centered, blocking fluid passage to wheel motors whenever

electric power is cut off to control valve. When forward or reverse solenoid is electrically energized, the pilot valve opens and admits pressure to one end of direction valve, shifting the valve and routing pressure fluid to front wheel motors in forward or reverse direction depending on which way valve is moved.

Torque valve is spring loaded in High Torque position. When Torque Pilot Valve (T) is not energized, flow through

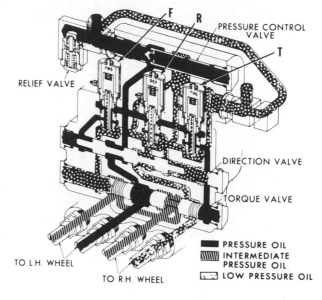

Fig. 23 — Cross-sectional view of Power Front Wheel Drive control valve showing component parts. Shaded areas indicate pressurized operating fluid with valve in forward, low torque position.

(Labels: F, R, PRESSURE CONTROL VALVE, T, RELIEF VALVE, DIRECTION VALVE, TORQUE VALVE, TO L.H. WHEEL, TO R.H. WHEEL)

Legend: PRESSURE OIL / INTERMEDIATE PRESSURE OIL / LOW PRESSURE OIL

Fig. 24 — Schematic view of wheel drive unit showing double reduction principle. "A" shows hydraulic motor, shaft, and primary sun gear which turn at the fastest speed. "B" shows primary planet carrier and secondary sun gear, which turn at intermediate speed. "C" shows fluid housing and wheel hub which turn at the slowest speed. "D" shows secondary planetary ring gear which is stationary when unit is engaged.

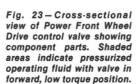

direction valve is delivered at equal pressure to each front wheel motor. In Forward position, when torque pilot valve is energized, pressure fluid enters piston end of torque valve and moves it against spring pressure to route full volume flow to left wheel motor. Discharge port of left wheel motor is connected through control valve to inlet port of right wheel motor and discharge port of right wheel motor to discharge port of valve.

In reverse position, when torque pilot valve is energized, pressure fluid enters piston end of torque valve and moves it against spring pressure to route full volume flow to right wheel motor. Discharge port of right wheel motor is connected through control valve to inlet port of left wheel motor and discharge port of left wheel motor to discharge port of valve.

Electrical switches control current to solenoids of control valves. Refer to Figs. 26, 27 or 28 for schematic.

24. WHEEL DRIVE UNIT. Refer to Fig. 24. The front wheel drive unit consists of a fixed displacement axial piston motor and primary sun gear (A), which drives a dual reduction planetary gear unit (B). The primary planet carrier is splined to a secondary sun gear which drives hub (C).

The planetary secondary ring gear (D) floats in its housing and is coupled to spindle carrier by hydraulic pressure whenever power is applied. The wheel manifold contains check valves to admit pressure to brake pistons and wheel motor pistons simultaneously. One planet brake piston contains a relief valve to prevent damage due to pressure buildup during neutral operation. In forward direction, pressure oil is routed through upper hoses to front wheel assembly, and return oil is directed through lower hoses to control valve and then to reservoir. For reverse direction, pressure oil is directed to lower hoses and return is through upper hoses. Refer to Fig. 25 for oil pressure flow through manifold. The upper check valve body has a 0.031 inch orifice which allows brake piston apply pressure to bleed to low pressure passage after controls are moved to neutral, or off position. A relief valve is provided in center of manifold to prevent damage due to excessive pressure buildup.

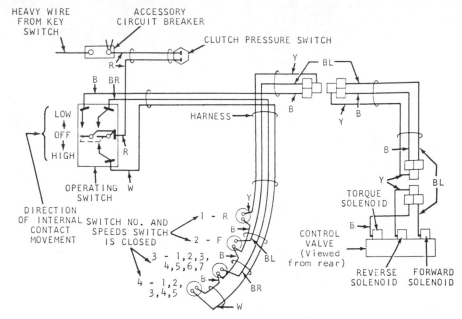

Fig. 26 — *Wiring diagram for Power Front Wheel Drive electrical system used on Power Shift models.*

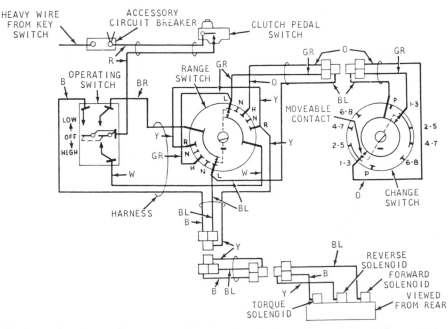

Fig. 27 — *Power Front Wheel Drive electrical system wiring diagram used on some early model Syncro-Range tractors.*

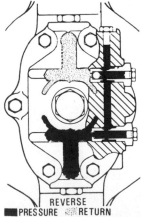

Fig. 25 — *Schematic view of pressure and return oil flow through wheel manifold. Brake piston port is pressurized in both directions.*

TROUBLESHOOTING
Models So Equipped

CAUTION: If engine is to be run while checking Power Front Wheel Drive, shut off hydraulic pump and disconnect wiring harness to solenoids at electrical connectors. This will prevent tractor movement if a short circuit or accidental switch movement should attempt to engage front wheel drive.

25. On all models, electrical circuits can be checked whenever key switch is turned on, but on models with Power Shift the clutch pressure switch must be by-passed using a jumper wire. Solenoids close with an audible click. With key switch on and clutch switch jumper wire installed, test as follows:

(1) Move selector switch from "OFF" to "LOW TORQUE" (lower) position; torque solenoid should engage with an audible click.

(2) Move selector switch back to "OFF" position and Gear or Speed selector lever to 1st or 2nd forward gear; then, move selector switch to "HIGH TORQUE" (upper) position. Forward solenoid should engage with an audible click.

(3) Repeat test 2 except with Gear or Speed selector lever in 1st or 2nd reverse gear. Reverse solenoid should engage with an audible click.

NOTE: HIGH TORQUE (upper) position of selector switch is used to test forward and reverse solenoids to keep from energizing torque solenoid.

If any solenoids fail to engage at proper time, continuity checks of circuits should be conducted to determine cause. Refer to Figs. 26, 27 or 28, for wiring diagrams of power front drive units. Check red wire from clutch pressure switch to circuit breaker and make sure circuit breaker is not open.

Internal hydraulic leakage or failure, because of the closed center system, will usually be signalled by heat or noise.

ELECTRICAL SYSTEM
Models So Equipped

26. ADJUSTMENT. On Syncro-Range models, the two rotary switches must be synchronized as follows:

Remove console cover and be sure key is off, or remove from switch. Place shift lever in Second Gear.

Disconnect rod from Speed Change Switch and turn switch counter-clockwise against internal stop. Turn switch clockwise, two detents, then adjust control rod yoke if necessary until rod can be reconnected to switch arm.

NOTE: Be sure all shift lever supports and switch brackets are tight before making the above adjustments.

Disconnect rod from Speed Range Switch and turn switch counter-clockwise against internal stop. Turn switch clockwise one detent, then adjust control rod yoke as necessary to reconnect rod to switch arm.

On Power Shift models, disconnect wiring harness from the four transmission switches and check for continuity as follows:

TOP switch (No. 1)-closed in all reverse speeds.
2nd switch-closed in all forward speeds.
3rd switch-closed in 1st through 7th speeds.
BOTTOM switch (No. 4)-closed in 1st through 5th speeds. Refer to Fig. 26 for wiring diagram. If adjustment is necessary, remove switches and add or remove shim washers to make switches close ONLY as outlined above.

The clutch pressure switch should be checked for continuity, and should make contact as clutch engages.

OVERHAUL
Models So Equipped

27. PILOT VALVES. Solenoid coil (2 – Fig. 29) can be removed from pilot valve after removing nut (1), without disturbing hydraulic circuits. Pilot valves can be removed for inspection, renewal, or renewal of "O" rings after removing solenoid coil. Main hydraulic pump should be shut off and system pressure relieved before attempting removal of valve unit.

Valve spool and bushing are only available as an assembly (10). All other parts are available individually, and seals are provided in control valve seal kit assembly.

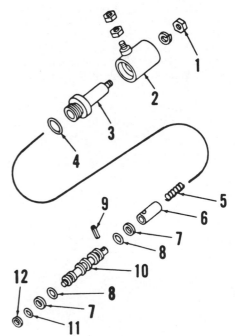

Fig. 29 — Exploded view of solenoid and pilot valve assembly, interchangeable in all three locations.

1. Nut	7. Back-up washer
2. Solenoid	8. "O" ring
3. Armature	9. Spring pin
4. "O" ring	10. Bushing & valve
5. Spring	11. "O" ring
6. Plunger	12. Back-up washer

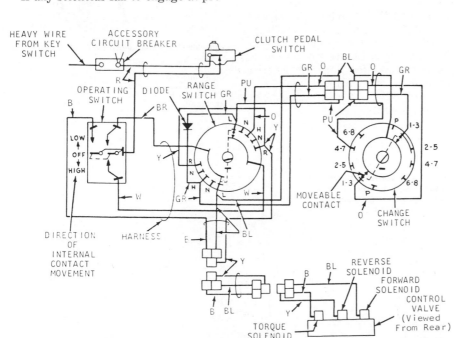

Fig. 28 — Power Front Wheel Drive electrical system wiring diagram used on late model Syncro-Range tractors.

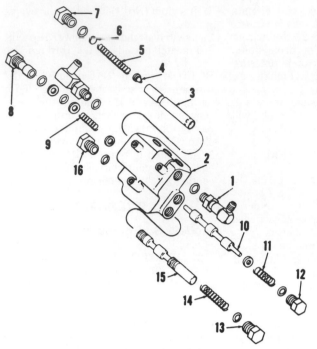

Drain wheel housing by turning plug (30 – Fig. 39) downward and removing plug, Housing capacity is approximately 9 quarts. Turn wheel for access to manifold cap screws and loosen screws evenly. Wheel motor spring should push manifold approximately 1/8-inch away from motor housing (spindle) as screws are loosened. Valve plate (1 – Fig. 32) and/or bearing plate (13 – Fig. 33) may remain with motor or be removed with manifold. Do not allow plates to drop as manifold is removed. Upper check valve body (17 – Fig. 32) has an orifice, and is longer than the lower, which has no orifice.

Install by reversing removal procedure. Manifold plate should easily slide on attaching bolts to within 1/8-inch of motor housing. Considerable pressure is required to compress motor spring, but compression should be even without binding. Tighten manifold cap screws to a torque of 35 ft.-lbs.

Fig. 30 – Exploded view of Power Front Wheel Drive control valve showing component parts.

1. Inlet elbow
2. Valve body
3. Priority valve
4. Orifice
5. Spring
6. Shim
7. Plug
8. Plug
9. Centering spring
10. Directional valve
11. Centering spring
12. Plug
13. Plug
14. High-torque spring
15. Torque valve
16. Plug

28. CONTROL VALVE. Refer to Fig. 30 for an exploded view of control valve assembly. Valve spools (3, 10 and 15) or body (2) are not available separately, but spools may be removed for inspection or renewal of other parts. All seals, including those on pilot valves (Fig. 29), are available in a valve sealing kit. Shims (6 – Fig. 30) control setting of priority valve (3). With fluid at operating temperature, operating pressure should be 1930-1970 psi at the following engine speeds; 3020-2100 rpm, 4020-1900 rpm, 4320-2150 rpm, 4520 and

4620-1600 rpm. On late models, a relief valve (Fig. 31) was added on inlet side of control valve housing to absorb pressure peaks and limit possible pressure build-up, resulting from a rapid change of direction, to 2700 psi. Pressure setting can be adjusted using shims (10).

Refer to Figs. 23, 30 and 31 and assemble by reversing disassembly procedure, making sure valve spools are installed correct end forward. New sealing rings should always be used when assembling control valve unit.

29. WHEEL MANIFOLD. Refer to Fig. 32 for an exploded view of wheel manifold and associated parts. Valves (7 through 12 and 13 through 17 – Fig. 32) can be removed for inspection or service without removing manifold plate. Plate can be removed without taking weight off wheel; proceed as follows:

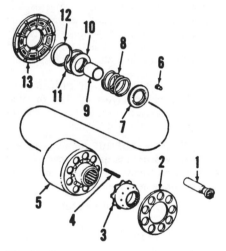

Fig. 33 – Exploded view of front wheel drive motor.

1. Piston
2. Slipper retainer
3. Ball guide
4. Springs
5. Cylinder block
6. Aligning dowel
7. Spring seat
8. Spring
9. Spring guide
10. Spring retainer
11. Retaining ring
12. Centering ring
13. Brass bearing plate

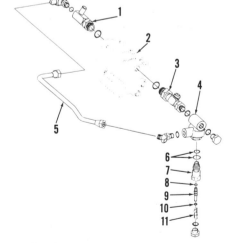

Fig. 31 – Exploded view of Power Front Wheel Drive relief valve and hydraulic connectors. Valve is not used on some early models.

1. Return fluid tee
2. Control valve
3. Pressure fluid tee
4. Relief valve housing
5. By-pass pipe
6. "O" rings
7. Relief valve body
8. Piston ring
9. Poppet
10. Shim
11. Spring

Fig. 32 – Exploded view of Power Front Wheel Drive manifold showing component parts.

1. Valve plate
2. Packing
3. Bearing cone
4. Bearing cup
5. Shims
6. Manifold
7. Snap ring
8. Retainer
9. Spring
10. Valve plug
11. "O" ring
12. Valve
13. Plug
14. Spring pin
15. Spring
16. Check valve ball
17. Valve body

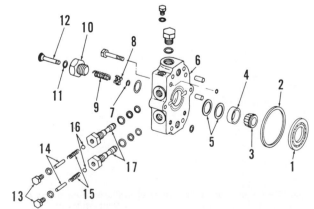

30. WHEEL MOTOR. The hydrostatic wheel motor is axial piston type shown exploded in Fig. 33. Wheel motor can be removed as outlined in paragraph 31, and overhauled as in paragraph 32.

31. REMOVE AND REINSTALL. To remove hydrostatic wheel motor, first remove manifold as outlined in paragraph 29 and primary planetary unit as in paragraph 33. If brass bearing plate (13–Fig. 33) remained with motor, insert a suitable wood dowel in one of the small round drain holes to use as a pry. NEVER use a screwdriver or steel wedge of any kind.

Grasp inner bearing cone (3–Fig. 32) and withdraw wheel motor (2–Fig. 39), swashplate (3), motor shaft (7) and outer bearing cone (8) as a unit from motor housing. If swashplate binds in housing or drags on locating pin, pressure can be applied at outer end by inserting a wood dowel or brass drift in spline end of shaft (7).

NOTE: Do not rotate assembly when out of housing, as damage may occur to motor parts while not supported by housing.

Install by reversing removal procedure, making sure thick side of swashplate ramp is toward rear of tractor and locating dowel pin enters appropriate notch in swashplate.

32. OVERHAUL. To overhaul removed wheel motor unit, use specially cut pieces of clean cardboard, sheet gasket material or plastic to protect polished inner face of cylinder block (5–Fig. 33), then use suitable pullers to remove inner bearing cone (3–Fig. 32). Motor shaft, outer bearing cone and swashplate can now be removed. (See preceding NOTE.)

Before disassembling motor, note splines on ball guide and cylinder block. If splines do not have a master spline, mark both surfaces for reassembly.

To disassemble motor, grasp outer diameter of slipper retainer (2–Fig. 33) and remove retainer and nine pistons (1). Remove ball guide (3) and withdraw nine slipper retainer springs (4). See Fig. 34 for method of removal and stacking of pistons for maximum cleanliness.

Wheel motor spring (8–Fig. 33) must be slightly compressed to remove or install retaining ring (11). Either of two methods is acceptable:

(1) Select a ⅜-inch bolt long enough to extend through cylinder block, nut and large flat washers. Install bolt through cylinder block as shown in Fig. 35 (washer on each end) and compress spring by tightening nut, then unseat and remove retaining ring (11–Fig. 33).

NOTE: Retaining ring should be pulled (not pried) out of its groove. Prying will probably cause a burr to be turned up on the lapped surface, resulting in oil leakage.

(2) As an alternate method, place cylinder block in a press on wood blocks, retaining ring up. Use a step plate on spring retainer and slightly compress spring to remove retaining ring.

When using either method, release pressure slowly after removing retaining ring using extreme care to keep from scratching or scoring polished, machined face of block.

Clean all parts and inspect for excessive wear or other damage. Check pistons and bores in cylinder block for linear scratches and excessive wear. Check cylinder block face for shiny streaks, indicating high pressure leakage between cylinder block and brass bearing plate. Inspect piston slippers for scratches, imbedded material or other damage. Light scratches in slippers can be removed by lapping. All nine slippers must be within 0.002 inch of each other in thickness.

Lubricate all parts with hydraulic fluid and reassemble by reversing disassembly procedure. Spring seat (7–Fig. 33) must be installed with beveled side away from cylinder block spring (8) and spring retainer (10) installed with retaining groove away from spring. When installing ball guide (3), align master spline of ball guide and cylinder block. On some motors, a prick punch mark is used on one outer tang of ball guide which aligns with a corresponding mark on cylinder block face.

Inspect steel valve plate (1–Fig. 32)

Fig. 36 – Inspect steel valve plates for wear or scoring in areas shown.

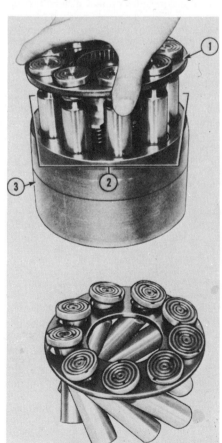

Fig. 34 – Method of removal and convenient stacking of pistons.

1. Retainer
2. Pistons
3. Cylinder block

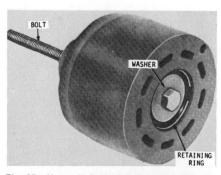

Fig. 35 – Use a ⅜-inch, full threaded bolt, flat washers and nut to disassemble motor spring as shown.

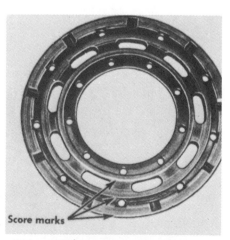

Fig. 37 – Inspect brass bearing plates as indicated.

and brass bearing plate (13 – Fig. 33) for excessive wear or scoring in areas shown in Figs. 36 and 37. Although both plates are available individually, it is good shop practice to renew both plates whenever one is damaged, to assure proper sealing necessary for efficient operation.

Before installing motor assembly and shaft in motor housing (4 – Fig. 39), check wheel hub bearings (10 and 11) for proper adjustment. Shims (13) behind outer bearing cup are available in 0.003 inch and 0.010 inch thickness to allow for an adjustment range of from 0.004 inch preload to 0.002 inch end play. Preload may be considered correct when a pull of 3 pounds at wheel hub is required to turn hub.

Install outer bearing cone (8) on motor shaft (7). Install motor assembly and swashplate (3) on motor shaft and press inner bearing cone (3 – Fig. 32) onto motor shaft. Thoroughly lubricate motor and shaft assembly and carefully install into motor housing (4 – Fig. 39), with swashplate (3) installed thick side toward rear of tractor and brass bearing plate (13 – Fig. 33) protecting lapped surface of cylinder block. Refer to Fig. 32. Lubricate valve plate (1) and install before oil manifold is bolted to housing. Spring (8 – Fig. 33) will resist as the manifold is installed so tighten evenly to avoid binding. Tighten manifold cap screws to a torque of 35 ft.-lbs. Motor shaft end play should be 0.006-0.012 inch and can be measured with a dial indicator by removing relief valve plug (10 – Fig. 32) and associated parts. Shims (5) are available in thicknesses of 0.003-0.010 inch and are used for adjustment if necessary. Remove bearing cup (4) and adjust by adding or removing shims as required.

33. PRIMARY PLANETARY UNIT. The primary planetary unit and secondary sun gear are shown exploded in

Fig. 38. Primary ring gear is carried in planetary housing (29 – Fig. 39).

Each planet pinion (7 – Fig. 38) rides on 17 loose needle rollers (6). Secondary sun gear (1) splines into hub of planet carrier (8), and serves as pilot for primary sun gear (2). Primary planetary unit may be withdrawn after removing retaining cover (10). Tighten cover retaining cap screws to a torque of 35 ft.-lbs. when installing.

34. SECONDARY PLANETARY AND DRIVE BRAKE UNITS. Refer to Fig. 39 for an exploded view of wheel drive unit. Secondary planet gears (28) are carried in planetary housing (29) which also contains primary ring gear. Secondary ring gear (22) floats in housing until hydraulic pressure is applied by the nine brake pistons (18).

Each planet pinion (28) rides on 22 loose needle rollers on shaft (27). Shaft (27) is keyed to housing (29) and carrier plate (24) by steel balls (26). Tighten cap screws retaining carrier plate (24) to a torque of 35 ft.-lbs. when unit is reassembled.

Brake backing plate (23) attaches to planetary piston housing (15) with 18 cap screws and encloses planetary ring gear (22), brake facings (21), brake pressure plate (19) and six separator springs (20). Backing plate must be removed for renewal of brake parts, including the nine brake pistons (18).

NOTE: One of the nine brake pistons should contain a bleed valve, and should be installed slotted end out in the second hole counter-clockwise from the word "UP", which is cast into the brake piston housing.

Tighten brake backing plate attaching screws to a torque of 35 ft.-lbs. when

unit is assembled. Mount a dial indicator on outboard side of brake backing plate (23) and check free play of planetary ring gear (22) at three places around internal teeth (inside) circumference. Free play should be at least 0.014 inch.

Tighten planetary brake piston housing retaining cap screws to a torque of 45 ft.-lbs. when unit is reassembled.

ENGINE

All Models

35. REMOVE AND REINSTALL. To remove engine and clutch assembly as a unit, first drain cooling system and, if engine is to be disassembled, drain oil pan. Remove muffler, hood, cowl, grille screens, engine side panels and control support covers. Discharge brake accumulators by opening the right brake bleed screw and depressing right brake pedal for about 60 seconds.

Make a clutch split as follows: Disconnect battery cables, tachometer cable, throttle linkage (diesel), hydraulic lines and clamps on right side of engine. Remove wiring harness clamps and disconnect wiring connectors. Disconnect engine temperature gage bulb, and on diesel engine, disconnect ether starting aid pipe. Unbolt starter relay mounting bracket and disconnect battery cable from starter. Disconnect choke cable and speed control rod on gasoline engine. On LP-Gas tractors, remove fuel withdrawal valve handles, then disconnect control rods at withdrawal valves and pull rods rearward. If equipped with cab, disconnect heater hoses. Remove

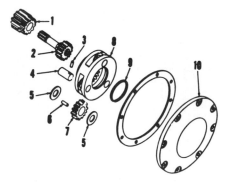

Fig. 38 — Exploded view of front wheel drive cover, planet carrier and associated parts.

1. Secondary sun gear	6. Bearing roller
2. Primary sun gear	7. Primary pinion
3. Dowel	8. Planet carrier
4. Pinion shaft	9. Snap ring
5. Thrust washer	10. Cover

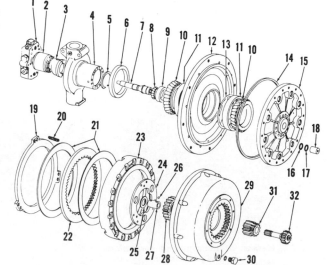

Fig. 39 — Exploded view of wheel drive unit showing component parts.

1. Wheel manifold
2. Motor
3. Swashplate
4. Motor housing
5. Snap ring
6. Oil seal
7. Motor shaft
8. Bearing cone
9. Bearing cup
10. Bearing cone
11. Bearing cup
12. Wheel hub
13. Shim
14. Packing
15. Piston housing
16. Quad ring
17. Back-up ring
18. Brake piston
19. Pressure plate
20. Separator spring
21. Brake facings
22. Planetary ring gear
23. Backing plate
24. Retainer
25. Thrust washer
26. Locking ball
27. Pinion shaft
28. Planet pinion
29. Planetary housing
30. Drain plug
31. Secondary sun gear
32. Primary sun gear

air conditioning compressor and hoses as an assembly, if equipped, and bend hoses back and place compressor in cab. Disconnect hydraulic pump inlet pipe. If equipped with Power Front Wheel Drive, remove front drain pipe, disconnect electrical connector, hydraulic pressure pipe and return hose, then unbolt and lower valve assembly from tractor. Suitably support both halves of tractor, remove cap screws securing engine to clutch housing and roll rear half of tractor away from engine.

Remove engine from front end as follows: Remove radiator hoses. Disconnect wiring harness from engine. On 4520 and 4620 models, remove power steering pipes. Remove air intake pipe from carburetor or intake manifold. On some diesel models, pre-cleaner and air cleaner are located above engine and must be removed. On all models, close fuel tank shut-off valve and disconnect fuel transfer pump inlet pipe. Disconnect fuel leak off pipe (diesel) or carburetor inlet pipe (gasoline). On LP-Gas engine, disconnect fuel line at fuel lock strainer. On all models, disconnect hydraulic pump coupling and support. Remove alternator and hydraulic pipe support bracket.

NOTE: Remove front weights and drain fuel tank. Support front of unit to prevent tipping.

Install engine lift brackets. Using a sling and suitable hoist, remove side frame to engine cap screws and slide engine out of side rails.

Reassemble tractor by reversing disassembly procedure. Tightening torques are as follows:
Hydraulic Pump Drive . . . . . . .25 ft.-lbs.
Hydraulic Pump Support . . . . .85 ft.-lbs.
Side Frame to Cylinder
   Block . . . . .250 ft.-lbs.
Cylinder Block to Clutch
   Housing . . . . .300 ft.-lbs.

CYLINDER HEAD
Model 3020

36. To remove cylinder head, drain cooling system and remove side panels, grille screens, muffler and hood. Remove air intake pipes up to intake manifold or carburetor. Remove thermostat by-pass hose, upper radiator hose and water manifold. Disconnect interfering linkage, then unbolt and remove intake and exhaust manifolds. On

diesel models, remove injector lines and injectors. On all models, remove vent tube, rocker arm cover, rocker arm assembly, then unbolt and remove cylinder head.

NOTE: Make sure cylinder liners are held down with bolts and washers if engine crankshaft is to be turned.

Install cylinder head gasket dry. Coat threads of cylinder head retaining cap screws with oil before installation, and be sure hardened flat washers are installed under head of all cylinder head retaining screws. Tighten cylinder head retaining screws evenly to 130 ft.-lbs. torque using sequence shown in Fig. 40. Tighten rocker arm clamp screws to 55 ft.-lbs. torque. Retorque cylinder head and readjust valve clearance after engine has run at 1900-2100 rpm for ½-hour. Use same sequence as shown in Fig. 40 and tighten screws to 130 ft.-lbs. torque. Refer to paragraph 40 for adjusting valve clearance.

Models 4000-4020

37. To remove cylinder head, drain cooling system and remove side panels, grille screens, muffler and hood. Remove air intake pipes up to intake manifold or carburetor. Remove thermostat by-pass tube (or tubes), upper radiator hose and water manifold. Disconnect interfering linkage and remove intake and exhaust manifolds. Remove injector lines and injectors from diesel engines. On all models, remove crankcase vent tube, rocker arm cover and rocker arm assembly, then unbolt and remove cylinder head.

NOTE: Make sure cylinder liners are held down with bolts and washers if engine crankshaft is to be turned.

On all models, install cylinder head gasket dry, but lightly oil threads of cylinder head retaining cap screws. Be sure to install a flat hardened washer on all cap screws. Refer to Fig. 41 and tighten center cap screw (No. 17) to 105 ft.-lbs. torque first to hold cylinder head in correct position during tightening sequence. Starting at No. 1 cap screw, tighten remainder of cap screws in sequence indicated to 105 ft.-lbs. torque. Then, retighten all cap screws in same sequence to 115 ft.-lbs. torque. Tighten rocker arm clamp bolts to 55 ft.-lbs. torque.

After engine is reassembled, run for

about 30 minutes. Retorque the cylinder head cap screws (engine hot) to 130 ft.-lbs. Readjust valve clearance as outlined in paragraph 40.

Models 4320-4520-4620

38. To remove cylinder head, drain cooling system and remove side panels, grille screens, muffler and hood. Relieve hydraulic pressure by pumping brake pedals several times. Remove air intake pipes up as far as turbocharger. On some models, air cleaner is mounted above engine and must be removed. Remove water manifold, by-pass pipe, thermostat housing and upper water hose. Disconnect turbocharger oil pipes and remove intake manifold; then unbolt and remove turbocharger and exhaust manifold as a unit.

Remove alternator and fan blades, then unbolt water pump and tip it forward out of the way. Remove injector lines and injectors. Remove ventilator tube, rocker arm cover, rocker arm assembly and push rods, then unbolt and remove cylinder head.

NOTE: Make sure cylinder liners are held down with bolts and washers if crankshaft is to be turned.

Install cylinder head gasket dry, but lightly oil threads of retaining cap screws. Be sure to install a hardened flat washer on all cap screws. Refer to Fig. 41 and tighten center cap screw (No. 17) to 105 ft.-lbs. torque first to hold cylinder head in correct position during tightening sequence. Starting at No. 1 cap screw, tighten remainder of cap screws in sequence indicated to 105 ft.-lbs. torque. Then, retighten all cap screws in same sequence to 115 ft.-lbs. torque. Tighten rocker arm clamp screws to 55 ft.-lbs. torque.

NOTE: NEVER run engine with turbocharger oil lines disconnected.

Complete assembly to point of permitting engine to run. Operate engine for about 30 minutes. then retorque all cylinder head cap screws (engine hot) to 130 ft.-lbs. Readjust valve clearance as outlined in paragraph 40.

VALVES AND SEATS
All Models

39. Some models are originally equipped with hardened steel seat inserts which are pressed into machined bores

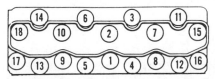

Fig. 40—When installing cylinder head on 3020 Models, tighten cylinder head cap screws to a torque of 130 ft-lbs. using sequence shown.

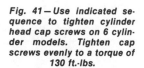

Fig. 41—Use indicated sequence to tighten cylinder head cap screws on 6 cylinder models. Tighten cap screws evenly to a torque of 130 ft.-lbs.

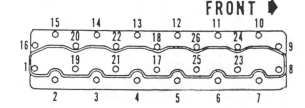

in cylinder head. When installing new inserts, chill insert in dry ice for easier installation. Measure distance valve head is recessed below cylinder head surface and compare to following specification limits.

3020 and 4020
Intake . . . . 0.036-0.050 in.
Exhaust . . . . 0.054-0.068 in.

4320
Intake . . . . 0.036-0.050 in.
Exhaust . . . . 0.042-0.056 in.

4520 (Engine S.N. E245757 and Below)
Intake . . . . 0.026-0.040 in.
Exhaust . . . . 0.042-0.056 in.

4520 (Engine S.N. 245758 and Above)
Intake . . . . 0.036-0.050 in.
Exhaust . . . . 0.042-0.056 in.

4620
Intake . . . . 0.036-0.050 in.
Exhaust . . . . 0.042-0.056 in.

Intake and exhaust valve stem diameter is 0.3715-0.3725 inch and valve guide bore is 0.3745-0.3755 inch on all models. Recommended stem to guide clearance is 0.002-0.004 inch. Guides can be knurled to provide correct clearance if clearance is less than 0.008 inch. If clearance exceeds 0.008 inch, bores in guides should be reamed to correct oversize and new valves with larger (oversize) stem installed. Be sure to reseat valve after guide has been knurled or reamed oversize. Recommended valve face and valve seat angles are as follows:

3020, 4000, 4020 and 4520 (Prior to Engine S.N. 245758)
Valve Face Angle
Intake . . . . 44½ degrees
Exhaust . . . . 44½ degrees

Valve seat angle
Intake . . . . 45 degrees
Exhaust . . . . 45 degrees

4320, 4520 (Engines S.N. 245758 and After) and 4620
Valve Face Angle
Intake . . . . 29½ degrees
Exhaust . . . . 44½ degrees
Valve Seat Angle
Intake . . . . 30 degrees
Exhaust . . . . 45 degrees
Valve tappet gap should be adjusted using procedure outlined in paragraph 40.

TAPPET GAP ADJUSTMENT

All Models

40. The two-position method of valve tappet gap adjustment is recommended. Refer to Fig. 42 for 3020 models or Fig. 43 for all other models and proceed as follows:

Turn engine crankshaft by hand until number 1 and 4 cylinders on 3020 models or number 1 and 6 cylinders on all other models are at TDC. Check valves to determine whether front or rear cylinder is at top of compression stroke (exhaust valve on adjacent cylinder will be partly open). Use appropriate diagram (Fig. 42 or 43) and adjust valves indicated; then rotate crankshaft one complete turn and adjust remainder of valves. Recommended valve tappet clearance is as follows:

3020, 4000 and 4020
Intake (Gasoline and
  LP-Gas) . . . . 0.015 in.
Exhaust (Gasoline and
  LP-Gas) . . . . 0.028 in.
Intake (Diesel) . . . . 0.018 in.
Exhaust (Diesel) . . . . 0.018 in.

4320 and 4620
Intake . . . . 0.018 in.
Exhaust . . . . 0.028 in.

4520 (Prior to Engine S.N. 303519)
Intake . . . . 0.018 in.
Exhaust . . . . 0.022 in.

4520 (Engine S.N. 303519 and After)
Intake . . . . 0.018 in.
Exhaust . . . . 0.028 in.

VALVE ROTATORS

All Models

41. Positive type valve rotators are originally installed only on exhaust valves of 3020, 4000 and 4020 gasoline and diesel models. Models 4320, 4520 and 4620 are originally equipped with rotators on all (intake and exhaust) valves.

Normal service consists of renewing complete unit. Rotator can be considered satisfactory if it turns freely in one direction.

VALVE GUIDES AND SPRINGS

All Models

42. Valve guide bores are an integral part of cylinder head. Standard bore diameter is 0.3745-0.3755 inch and normal stem to guide clearance is 0.002-0.004 inch. If clearance is 0.006-0.008 inch, the guide can be knurled; however, if clearance exceeds 0.008 inch, ream guide to correct oversize and install valve with oversize stem.

Intake and exhaust valve springs are interchangeable and may be installed either end up. Renew any spring which is distorted, rusted or discolored, or does not meet test specifications which follow:
Free Length (approx.) . . . . 2.12 in.
Lbs. Test at Length (Inches)
  Closed . . . . 54-62 at 1.81 in.
  Open . . . . 133-153 at 1.36 in.

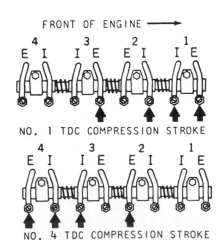

FRONT OF ENGINE →

NO. 1 TDC COMPRESSION STROKE

NO. 4 TDC COMPRESSION STROKE

Fig. 42—Adjust valves on 3020 models as indicated by arrows in upper half of Fig. when No. 1 piston is at TDC on compression stroke. Refer to lower half of Fig. for remainder of valves.

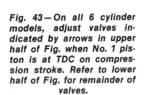

Fig. 43—On all 6 cylinder models, adjust valves indicated by arrows in upper half of Fig. when No. 1 piston is at TDC on compression stroke. Refer to lower half of Fig. for remainder of valves.

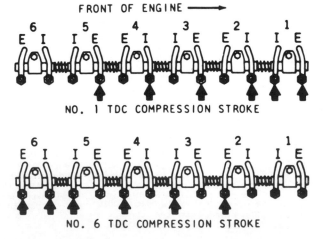

FRONT OF ENGINE →

NO. 1 TDC COMPRESSION STROKE

NO. 6 TDC COMPRESSION STROKE

ROCKER ARMS

All Models

43. The rocker arm shaft attaches to bosses which are cast into cylinder head and is held in place by clamps. Shaft rotation is prevented by a spring pin in cylinder head which enters a hole in shaft for positive positioning of lubrication passages.

Rocker arms are right-hand and left-hand assemblies. Recommended clearance of rocker arms to shaft is 0.0005-0.0035 inch. Bushings are not available; if clearance is excessive, renew rocker arms and/or shaft.

When reassembling, make sure spring pin aligns with locating hole in shaft, tighten clamp screws to 55 ft.-lbs. torque, then adjust tappet gap as outlined in paragraph 40.

CAM FOLLOWERS (TAPPETS)

All Models

44. The mushroom type cam followers can be removed from below after removing camshaft as outlined in paragraph 47. The cam followers operate in unbushed bores in engine block and are available in standard size only.

TIMING GEAR COVER AND CRANKSHAFT FRONT OIL SEAL

All Models

45. To remove timing gear cover, first drain cooling system and remove hood, grille screens and engine side panels. Remove radiator and fan shroud from left side after disconnecting oil cooler from radiator.

Remove pressure and return lines from hydraulic pump, disconnect pump drive shaft and coupler and remove pump and support. Loosen fan belts, remove crankshaft damper pulley using a suitable puller, remove cover retaining cap screws and lift off cover. The lip type front oil seal is supplied in a kit which also includes a steel wear sleeve which is pressed on crankshaft in front of gear (Fig. 44). Score old sleeve lightly with a blunt chisel and pry sleeve from shaft. Coat inner surface of new sleeve with a non-hardening sealant and install with a suitable screw-type installer such as JDE-3. Install oil seal in cover from inside. Sealing lip should be toward the rear and seal should bottom in its bore.

When installing timing gear cover, tighten retaining cap screws to a torque of 30 ft.-lbs. and damper pulley retaining cap screw to 150 ft.-lbs. Complete assembly by reversing disassembly procedure.

TIMING GEARS

All Models

46. Gears are available in standard size only. If backlash is excessive, renew parts concerned. Before removing gears, set No. 1 piston at TDC on compression stroke to aid timing on reassembly.

When renewing crankshaft gear, heat new gear to approximately 350 degrees F. using a hot plate or oven and install with a press or JDH-7 driver with timing mark to front. On all models, No. 1 cylinder must be at TDC on compression stroke when installing gears and aligning timing marks. On 3020, 4000, 4020 and 4520 (engine S.N. E303518 and below) models, "V" mark on camshaft gear must align with "V" mark on crankshaft gear (Fig. 45). On 4320, 4520 (engine S.N. E303519 and above) and 4630 models, align "V" mark on crankshaft gear with "V" mark on camshaft gear and "V" mark on injection pump gear with "V" mark on injection pump drive gear (Fig. 46).

CAMSHAFT AND BEARINGS

All Models

47. To remove camshaft, first split tractor at front of engine as follows: Drain cooling system, remove air stack (if equipped) and muffler, grille screens, side shields, cowl and hood. Disconnect battery ground cable. If equipped with air conditioner, unbolt and remove compressor and hoses as a unit. Disconnect radiator hoses and heater hoses (if equipped), steering and hydraulic tubes and fuel gage sender wire. Remove hydraulic tube clamps and separate wiring harness from tubes as needed. Disconnect hydraulic pump inlet and pressure pipes and pump oil seal drain tube. Shut off fuel valve and disconnect fuel inlet pipe. Disconnect fuel leak off line (diesel). Remove hydraulic pump drive coupler and detach pump support from engine. Remove air intake pipe and air cleaner assembly (if necessary). On LP-Gas engines, disconnect fuel control rods from withdrawal valve and slide rods rearward, and disconnect fuel line from withdrawal valve at fuel lock strainer. If equipped with Power Front Wheel Drive, remove front drain pipe and disconnect electrical connector, hydraulic pressure pipe and return hose. Unbolt and lower valve from tractor. Remove rocker arm cover and rocker arm assembly.

NOTE: Remove front weights and drain fuel tank to prevent front of tractor from tipping.

Install JDE-63 engine lift brackets and JDG-1 sling and front and rear support stands. Support front end with suitable hoist. Remove side frame to engine cap screws and separate front end from engine.

Fig. 44—Score oil seal wear sleeve as shown using a blunt chisel, then pry from shaft.

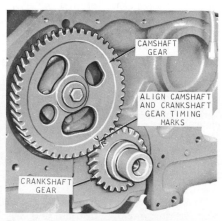

Fig. 45—Timing marks on camshaft gear and crankshaft gear must align as shown. 3020, 4000 4020 and early 4520 models use this gear arrangement.

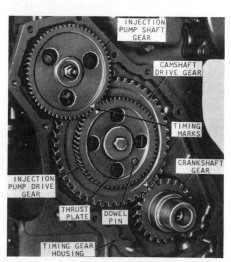

Fig. 46—When installing timing gears on 4320, late 4520 and 4620 models, align timing marks as shown.

NOTE: Before disassembly, set No. 1 piston on TDC on compression stroke to aid timing on reassembly.

Remove crankshaft damper pulley using a suitable puller. Remove timing gear cover. Remove oil pan and oil pump. Remove fuel supply pump (gasoline and diesel engines) and distributor (gasoline and LP-Gas engines). On 4320, late 4520 and 4620 models, remove fuel injection pump gear (Fig. 46).

Remove push rods, then raise and secure cam followers away from camshaft using magnetic holders or other suitable means. Before removing camshaft, check end play. Maximum end play is 0.015 inch. Excessive end play indicates worn thrust washer. Working through openings in camshaft gear, remove cap screws securing thrust washer to engine block, then withdraw camshaft and gear assembly from engine.

All models have 2.3745-2.3755 inch camshaft journals. All models should have a clearance of 0.002-0.005 inch in bushings. The presized camshaft bushings are interchangeable. To install bushings after camshaft is out, detach cylinder block from clutch housing, remove clutch, flywheel and camshaft bore plug, then pull bushings into block bores using a piloted puller. Make sure oil supply holes in block are aligned with holes in bushings. The elongated hole in bushing goes to the top.

Camshaft end play of 0.0025-0.0085 inch is controlled by thickness of camshaft thrust plate. End play can be measured with a dial indicator when camshaft is installed, or with a feeler gage when camshaft unit is out. Thrust plate thickness is 0.187-0.189 inch on all models.

Camshaft gear and injector pump drive gear (if so equipped) can be removed with a press when camshaft is out, after removing retaining cap screw and washer. Align key slot in gear with Woodruff key in shaft, make sure thrust

plate and spacer are installed and press gear on shaft until it bottoms. Tighten gear retaining cap screw to a torque of 85 ft.-lbs. If camshaft is replaced, cam followers should also be replaced.

Reassemble by reversing removal procedure. Align timing marks as outlined in paragraph 46. Tighten thrust plate retaining cap screws to a torque of 20 ft.-lbs.

ROD AND PISTON UNITS

All Models

48. Connecting rod and piston units are removed from above after removing cylinder head, oil pan and rod bearing caps. When reinstalling, correlation numbers, small and large slots and tangs on rod and cap must be in register. Rods and head of pistons are stamped "FRONT" for proper installation. Tighten connecting rod cap screws to a torque of 85-95 ft.-lbs. on spark ignition models and 100-110 ft.-lbs. on diesel models.

NOTE: Do not rotate crankshaft with head removed or attempt to remove rod and piston units without first bolting liners down using washers and short cap screws as shown in Fig. 47.

PISTONS, RINGS AND SLEEVES

All Models

49. **PISTONS AND RINGS.** All models are equipped with aluminum alloy, cam ground pistons which use two compression rings and one oil control ring. All rings are located above the piston pin.

Spark ignition engines use conventional rectangular rings. Piston ring side clearance in groove should be less than 0.010 inch. On diesel engines, all models use a keystone type piston ring in top groove and some models also use keystone ring in second groove. Special ring groove wear gages JDE55 (4320, 4520 and 4620) and JDE62 (3020, 4000 and 4020) are used to check keystone ring groove wear. On models with rectangular second compression ring, a feeler gage can be used to check ring side clearance in groove, which should be less than 0.010 inch.

The manufacturer recommends cleaning pistons using Immersion Solvent "D-Part", and Hydra-Jet Rinse Gun or Glass Bead Blasting Machine.

Coat pistons and liners with engine oil, stagger ring gaps on pistons, and use JDE-57 ring compressor or other suitable ring compressor when installing pistons in sleeves.

50. **SLEEVES, PACKING AND "O" RINGS.** The renewable, wet type cylinder sleeves are available in standard size only. However, some sleeves and pistons are selectively fitted. Pistons marked with green paint are matched with sleeves stamped "LO" or "LH". Pistons marked with yellow paint are matched to sleeves stamped "HO" or "HH".

Sleeve flange at upper edge is sealed by cylinder head gasket. Sleeves are sealed at lower edge by packing shown in Fig. 48. Sleeves normally require loosening using a sleeve puller, after which they can be withdrawn by hand. Out-of-round or taper should not exceed 0.005 inch. If sleeve is to be reused, it should be deglazed using a normal cross-hatch pattern.

When reinstalling sleeves, first make sure sleeve and block bore are absolutely clean and dry. Carefully remove all rust and scale from seating surfaces and packing grooves in block and from areas of water jacket where loose scale might interfere with sleeve or packing installation. If sleeves are being reused, buff rust and scale from outside of sleeve.

Install sleeve without seals and secure with cap screws and washers, then measure standout. Check sleeve standout at several locations around sleeve. Also check to be sure sleeve will slip fully into bore without force. If sleeve cannot be pushed down by hand, recheck for scale or burrs. Sleeve standout should be 0.001-0.004 inch. Check fit of other sleeves if standout is less than specified. If standout is excessive, check for scale or burrs, then, if necessary, select another sleeve. After matching sleeves to all the bores, mark the sleeves; then

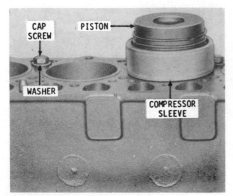

Fig. 47—Hold sleeves in position using a cap screw and washer as shown when cylinder head is removed.

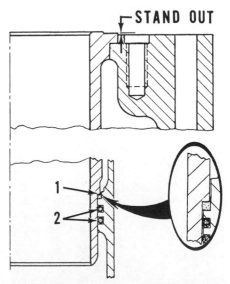

Fig. 48—Cross-section of cylinder sleeve showing square section packing (1) and round "O" rings (2). Refer to text for correct installation of cylinder sleeves.

refer to following paragraph for packing and sleeve installation.

Apply liquid (lubricating) soap such as part number AR54749 to square section ring (1 – Fig. 48) and install over lower end of cylinder sleeve. Slide square section ring up against shoulder on sleeve, make sure ring is not twisted and longer sides are parallel with side of sleeve as shown. Apply liquid soap to round section "O" rings (2) and install in grooves in cylinder block. If some of the "O" rings are red and some are black, red "O" rings should be installed in top groove in block and black "O" rings should be in lower groove. Be sure "O" rings are completely seated in grooves so installing sleeve will not damage the "O" rings. Observe previously affixed mark indicating correct cylinder location, then install sleeves carefully into correct cylinder block bore. Work sleeve gently into position by hand until it is finally necessary to tap sleeve into position using a hardwood block and hammer.

51. **SPECIFICATIONS.** Specifications of pistons and sleeves are as follows:

3020, 4000 and 4020 Gasoline and LP-Gas
Piston Skirt
  Diameter*.........4.2455-4.2465 in.
Cylinder Bore
  Diameter..........4.2493-4.2507 in.
Piston Skirt* to Cylinder
  Clearance ....0.0028-0.0052 in.
Piston Pin Diameter .. 1.2497-1.2500 in.
Piston Pin Bore in
  Piston ....1.2503-1.2509 in.
*Measured at bottom of skirt, right angle to pin.

3020 Diesel
Piston Skirt
  Diameter*.........4.2450-4.2460 in.
Cylinder Bore
  Diameter..........4.2493-4.2507 in.
Piston Skirt* To Cylinder
  Clearance ....0.0033-0.0057 in.
Piston Pin Diameter .. 1.4997-1.5003 in.
Piston Pin Bore in
  Piston ....1.5003-1.5009 in.
*Measured at bottom of skirt, right angle to pin.

4000 and 4020 Diesel
Piston Skirt Diameter*
  Original (not
    marked) ....4.2450-4.2460 in.
  Marked green ......4.2450-4.2455 in.
  Marked yellow .....4.2455-4.2460 in.
Cylinder Bore Diameter
  Original (not
    marked) ....4.2493-4.2507 in.
  Marked "LO" ......4.2500-4.2507 in.
  Marked "HO" ......4.2507-4.2514 in.

Piston Skirt* to Cylinder
  Clearance
  Original (not
    marked) ....0.0033-0.0057 in.
  Marked green &
    "LO" ....0.0045-0.0057 in.
  Marked yellow &
    "HO" ....0.0047-0.0059 in.
Piston Pin Diameter .. 1.4997-1.5003 in.
Piston Pin Bore in
  Piston ....1.5003-1.5009 in.
*Measured at bottom of skirt, right angle to pin.

4320, 4520 and 4620
Piston Skirt Diameter*
  Marked green ......4.2464-4.2471 in.
  Marked yellow .....4.2471-4.2478 in.
Cylinder Bore Diameter
  Marked "LH" ......4.2490-4.2500 in.
  Marked "HH" ......4.2500-4.2510 in.
Piston Skirt* to Cylinder
  Clearance
  Marked green &
    "LH" ....0.0019-0.0036 in.
  Marked yellow &
    "HH" ....0.0022-0.0039 in.
Piston Pin Diameter .. 1.6247-1.6253 in.
Piston Pin Bore in
  Piston ....1.6253-1.6259 in.
*Measured at bottom of skirt, right angle to pin.

PISTON PINS

All Models

52. The full floating type piston pins are retained in piston bosses by snap rings. Pins are often available in oversizes as well as standard. Check parts source for availability.

The recommended fit of piston pins is a hand push fit in piston bores and a slip fit in connecting rod bushings. Standard diameter and clearances are as follows:
3020, 4000 and 4020 Gasoline and LP-Gas
Piston Pin Diameter
  Standard..........1.2497-1.2500 in.
  Wear Limit ....1.2492 in.
Piston Pin Bore in Connecting
  Rod ....1.251-1.252 in.
Piston Pin Bore in Piston
  Standard..........1.2503-1.2509 in.
  Wear Limit ....1.2519 in.

3020, 4000 and 4020 Diesel
Piston Pin Diameter
  Standard..........1.4997-1.5003 in.
  Wear Limit ....1.4992 in.
Piston Pin Bore in Connecting
  Rod ....1.501-1.502 in.
Piston Pin Bore in Piston
  Standard..........1.5003-1.5009 in.
  Wear Limit ....1.5019 in.

4320, 4520 and 4620 Diesel
Piston Pin Diameter
  Standard..........1.6247-1.6253 in.
  Wear Limit ....1.6242 in.
Piston Pin Bore in Connecting
  Rod ....1.6260-1.6270 in.
Piston Pin Bore in Piston
  Standard..........1.6253-1.6259 in.
  Wear Limit ....1.6269 in.

CONNECTING RODS AND BEARINGS

All Models

53. Connecting rod bearings are steel-backed inserts. Bearings are available in standard size as well as undersizes of 0.002, 0.010, 0.020 and 0.030 inch. Refer to paragraph 52 for specifications concerning the piston pin bushing.

Mating surfaces of rod and cap have milled tongues and grooves which positively locate cap and prevent it from being reversed during installation. Connecting rods are marked "FRONT" for proper installation. Check connecting rods, bearings and crankpin journals for excessive taper and against values which follow:

3020, 4000 and 4020 Gasoline and LP-Gas
Crankpin Std.
  Diameter..........2.9985-2.9995 in.
Crankpin to Rod Bearing
  Diametral Clearance
  Desired ....0.0015-0.0045 in.
  Wear Limit ....0.0065 in.
Rod Bolt Torque ....85-95 ft.-lbs.

3020, 4000, 4020, 4320, 4520 and 4620 Diesel
Crankpin Std.
  Diameter..........2.9982-2.9992 in.
Crankpin to Rod Bearing
  Diametral Clearance
  Desired ....0.0015-0.0045 in.
  Wear Limit ....0.0065 in.
Rod Bolt Torque ....100-110 ft.-lbs.

CRANKSHAFT AND BEARINGS

All Models

54. The crankshaft is forged steel and dynamically balanced. On 3020 spark ignition engine, three main bearings are used with center bearing taking end thrust. On 3020 diesel engine, five main bearings are used and third bearing from rear is thrust bearing. On 4000 and 4020 spark ignition engine, four main bearings are used with end thrust taken by second bearing from rear. On 4000, 4020, 4320, 4520 and 4620 diesel engines, seven main bearings are used and crankshaft end thrust is taken by third bearing from rear.

Upper and lower bearing shells may

be interchangeable, with both halves containing oil holes; however, if only one half of bearing has a hole, be sure that half is installed in the block to insure proper lubrication.

All main bearing caps can be removed from below after removing oil pan and oil pump. Identify caps so they can be reinstalled on main bearing bosses from which they were removed. When renewing bearings, make sure locating lug on bearing shell is aligned with milled slot in cap and block bore. After caps are loosely installed, pump crankshaft forward and rearward to align thrust flanges; then tighten main bearing cap screws to a torque of 150 ft.-lbs. Main bearings are available in undersizes of 0.002, 0.010, 0.020 and 0.030 inch. The thrust bearing is available in all undersizes with standard flange width and in 0.010 inch undersize and 0.007 inch oversize flange width. Thrust flange thickness should be sufficient to permit crankshaft end play within desired limits.

To remove crankshaft, first remove engine as outlined in paragraph 35 and proceed as follows: Remove flywheel and crankshaft rear oil seal retainer. Remove crankshaft pulley, timing gear cover, oil pan and oil pump. On 3020 models, unbolt and remove engine balancer. On all models, remove rod and main bearing caps and lift out crankshaft. The hardened crankshaft front and rear wear sleeves are a press fit on flywheel flange and nose of crankshaft, and are both renewable. When renewing, refer to paragraph 45 for front wear sleeve, and paragraph 56 for rear sleeve.

Examine crankshaft damper pulley for signs of eccentricity, wobble or damage at attaching point. If damper has failed, check crankshaft for cracks. The manufacturer recommends on diesel engines that the damper be replaced, regardless of condition, at time of major engine repair.

Check crankshaft and bearings against the following values:

All Models

Crankshaft End Play –
Desired . . . . 0.0025-0.0086 in.
Wear Limit . . . . 0.0235 in.
Main Bearing Journal –
Standard Diameter 3.3715-3.3725 in.
Desired Clearance in
Bearing . . . . 0.0017-0.0047 in.
Wear Limit . . . . 0.0077 in.
Regrind if Taper
Exceeds . . . . . 0.001 in./1 in. length
Regrind if Out-of-Round
Exceeds . . . . 0.0040 in.
Connecting Rod Journal –
Standard Diameter,
Gasoline & LP-Gas 2.9985-2.9995 in.
Diesel . . . . 2.9982-2.9992 in.
Desired Clearance in
Bearing . . . . 0.0015-0.0045 in.
Wear Limit . . . . 0.0065 in.
Main Cap Torque . . . . 150 ft.-lbs.
Rod Bolt Torque
Gasoline & LP-Gas . . . . . . 85-95 ft.-lbs.
Diesel . . . . 100-110 ft.-lbs.
Crankshaft Damper or Pulley
to Crankshaft Torque . . . . . 150 ft.-lbs.
Flywheel to Crankshaft
Torque . . . . 85 ft.-lbs.

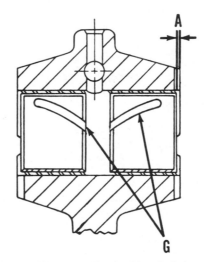

Fig. 50 — When installing bushings in balancer frame, make sure end of oil grooves (G) are together and bushing is 3/64-inch below bore surface as shown at (A).

ENGINE BALANCER
Model 3020

55. A crankshaft driven engine balancer is used on four cylinder engines to reduce vibration. The balancer assembly is bolted to bottom of engine block within the oil pan.

When removing balancer, be careful not to damage oil pump intake pipe. To disassemble, remove oil shield plates (5 – Fig. 49) and snap rings (1), then press shafts out of gears; support at driven gear (8) or balance weight (9) to prevent damage to housing. To renew bushings (3), use JDE-16 or other suitable driver, and install bushings with open end of oil grooves (G-Fig. 50) toward inside of housing and edge of bushing recessed 3/64-inch below housing surface as shown at A.

To assemble, reverse disassembly procedure. Align timing marks as shown in Fig. 51 when installing gears and shafts in housing. Turn crankshaft so No. 1 and 4 pistons are at bottom center, then install balancer assembly aligning timing marks on drive gear and driven gear as shown in Fig. 52. Tighten cap screws to 85 ft.-lbs. torque.

To renew crankshaft drive gear, crankshaft must be removed as outlined in paragraph 54. Carefully grind off old welds and dress weld surfaces on crankshaft flange to conform with rest of flange surface. Install new gear on crankshaft with chamfered edge towards front of shaft. Press gear on until front face of gear is two inches from front face of number two connecting rod journal. Place in lathe or V-blocks and check gear straightness, then tack weld gear in four places as shown in Fig. 53.

NOTE: Tack-weld length must not exceed 3/8-inch. DO NOT apply more heat than necessary to make weld. Protect gear teeth and crankshaft journals from weld splatter.

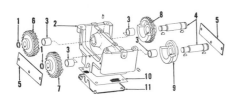

Fig. 49 — Exploded view of engine balancer used on 3020 models.

1. Snap ring
2. Housing
3. Bushings
4. Shafts
5. Plates
6. Balancer gear
7. Balancer gear
8. Driven gear & weight
9. Balancer weight
10. Screen
11. Cover

Fig. 51 — Align balancer gear timing marks when reinstalling gears and shafts.

Fig. 52 — When installing balancer assembly, timing marks on crankshaft gear and balancer driven gear must be aligned as shown.

Use ⅛-inch, E7018 welding rod at approximately 100 amperes current, and weld gear in place as shown in Fig. 53. Beads must be equal in length and approximately one inch long. Clean slag from beads.

CRANKSHAFT REAR OIL SEAL

All Models

56. The crankshaft rear oil seal is contained in a retainer plate which is attached to rear face of cylinder block by cap screws. Seal is available only in a kit which also includes the steel wear sleeve. To renew seal, first detach (split) engine from clutch housing as outlined in paragraph 120 and remove clutch and flywheel.

Unbolt and remove oil seal retainer plate and score wear sleeve lightly with a dull chisel, then pry wear sleeve from crankshaft using a screwdriver or pry bar. Drive or press seal from retainer being careful not to damage retainer plate. Install seal with closed side to rear of engine. Install wear sleeve with a suitable driver so rear edge is flush with rear face of crankshaft flange. Check sealing face runout with a dial indicator. Runout should not exceed 0.006 inch. Tighten screws attaching seal retainer to 20 ft.-lbs. torque when correctly centered.

FLYWHEEL

All Models

57. Flywheel is doweled to crankshaft flange and retained by four cap screws. Flywheel can be removed by using forcing screws in tapped holes provided.

To install flywheel ring gear, heat gear evenly to approximately 300 degrees F. and position gear so chamfered end of gear teeth face toward front of engine. When installing flywheel, align dowel hole in flywheel over

dowel in flange, coat threads of retaining cap screws with sealant and tighten evenly to a torque of 85 ft.-lbs.

OIL PAN

All Models

58. Engine oil pan can be removed without interference from other components. Drain engine oil, remove oil filter access plate and filter from side of oil pan (3020, 4000 and 4020 models). Use a floor jack or other lifting means to remove and install the cast pan. When installing, tighten the ⅜-inch cap screws to a torque of 35 ft.-lbs. and ½-inch cap screws to a torque of 85 ft.-lbs.

OIL PUMP

All Models

59. **REMOVE AND REINSTALL.** To remove oil pump, first remove oil pan as outlined in paragraph 58, turn crankshaft until No. 1 piston is at TDC on compression stroke; then unbolt and remove oil pump. Correct distributor timing on spark ignition models depends on proper installation of oil pump. When installing pump, remove tachometer drive housing and drive gear. If crankshaft was properly timed (TDC-1) when oil pump was removed and crankshaft has not been turned, it may not be necessary to retime distributor. If drive slots will not engage, distributor must be removed or loosened. Refer to Fig. 54. Bottom of tooth which aligns with drive slot must align with a small casting mark just to the side of oil hole in housing as shown.

When pump is installed, "V" mark should be toward crankshaft and drive slot approximately 15 degrees from parallel with crankshaft center line. On all models, tighten pump to block attaching screws to 20-25 ft.-lbs. torque. On non-diesel models, refer to paragraph 113 and retime ignition distributor.

60. **OVERHAUL.** To disassemble oil pump, remove pump, cover (1 – Fig. 55 or 56) and idler gear (2). Examine pump

gears and cover for wear or scoring. Gears are available as a matched set only. Check for wear or excessive looseness at drive gear end.

Driven gear (4) on all models is retained to drive shaft (7) by a press fit on shaft. Some early model pumps utilize a Woodruff key (6) in oil pump drive shaft and driven gear. To prevent damage to pump housing, refer to Fig. 57 and press shaft (2) from gear (3) as follows: Support pump housing and press shaft ¼-inch from gear, then place two ¼-inch shims (1) between gear and housing. Press shaft another ¼-inch and place two ½-inch shims between gear and housing. Press shaft another ½-inch and

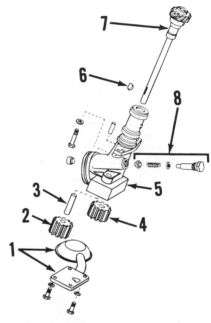

Fig. 55 — Exploded view of engine oil pump used on 3020, 4000 and 4020 models.

1. Pump cover & intake
2. Idler gear
3. Idler shaft
4. Driven gear
5. Pump housing
6. Woodruff key (some models)
7. Drive gear & shaft
8. Filter by-pass valve

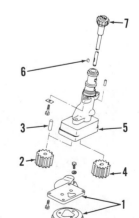

Fig. 56 — Exploded view of engine oil pump used in 4320, 4520 and 4620 models. Refer to Fig. 55 for legend.

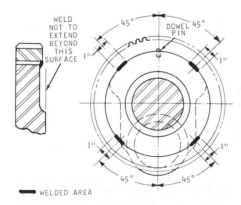

Fig. 53 — Engine balancer drive gear is located on crankshaft by a dowel pin. When renewing gear, weld at four points as outlined in text.

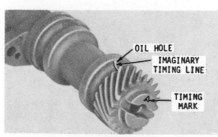

OIL HOLE
IMAGINARY TIMING LINE
TIMING MARK

Fig. 54 — Oil pump must be properly timed when installing. Refer to text.

gear should be free from shaft.

When reassembling, press drive gear on shaft until gear is flush with mating surface of housing. Driven gear and idler must spin freely and cover must not bind gears.

PRESSURE REGULATOR VALVES

Models 3020-4000-4020

61. The oil pressure regulator valve (Fig. 58) is housed in speed-hour meter cover, filter by-pass valve (8 – Fig. 55) is located in pump housing and oil cooler by-pass valve (Fig. 60) is located in coolant outlet body.

With engine at normal operating temperature, correct regulated oil pressure should be 25-40 psi at 1900 rpm. Pressure can be adjusted by adding or removing shims. One shim will change regulated pressure by about 5 psi. Oil filter by-pass valve operates on pressure differential of approximately 15 psi. Filter by-pass valve opens for cold starting, or if filter becomes plugged, to assure continuing lubrication. Oil cooler by-pass valve operates on pressure differential of 15 psi and opens for cold starting, or in case of cooler restriction.

Models 4320-4520-4620

62. The oil pressure regulator valve (1 – Fig. 59), filter by-pass valve (3) and oil cooler by-pass valve (4) are located in oil filter housing.

With engine at normal operating temperature, correct regulated pressure should be 40-50 psi at 1900 rpm. To adjust pressure, remove filter housing (5)

and add or subtract shims (2) at regulating valve spring. One shim will change regulated pressure by about 5 psi. Oil filter by-pass valve operates on pressure differential of approximately 25 psi. Oil cooler by-pass valve operates on pressure differential of approximately 15 psi.

OIL COOLER

All Models

63. All models are equipped with an engine oil cooler of the type shown in Figs. 60 or 61. Some variation from coolers shown may be noted on some models. Oil is cooled by engine coolant circulating through tubes in cooler body. Disassembly for cleaning or other service is normally not required except in cases of contamination of cooling or lubrication system.

AIR INTAKE SYSTEM

All Models

64. Air used to operate engine enters through a pre-cleaner which separates

dust from air. The air then passes through a dual element dry type filter into carburetor, manifold or turbocharger; then on into the engine. A vacuum switch connects to a warning indicator lamp which lights to warn operator if filters are restricted.

Because of the balanced system and large volume of air demanded (especially by a turbocharged diesel engine), it is of utmost importance that only approved parts which are in good condition be used.

To check for air cleaner restriction, install a "T" fitting, an elbow and the vacuum switch, where switch was originally installed, and connect a JDST-11 water vacuum gage in "T" fitting.

Normal vacuum at port for switch with clean filters and engine operating at full load is as follows:

Model	Inches of Water
3020	
Spark Ignition	3
Diesel	4
4020	9
4320	8
4520	8
4620	8

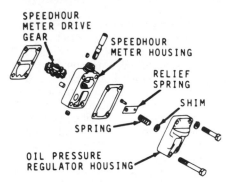

Fig. 58—Exploded view of speed-hour meter drive housing and oil pressure regulator assembly used on 3020, 4000 and 4020 models.

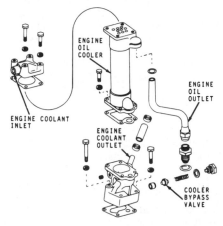

Fig. 60—Exploded view of typical engine oil cooler used on 3020, 4000 and 4020 models. Cooler by-pass valve is located in coolant outlet body.

Fig. 57—When pressing oil pump drive shaft (2) out of driven gear (3), use shims (1) under gear as shown to prevent Woodruff key from damaging pump housing. Refer to text for procedure.

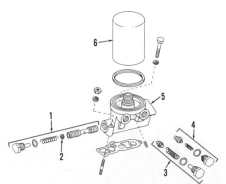

Fig. 59—Exploded view of oil filter housing used on 4320, 4520 and 4620. Housing contains oil pressure regulator valve (1), oil filter by-pass valve (3) and oil cooler by-pass valve (4).

1. Oil pressure regulator valve
2. Shim
3. Oil filter by-pass valve
4. Oil cooler by-pass valve
5. Oil filter housing
6. Oil filter

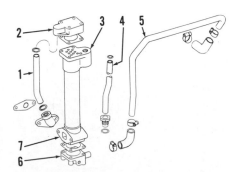

Fig. 61—Exploded view of typical engine oil cooler used on 4320, 4520 and 4620 models.

1. Water inlet tube
2. Cooler top cover
3. Oil cooler body
4. Oil outlet tube
5. Water outlet tube
6. Cooler bottom cover
7. Oil inlet

Warning switch closes at a vacuum of 24-26 inches of water (approximately 1.8 inches Hg) and vacuum should never be permitted to be higher than 25 inches of water on any model.

On turbocharged models, intake manifold pressure, checked at the ⅜-inch pipe plug near the inlet tube on intake manifold at 2200 rpm, full load, should be 17-19 psi. A low manifold air pressure may indicate intake manifold air leaks, restricted air intake or air cleaner, exhaust leaks or a malfunctioning turbocharger. Refer to paragraph 71 for service to intercooler located in intake manifold on models so equipped.

TURBOCHARGER

OPERATION

Models 4320-4520-4620

65. The exhaust driven turbocharger supplies air to intake manifold at above normal atmospheric pressure. The additional air entering the combustion chamber permits an increase in the amount of fuel burned, and increased power output over an engine of comparable size not so equipped.

Because a turbocharger compresses incoming air, heat of compression causes the air to expand and become less dense than it would be at a lower temperature. Models 4520 (engine S.N. 303519 and above) and 4620 are equipped with an intercooler which lowers the temperature of intake air which, in turn, increases the density of intake air. This permits the use of more fuel and results in more power.

The turbocharger contains a rotating shaft which carries an exhaust turbine wheel on one end and a centrifugal air compressor on the other. The rotating member is precisely balanced and capable of rotative speeds up to 100,000 rpm. Bearings are floating sleeve type, and unit is lubricated and cooled by a flow of engine oil under pressure. Exchange turbocharger units are available, or a qualified technician can overhaul unit if parts are available.

SERVICE

Models 4320-4520-4620

66. Schwitzer turbocharger Model 3LD or AiResearch Model TO4 units have been used. Be sure to install correct replacement unit if turbocharger is exchanged. The turbocharger consists of the following three main sections. The turbine, bearing housing and compressor.

Engine oil taken directly from clean oil side of engine oil filters, is circulated

through the bearing housing. This oil lubricates the sleeve type bearings and also acts as a heat barrier between the hot turbine and compressor. The oil seals used at each end of the shaft are piston ring type. When servicing turbocharger, extreme care must be taken to avoid damaging any of the parts.

CAUTION: DO NOT operate turbocharger without adequate lubrication. When turbocharger is first installed, or engine has not been run for a month or more, or new oil filter has been installed, turn engine over with starter with fuel cutoff wire disconnected until oil pressure indicator light goes out; reconnect wire and start engine. Run engine at slow idle speed for at least two minutes before opening throttle or putting engine under load.

Some other precautions to be observed in operating and servicing a turbocharged engine are as follows:

Do not operate at wide-open throttle immediately after starting. Allow engine to idle until turbocharger slows down before stopping engine. This will insure adequate lubrication to shaft bearings at all times.

Because of increased air flow, care of air cleaner and connections is of added importance. Check the system and condition of restriction indicator whenever

tractor is serviced. Make sure exhaust pipe opening is closed and air cleaner connected whenever tractor is transported, to keep turbocharger from turning due to air pressure. If exhaust outlet is equipped with weathercap, tape cap closed. If weathercap is missing, use tape to close exhaust outlet pipe.

67. **REMOVE AND REINSTALL.** To remove turbocharger, first remove muffler, hood, exhaust elbow and adapter. Remove oil lines and intake hose connections, then unbolt and remove turbocharger from exhaust manifold.

To inspect removed turbocharger unit, examine turbine wheel and compressor impeller for blade damage, looking through housing and end openings. Using a dial indicator with plunger extension, check radial bearing play through outlet port while moving both ends of turbine shaft equally. Check shaft end play with dial indicator working from either end. If end or side play exceeds 0.009 inch, or if any blades are broken or damaged, renew or overhaul unit.

When installing, attach turbocharger to manifold using a new gasket and tighten stud nuts to a torque of 35 ft.-lbs. Install inlet oil line, outlet line, adapter and exhaust elbow after first making sure parts are perfectly aligned.

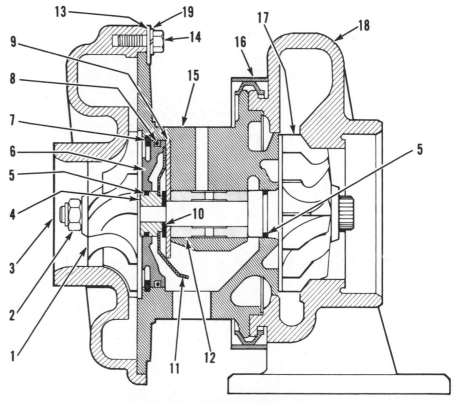

Fig. 62 — Cross-sectional view of Schwitzer Model 3LD turbocharger showing location of component parts. Refer to legend under Fig. 63.

Adapter must have a minimum of 1/16-inch end play and be free to rotate. Undue stress on turbocharger at installation may cause bearing failure. Prime turbocharger as outlined in paragraph 66.

Schwitzer Model 3LD

68. OVERHAUL. Remove turbocharger as outlined in paragraph 67. Before disassembling, place a row of light punch marks across compressor cover, bearing housing and turbine housing to aid in reassembly. Clamp turbocharger mounting flange (exhaust inlet) in a vise and remove cap screws (14 – Fig. 63), lockwashers and clamp plates (13). Remove compressor cover (3). Remove nut from clamp ring (16), expand clamp ring and remove bearing housing assembly (15) from turbine housing (18).

CAUTION: Never allow weight of bearing housing assembly to rest on either turbine or compressor wheel vanes. Lay bearing housing assembly on a bench so turbine shaft is horizontal.

Remove locknut (2) and slip compressor wheel (1) from end of shaft. Withdraw turbine wheel and shaft (17) from bearing housing. Place bearing housing on bench with compressor side up. Remove snap ring (7), then using two screwdrivers, lift flinger plate insert (6) from bearing housing. Push spacer sleeve (4) from the insert. Remove oil deflector (11), thrust ring (10), thrust plate (9) and bearing (12). Remove "O" ring (8) from flinger plate insert (6) and remove both seal rings (5) from spacer sleeve and turbine shaft.

Soak all parts in Bendix metal cleaner or equivalent and use a soft brush, plastic blade or compressed air to remove carbon deposits. CAUTION: Do not use wire brush, steel scraper or caustic solution for cleaning, as this will damage turbocharger parts. Glass bead dry blast may be used for cleaning if air pressure does not exceed 40 psi and all traces of glass beads are rinsed out before assembling.

Inspect turbine wheel and compressor wheel for broken or distorted vanes. DO NOT attempt to straighten bent vanes. Check bearing bore in bearing housing, floating bearing (12) and turbine shaft for excessive wear or scoring. Inspect flinger plate insert (6), flinger sleeve (4), oil deflector (11), thrust plate (9) and thrust ring (10) for excessive wear or other damage. Refer to Fig. 63 and the following for specifications of new parts.

Bearing Bore in
   Housing (15) . . . . 0.8750-0.8755 in.
Bearing (12) Length . . . . 1.484-1.485 in.
Flinger Sleeve (4)
   Length . . . . 0.517-0.519 in.
Piston Ring Grooves . . . . 0.064-0.065 in.
Shaft (17) Concentricity –
   Maximum Run-out . . . . 0.001 in.
Shaft (17) Diameter . . . 0.4800-0.4803 in.
Shaft (17) Shoulder
   Length . . . . 1.595-1.596 in.
Thrust Plate (9)
   Thickness . . . . 0.107-0.108 in.
Assembled Bearing Clearances –
   Axial End Play
    of Shaft . . . . 0.003-0.008 in.
Radial Clearance Measured at
   Exhaust Turbine
    Blades . . . . 0.018-0.049 in.

Renew all damaged parts and use new "O" ring (8) and seal rings (5) when reassembling. The seal ring used on turbine shaft is copper plated and is larger in diameter than seal ring used on spacer sleeve. Refer to Figs. 62 and 63 when reassembling.

Install seal ring on turbine shaft, lubricate seal ring and install turbine wheel and shaft in bearing housing. Lubricate I.D. and O.D. of bearing (12), install bearing over end of turbine shaft and into bearing housing. Lubricate both sides of thrust plate (9) and install plate (bronze side out) on the aligning dowels. Install thrust ring (10) and oil deflector (11), making certain holes in deflector are positioned over dowel pins. Install new seal ring on spacer sleeve (4), lubricate seal ring and press spacer sleeve into flinger plate insert (6). Position new "O" ring (8) on insert, lubricate "O" ring

and install insert and spacer sleeve assembly in bearing housing, then secure with snap ring (7). Place compressor wheel on turbine shaft, coat threads and back side of nut (2) with graphite grease, or equivalent, then install and tighten nut to 156 in.-lbs. torque. Assemble bearing housing to turbine housing and align punch marks. Install clamp ring, apply graphite grease to threads, install nut and tighten to 120 in.-lbs. torque. Apply a light coat of graphite grease around machined flange of compressor cover (3). Install compressor cover, align punch marks and secure cover with cap screws, washers and clamp plates. Tighten cap screws evenly to 60 in.-lbs. torque. Fill oil inlet with engine oil and turn turbine shaft by hand to lubricate bearing and thrust plate.

Check rotating unit for free rotation within the housings. Cover all openings until turbocharger is reinstalled.

Use a new gasket and install and prime turbocharger as outlined in paragraph 66.

Turbocharger oil supply pressure at 2200 engine rpm should be within 10 psi of engine oil pressure but never less than 25 psi. Minimum return oil flow from turbocharger is ½-gpm at 2200 engine rpm.

AiResearch Turbocharger

69. INSPECTION. Remove turbocharger unit as outlined in paragraph 67. To inspect removed turbocharger unit, examine turbine wheel and compressor impeller for blade damage, looking through housing openings. Using a dial indicator with plunger extension, check radial bearing play through outlet oil port while moving both ends of turbine shaft equally. Radial bearing clearance should not exceed 0.009 inch. Check shaft end play with dial indicator working from either end. If end play exceeds 0.009 inch or if any blades are broken or damaged, overhaul or renew turbocharger unit.

70. OVERHAUL. Mark across compressor housing (1 – Fig. 64), center housing (13) and turbine housing (17) to aid alignment when assembling.

CAUTION: Do not rest weight of any parts on impeller or turbine blades. Weight of only the turbocharger unit is enough to damage blades.

Remove lock plates and clamp plates (6) from compressor housing (1) and remove housing, some models use a clamp (1A) to secure housing to backplate. Remove lock plates and clamp plates (6) from turbine housing (17) and remove housing. Hold turbine shaft from turning using suitable holding fixture for

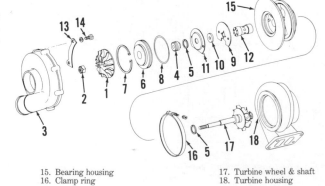

Fig. 63 – Exploded view of Schwitzer Model 3LD turbocharger used on some models.

1. Compressor wheel
2. Locknut
3. Compressor cover
4. Spacer sleeve
5. Seal rings (not identical)
6. Insert
7. Snap ring
8. "O" ring
9. Thrust plate
10. Thrust ring
11. Oil deflector
12. Bearing
13. Clamp plate
14. Cap screw
15. Bearing housing
16. Clamp ring
17. Turbine wheel & shaft
18. Turbine housing

turbine wheel (16) and remove locknut (2).

NOTE: Use a "T" handle or double universal socket to remove locknut in order to prevent bending turbine shaft.

Lift compressor impeller (3) off, then remove center housing from turbine shaft while holding shroud (15) onto center housing. Remove backplate (4), thrust bearing (9) and thrust collar (8). Carefully remove bearing retainers (10) from ends and withdraw bearings (11). Spring (5) is available as an assembly with backplate (4). Refer to Fig. 65.

CAUTION: Be careful not to damage bearings or surface of center housing when removing retainers. The center two retainers do not have to be removed unless damaged or unseated. Always renew bearing retainers if removed from grooves in housing.

Clean all parts in a cleaning solution which is not harmful to aluminum. A stiff brush and plastic or wood scraper should be used after deposits have softened. When cleaning, use extreme caution to prevent parts from being nicked, scratched or bent.

Inspect bearing bores in center housing (13 – Fig. 64) for scored surfaces, out-of-round or excessive wear. Make certain bore in center housing is not grooved in area where seal ring (18) rides. Compressor impeller (3) must not show signs of rubbing with either compressor housing (1) or backplate (4). Impeller should have 0.0002 inch tight to 0.0004 inch loose fit on turbine shaft. Make certain impeller blades are not bent, chipped, cracked or eroded. Oil passages in thrust collar (8) must be clean and thrust faces must not be warped or scored. Ring groove

shoulders must not have step wear. Inspect turbine shroud (15) for evidence of turbine wheel rubbing. Turbine wheel (16) should not show evidence of rubbing and vanes must not be bent, cracked, nicked, or eroded. Turbine wheel shaft must not show signs of scoring, scratching or overheating. Groove in shaft for seal ring (18) must not be stepped. If turbine shaft journals are damaged, undersized bearings of 0.005 and 0.010 inch may be ordered, and shaft reconditioned if so desired. Check shaft end play and radial clearance when assembling.

If bearing inner retainers (10) were removed, install new retainers. Oil bearings and install in center housing and install outer retainers. Position shroud (15) on turbine shaft (16) and install seal ring (18) in groove. Apply a light, even coat of engine oil to shaft journals, compress seal ring (18) with a thin strong tool such as a dental pick and install center housing (13). Install new seal ring (7) in groove of thrust collar (8), then install thrust bearing so smooth side of bearing (9) is toward seal ring (7) end of collar. Install thrust bearing and collar assembly over shaft, making certain pins in center housing engage holes in thrust bearing. Install new rubber seal ring (12), make certain spring (5) is positioned in backplate (4), then install backplate making certain seal ring (7) is not damaged. Seal ring will be less likely broken if open end of ring is installed in bore of backplate first. Install lock plates (14) and screws, tightening screws to 75-90 in.-lbs. torque. Install compressor impeller (3) and make certain impeller is seated against thrust collar (8). Install locknut (2) to 18-20 in.-lbs. torque, then use a "T" handle or double universal joint socket to turn locknut an additional 90° in order to stretch turbine

shaft the necessary 0.0055-0.0065 inch for proper tension.

CAUTION: If "T" handle or double universal joint socket is not used, shaft may be bent when tightening nut (2).

Install turbine housing (17) with clamp plates (6) next to housing, tighten screws to 100-130 in.-lbs., then bend lock plates up around screw heads.

Check shaft end play and radial play at this point of assembly. If shaft end play exceeds 0.009 inch, thrust collar (8) and/or thrust bearing (9) is worn excessively. End play of less than 0.003 inch indicates incomplete cleaning (carbon not all removed) or dirty assembly and unit should be disassembled and cleaned.

If turbine shaft radial play exceeds 0.009 inch, unit should be disassembled and bearings, shaft and/or center housing should be renewed, or shaft should be reconditioned and undersized bearings installed. Center housing bearing bore should not exceed 0.6228 inch and seal bore should not exceed 0.703 inch diameter. Shaft journal diameter should not be less than 0.3994 inch and seal hub diameter should not be less than 0.681 inch. Maximum permissible limits of all of these parts may result in radial play which is not acceptable.

Install compressor housing (1) and diffuser (1C) if so equipped, but do not tighten until unit is installed, so perfect alignment can be made. Fill reservoir with engine oil and protect all openings of turbocharger until unit is installed on tractor. After intake hoses are all assembled, tighten clamp plates (6) to 110-130 in.-lbs. torque to compressor housing (1), and complete assembly. Prime turbocharger as outlined in paragraph 66.

INTERCOOLER

Models So Equipped

71. The intake manifold on 4520 (engine S.N. 303519 and above) and

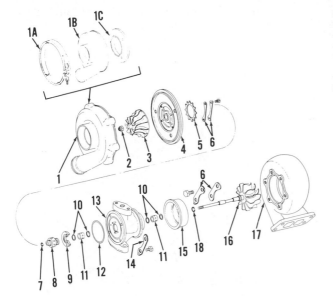

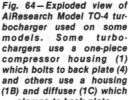

Fig. 64—Exploded view of AiResearch Model TO-4 turbocharger used on some models. Some turbochargers use a one-piece compressor housing (1) which bolts to back plate (4) and others use a housing (1B) and diffuser (1C) which clamps to back plate.

1A. Clamp
1B. Compressor housing
1C. Diffuser
1. Compressor housing
2. Locknut
3. Compressor impeller
4. Back plate
5. Spring
6. Clamp plates
7. Seal ring
8. Thrust collar
9. Thrust bearing
10. Retaining rings
11. Bearings
12. "O" ring
13. Center housing
14. Clamp plate
15. Shroud
16. Turbine shaft & wheel
17. Turbine housing
18. Seal ring

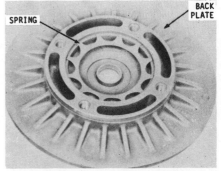

Fig. 65—AiResearch model back plate and spring must be renewed as an assembly if damaged.

4620 models contains an intercooler (3—Fig. 66) to lower intake air temperature. Coolant from engine cooling system flows through intercooler core and heat from compressed (turbocharged) intake air is conducted into engine coolant. The intercooler can lower the temperature of intake air as much as 80-90 degrees F., which will make intake air more dense and permit more air to be delivered to engine cylinders.

Since coolant from the radiator is circulated through intercooler core, a leak in the core could cause serious damage to the engine by allowing coolant into combustion area.

To remove intercooler, drain cooling system, remove muffler, and hood. Remove turbocharger as outlined in paragraph 67. Disconnect both water hoses at intercooler and ether starting aid pipe. Remove both adapter plates at hose connections to intercooler. Remove cap screws from underneath intake manifold, lift off manifold cover and intercooler core.

Test and repair procedures are much the same as for a radiator. The intercooler can be pressurized with air (20-25 psi) then submerged in water to check for leaks. Repair or renew aluminum intercooler core as necessary.

Reassemble in reverse order of disassembly using all new gaskets and prime turbocharger as outlined in paragraph 66.

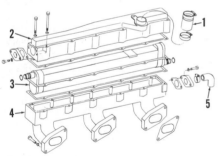

Fig. 66 — Exploded view of intercooler and intake manifold used on late 4520 models and all 4620 models.

1. Air inlet hose
2. Intake manifold cover
3. Intercooler
4. Intake manifold
5. Water inlet pipe

Fig. 67 — View of Marvel-Schebler carburetor showing idle mixture adjusting screw and load adjusting screw.

GASOLINE FUEL SYSTEM

CARBURETOR

Models 3020-4000-4020

Marvel-Schebler USX or Zenith Series 69 carburetors are used. Refer to appropriate following paragraphs.

72. MARVEL-SCHEBLER. All models are equipped with a vacuum actuated, diaphragm type accelerator pump, idle and load mixture adjustment needles and an idle speed adjusting screw. Suggested initial settings are 2½ turns open from closed position for load mixture adjustment needle and 1½ turns open for idle mixture adjustment

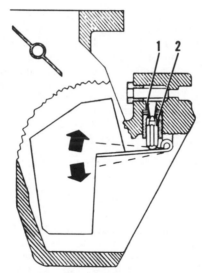

Fig. 68 — To check float adjustment, remove inlet needle (2) and reinstall float. Mark limits of float travel as indicated by dotted lines. With needle reinstalled, float should rest midway between the two marks. Float level can be lowered by adding a gasket (1).

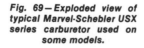

Fig. 69 — Exploded view of typical Marvel-Schebler USX series carburetor used on some models.

1. Accelerator pump assy.
2. Float
3. Idle adjusting needle
4. Fuel inlet strainer
5. Throttle shaft
6. Choke shaft
7. Fuel shut-off solenoid
8. Economizer jet
9. Accelerator jet
10. Discharge nozzle
11. Idle jet
12. Float valve & seat
13. Load needle
14. Load adjusting screw

needle. Refer to Fig. 67. Back out idle speed stop screw until it clears stop, then turn screw in 1⅛ turns after it contacts stop and begins to open throttle butterfly. Adjust speed and throttle linkage as outlined in paragraph 103.

A definite measurement of float level is not given, but float should close needle valve about midway between limit stops in fuel bowl. Refer to Fig. 68. To check float setting after removing bowl cover, first remove float and lift out inlet needle (2). Reinstall float and, using a pencil and marking on wall of fuel bowl, mark position of float lever with carburetor inverted and right side up, as shown by broken lines. Reinstall inlet needle (2); reinstall float and check float position with carburetor body inverted. Float should rest midway between marked positons as shown. If it does not, adjust by bending float lever or installing extra gasket (1) beneath inlet needle seat.

The accelerator pump discharge jet (9 — Fig. 69) screws into carburetor body and is ported through side of venturi. Check ball must unseat with minimum pressure but seal completely when ball is seated. A tapered weight, instead of a spring, seats the ball so threaded end of jet must be down when checking ball seating.

On all models, accelerator pump check valve is contained in diaphragm housing as shown in Fig. 70. Check ball (C) and spring can be inspected or renewed after removing seat (S). Distance (D) from end of diaphragm stop to gasket surface of housing should measure 29/64-inch. Diaphragm stop pin can be driven into or out of housing if distance (D) is not correct.

73. ZENITH. The Zenith, Model 69 carburetor used on some engines is equipped with a vacuum controlled accelerator pump. Refer to Fig. 71. Initial setting of idle mixture adjusting needle is 1 turn open; load mixture adjusting needle, 1¾ turns open. Adjust speed and

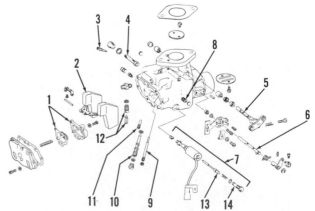

throttle linkage as outlined in paragraph 103. Make final mixture adjustments with engine at operating temperature.

Float setting is 1-19/32 inches when measured from gasket surface of carburetor body to farthest edge of float with needle valve closed. To adjust float, bend levers next to float body.

The accelerator pump piston and spring assembly can be withdrawn from top of bowl chamber after removing upper body. Both pistons of accelerator pump assembly should fit cylinder with a minimum of 0.001 inch or a maximum of 0.003 inch clearance in cylinder. Examine cylinder for scoring or deep scratches, and pistons for wear, scoring or other damage. A check valve is located in bottom of accelerator pump cylinder bore. Check valve is a drive fit and seats on a stepped shoulder in bore. Removal of check valve requires the use of a special tool (Zenith Part No. C161-15) which screws into check valve enabling valve to be withdrawn from top of cylinder bore. Check valve will be damaged in removal, and must be renewed if removed. Install check valve until it bottoms, using Zenith Tool C161-197.

GASOLINE FUEL PUMP AND FILTERS

Models So Equipped

74. A sediment bowl is located at rear of gasoline tank, and should be removed and strainer cleaned with each tune up, or when water is observed in bowl.

An in-line gasoline filter is located between fuel pump and carburetor. Renew filter at each tune up, and note arrow, which indicates direction of flow. If water is found in system, it may also be necessary to drain gasoline tank and carburetor, and clean sediment bowl.

The gasoline fuel pump is camshaft operated, and must be renewed if found to be faulty. Before renewing pump, check fuel pressure, which should be 3½-4½ psi at slow idle, and be sure sediment bowl or line to pump is not restricted.

LP-GAS SYSTEM

Models 3020-4000-4020

75. The pressure fuel tank is equipped with filter, vapor return, pressure relief, bleed and liquid and vapor withdrawal valves which can only be serviced as complete assemblies. Before renewal is attempted on any of these units, fuel tank must be completely exhausted of fuel.

If little fuel remains in tank, allow engine to run until fuel is exhausted, then open bleed valve and allow any remaining pressure to escape. If a considerable quantity of fuel is in tank consult an LP-Gas dealer about pumping out and saving the fuel.

The fuel gage sending unit consists of a magnetic sender unit which can be renewed at any time, and a float unit

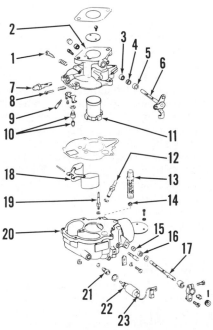

Fig. 71 — Exploded view of typical Zenith Model 69 carburetor used on some models.

1. Idle adjusting needle	12. Discharge nozzle
2. Throttle body	13. Accelerator pump
3. Needle bearing (2)	14. Check valve
4. Packing	15. Packing
5. Retainer	16. Retainer
6. Throttle shaft	17. Choke shaft
7. Fuel inlet strainer	18. Float
8. Throttle adjusting screw	19. Accelerator jet
9. Idle jet	20. Lower body
10. Float valve & seat	21. Load seat
11. Venturi	22. Fuel shut-off solenoid
	23. Load adjusting screw

which can only be renewed if tank is completely empty.

The safety relief valve is set to open at 312 psi to protect fuel tank against excessive pressures. This pressure should never be adjusted. If relief valve is faulty or inoperative, renew unit.

UL regulations prohibit any welding or repair on LP-Gas tanks. In the event of defect or damage, tank must be renewed rather than repaired.

Fuel lines and components may be removed at any time without emptying tank if liquid and vapor withdrawal valves are closed and engine allowed to run until fuel is exhausted in lines and filter.

76. **ADJUSTMENT.** Before starting engine for the first time after installing a new or overhauled carburetor, make the following preliminary adjustments:

Refer to Fig. 72 and turn load adjusting screw approximately 3½-turns open. Turn throttle stop screw ½-turn from closed position.

Start engine and bring to operating temperature. Place hand throttle in slow idle position and adjust throttle stop screw to obtain slow idle speed of 800 rpm. Loosen locknuts on drag link and adjust drag link to point of smoothest idle.

The load adjustment should properly be set under load. If a dynamometer is available, adjust load needle for best performance under rated load. If a dynamometer is not available, a reasonably satisfactory shop adjustment can be obtained as follows: Disconnect three spark plug wires (four cylinder models) or four spark plug wires (six cylinder models). Be sure to ground removed wires. Start engine and open hand throttle to first stop position. Adjust carburetor by opening load needle until highest engine speed is obtained, then close needle until speed just begins to drop. Shut off ignition, reinstall removed spark plug wires, then recheck engine low idle.

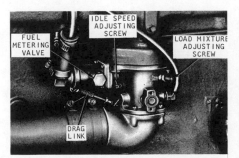

Fig. 72 — LP-Gas models use single barrel carburetor shown. Refer to text for carburetor adjustment.

Fig. 70 — Cross-section view of diaphram housing showing inlet check valve (C) and seat (S). Refer to text for recommended dimension (D).

77. TROUBLESHOOTING. If engine fails to start and if trouble is determined to be in the fuel system, set carburetor adjustments as outlined in paragraph 76. Check air cleaner for plugging and make sure air cleaner is properly serviced.

The following paragraphs list the more common troubles that can be attributed directly to fuel system. Many of the same troubles can be caused by malfunction of ignition system or of the valves and rings; therefore, check ignition system and compression also, when diagnosing trouble.

HARD STARTING could be caused by:

a. Improperly blended fuel.

b. Excess-flow valve in withdrawal valve closed. Close withdrawal valve to reset, then open slowly.

c. Incorrect starting procedure.

d. Plugged fuel filter or lines.

e. Liquid fuel in lines.

f. Automatic fuel shut-off not operating properly. Check solenoid and valve.

g. Plugged vent hole on converter.

h. Defective low pressure diaphragm in converter.

i. High pressure valve stuck or valve spring broken in converter.

LACK OF POWER could be caused by:

a. Throttle not opening properly due to maladjusted, bent or broken linkage.

b. Plugged vent hole in converter.

c. Clogged fuel strainer or lines.

d. Excess-flow valves closed.

e. Sticking high pressure valve in converter.

f. Restricted low pressure valve in converter.

g. Defective diaphragms or converter adjustment.

h. Engine not up to operating temperature.

i. Improperly adjusted carburetor.

j. Air leaks in carburetor fuel line or carburetor or manifold gaskets.

k. Clogged air filter.

POOR FUEL ECONOMY could be caused by:

a. Improperly adjusted carburetor or converter.

b. Leaks in fuel lines or tank.

c. Sticking converter valves.

d. Lack of power from any of the causes outlined above.

CONVERTER FREEZING UP could be caused by:

a. Running on liquid fuel before engine is warm.

b. Water circulating backwards through converter.

Leaks or restrictions in fuel system can sometimes be detected by frost forming at point of restriction. Check for frost at fuel filter and withdrawal valves on all complaints of lack of power or hard starting.

CARBURETOR

78. LP-Gas models use single barrel updraft carburetor shown in Fig. 72 and Fig. 73. When overhauling carburetor, check throttle shaft (13), needle bearings (6) and throttle disc (14) for wear and renew as needed. No special tools are required to replace needle bearings. Adjust as outlined in paragraph 76.

FUEL STRAINER AND SHUT-OFF VALVE

79. Fuel must pass through the strainer before reaching the converter. The strainer contains a filter element, consisting of a felt pad, chamois disc, and brass screens which remove all solids and gum from the fuel. A solenoid operated, automatic shut-off valve is located on top of strainer. A spring, plus system pressure keeps this valve closed when ignition switch is turned off.

If strainer is excessively cold or shows frost, it is probably clogged. To clean strainer, close both withdrawal valves and run engine until fuel is exhausted. Remove plug (15 – Fig. JD74) from bottom of strainer and open liquid with-

drawal valve momentarily. Pressure from fuel tank will blow out any accumulation of dirt.

To remove strainer assembly, close both withdrawal valves and run engine to exhaust any fuel, disconnect fuel lines and lead-in wire, then unbolt and remove unit.

Remove screws retaining cover (6) to strainer body (14) and lift off cover assembly. Remove filter pack (7) by prying out retaining ring (8). Filter pack can be cleaned in a volatile solvent and air dried. Reinstall filter pack with chamois disc toward the top.

Disconnect wire and lift off case (1) from shut-off valve; then lift off coil (2). Remove plunger housing (3) and lift out plunger and spring. Inspect and renew any damaged parts.

Assemble plunger housing by reversing disassembly procedure, then test by connecting unit to a battery. An audible "click" will be heard when solenoid opens valve.

After strainer is installed, turn on vapor withdrawal valve and use a soap solution to test for leaks at connections.

CONVERTER (REGULATOR)

80. **OPERATION.** Fuel enters converter as a liquid at tank pressure, through high pressure valve (6 – Fig. 75). The converter is connected to engine cooling system which supplies heat required for vaporization. Fuel pressure

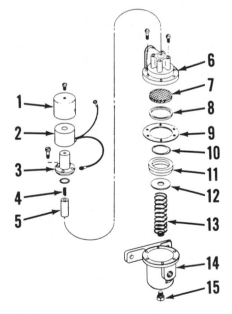

Fig. 73 – Exploded view of LP-Gas carburetor.

1. Air horn
2. Spray bar
3. Load adjustment screw
4. Cup
5. Packing (2)
6. Bearing (2)
7. Carburetor body
8. Metering valve body
9. Metering valve
10. Seal
11. Metering valve lever
12. Drag link
13. Throttle shaft
14. Throttle disk
15. Throttle lever
16. Throttle stop

Fig. 74 – Exploded view of fuel strainer and shut-off valve used on LP-Gas models.

1. Case	9. Gasket
2. Coil	10. "O" ring
3. Plunger housing	11. Magnet
4. Spring	12. Washer
5. Plunger	13. Spring
6. Strainer cover	14. Strainer body
7. Filter	15. Drain plug
8. Retaining ring	

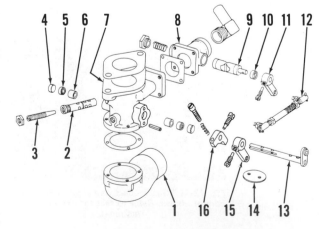

is reduced in the converter by means of regulating valves, to approximately atmospheric pressure required for engine operation. The plug (8) in bottom of converter is provided to drain out coolant compartment if radiator is drained in cold weather.

81. **R&R AND OVERHAUL.** To remove converter, first close both withdrawal valves and run engine until fuel is exhausted from converter and lines. Drain cooling system, remove plug (8 – Fig. 75) and allow coolant to drain from converter. Disconnect coolant lines and fuel lines from converter, remove attaching bolts and lift off converter assembly.

The high pressure valve (4 through 7) can be removed for service without removing converter from tractor. To remove, close both withdrawal valves and exhaust fuel by running engine. Disconnect inlet fuel line, then remove screws retaining inlet cover (4). The entire valve assembly will be removed with cover. Discard gasket (5), and examine seating surface of valve (6). To remove valve from lever (7), remove spring lock and withdraw valve. Renew valve (6) if seat is worn, cut or ridged. Clean metal parts in a suitable solvent and reassemble by reversing disassembly procedure, using a new gasket.

To disassemble converter after it is removed from tractor, remove end covers (1 and 24), and high pressure diaphragm cover (19). Examine diaphragms (16 and 23) for cracks or pin holes. Spring (18) is calibrated to maintain a converter pressure of 6 psi. Renew spring if it is bent, rusty, or has taken a permanent set. Remove pin retaining low pressure valve lever (12) and

remove lever and spring. Examine seating surface of valve seat (11) and examine spring (9) for rust, broken coils or loss of tension.

Reassemble by reversing disassembly procedure, using Fig. 75 as a guide. Always use new gaskets and make sure diaphragms are not wrinkled in assembly. When assembling low pressure valves, use a straight edge and rule to measure assembled distance between free end of low pressure lever (12) and gasket surface of converter body (3) as shown in Fig. 76. Bend lever if necessary, to obtain specified 5/16-inch distance below gasket surface of body.

Install assembled converter; turn on vapor withdrawal valve and check for leaks, using a soap and water solution. Adjust carburetor as outlined in paragraph 76.

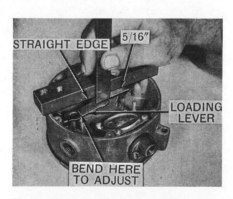

Fig. 76 – *Use a straightedge and scale to adjust low pressure lever. Bend lever, if necessary, to obtain dimension shown.*

DIESEL FUEL SYSTEM

FUEL FILTER, PRIMARY PUMP AND LINES

Models So Equipped

82. A single, two stage fuel filter contained in a sediment bowl is used on all late model tractors (Fig. 77). Some early models used dual canister type filters (Fig. 78). All models use a fuel transfer pump between fuel tank and filter assembly.

83. **BLEEDING.** To bleed air from system, proceed as follows: On late model tractors, open bleed screw (Fig. 77) in filter body. Actuate lever on fuel

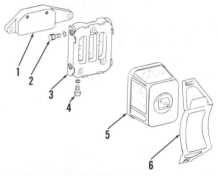

Fig. 77 – *Exploded view of single, renewable two-stage filter and sediment bowl element used on some models. Open bleed screw (2) and use transfer pump to pump air from system.*

1. Support	4. Drain screw
2. Bleed screw	5. Filter element
3. Filter body	6. Retaining spring

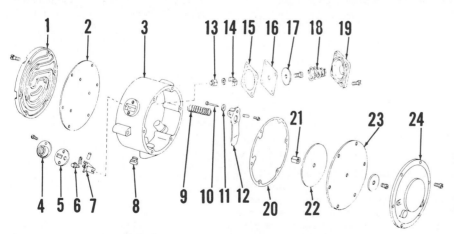

Fig. 75 – *Exploded view of LP-Gas converter (regulator).*

1. Rear cover	8. Plug	14. Damper	19. Cover
2. Gasket	9. Spring	15. Gasket	20. Gasket
3. Body	10. Valve seat pin	16. High pressure	21. Diaphragm button
4. Inlet cover	11. Low pressure valve	diaphragm	22. Plate
5. Gasket	12. Lever	17. Plate	23. Low pressure
6. Valve seat	13. Link	18. Spring	diaphragm
7. Lever			24. Front cover

Fig. 78 – *To bleed system with dual canister type fuel filters, open bleed screw (A) to first stage filter and use lift pump lever (P) to pump air from filter. Bleed second stage (upper) filter by opening bleed screw (B). Make sure lever (P) is in lower position before starting engine.*

transfer pump until air-free fuel flows out bleed screw opening, then close bleed screw. On early model tractors (Fig. 78), open bleed screw (A) and actuate lever on transfer pump until air-free fuel flows. Close bleed screw (A) and open bleed screw (B), then continue pumping until air is ejected from second stage filter. Close bleed screw. Make sure hand primer lever is in down position before attempting to start engine.

NOTE: If no resistance is felt when actuating pump lever, turn engine so camshaft is not on pump stroke.

On all models, loosen pressure line connections at injectors and, with throttle open, turn engine over with starter until fuel flows from all injector lines. Tighten connections and start engine. If engine will not start or misses, repeat bleeding procedure until system is free of trapped air.

INJECTION NOZZLES

Models 3020-4000-4020-4520 (Prior to Engine S.N. 303519)

Roosa Master pencil injector nozzles (9.5 mm) are used on these models. Two different types are used and should not be interchanged. Refer to Fig. 79 and Fig. 80 for identification.

Fig. 79 — Exploded view of Roosa-Master injector (9.5mm) used on 3020, 4000 and 4020 models prior to engine S.N. 280001 and some 4520 models prior to engine S.N. 303519.

1. Seal
2. Washer
3. Nozzle body
4. Nozzle valve
5. Spring seats
6. Spring
7. Nut
8. Pressure adjusting screw
9. Lift adjusting screw
10. Ball washer

84. TESTING AND LOCATING A FAULTY NOZZLE. If rough or uneven engine operation or misfiring indicate a faulty injector, defective unit can usually be located as follows:

With engine running at speed where malfunction is most noticeable (usually slow idle speed), loosen compression nut on high pressure line for each injector in turn, and listen for a change in engine performance. As in checking spark plugs, faulty unit is the one which, when its line is loosened, least affects running of the engine.

If a faulty nozzle is found and considerable time has elapsed since injectors have been serviced, it is recommended that all nozzles be removed and checked, or new or reconditioned units be installed. Refer to following paragraphs for removal and test procedure.

85. REMOVE AND REINSTALL. Wash injector, lines and surrounding area to remove any accumulation of dirt or foreign material. Remove muffler and hood. Disconnect leak-off pipe at return line fitting and at injection pump.

NOTE: If working on a nozzle near alternator, disconnect battery ground cables to prevent shorting a tool against alternator terminal.

Expand lower clamp on each leak-off boot and move clamp upward next to top clamp; then remove leak-off pipe and all boots as a unit.

Disconnect high-pressure line; remove nozzle clamp cap screw, clamp and spacer, then withdraw injector assembly.

NOTE: If injector cannot be easily withdrawn by hand, special OTC puller, JDE-38 will be required. DO NOT attempt to pry nozzle from its bore.

Before reinstalling injector nozzle, clean nozzle bore in cylinder head using OTC Tool JDE-39, then blow out foreign material with compressed air. Turn tool clockwise only when cleaning nozzle bore. Reverse rotation will dull tool.

Renew carbon seal at tip of injector body and seal washer at upper seat whenever injector has been removed. The protector cap can be used to push seal onto nozzle body, if cap is available.

NOTE: Nozzle tip may be cleaned of loose or flaky carbon using a brass wire brush. DO NOT use a brush, scraper or other abrasive on Teflon coated surface of nozzle body between the seals. The coating may become discolored by use, but discoloration is not harmful.

Insert the dry injector nozzle in its bore using a twisting motion. Tighten pressure line connection finger tight; then install hold-down clamp, spacer and

cap screw. Tighten cap screw to a torque of 20 ft.-lbs. Bleed injector if necessary, as outlined in paragraph 83, then tighten pressure line connection to approximately 35 ft.-lbs. Complete assembly by reversing disassembly procedure.

86. NOZZLE TEST. A complete job of testing and adjusting an injector requires the use of special test equipment. Only clean approved testing oil should be used in tester tank. The nozzle should be tested for opening pressure, seat leakage, back leakage and spray pattern. When tested, nozzle should open with a sharp popping or buzzing sound and cut off quickly at end of injection with a minimum of seat leakage and a controlled amount of back leakage.

Use tester to check injector as outlined in following paragraphs:

CAUTION: Fuel leaves nozzle tip with sufficient force to penetrate skin. Keep unprotected parts of body clear of nozzle spray when testing.

87. OPENING PRESSURE. Before conducting test, operate tester lever until fuel flows, then attach injector using No. 16492 Special Adapter. Close valve to tester gage and pump tester lever a few quick strokes to be sure nozzle valve is not plugged, that all spray holes are open and that possibilities are good that injector can be returned to service without overhaul.

Open valve to tester gage and operate lever slowly while observing gage reading. Compare opening pressure to the following specifications.

3020, 4000 and 4020
Roosa-Master No. 18270
New . . . . 2750-2850 psi
Used . . . . 2550-2650 psi
Orifice Diameter . . . . 0.011 in.

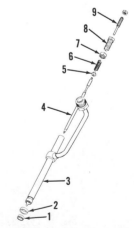

Fig. 80 — Exploded view of Roosa-Master injector used on 3020, 4000 and 4020 models with engine S.N. 280001 and after and on some 4520 models prior to engine S.N. 303519. Note that upper spring seat and ball washer are eliminated. Refer to Fig. 79 legend.

Roosa-Master No. 18646
New .... 2950-3050 psi
Used .... 2750-2850 psi
Orifice Diameter .... 0.012 in.
Roosa-Master No. 18767
New .... 3150-3250 psi
Used .... 2950-3050 psi
Orifice Diameter .... 0.012 in.

4520 (Prior to Engine S.N. 303519)
Roosa-Master No. 18660
New .... 2950-3050 psi
Used .... 2750-2850 psi
Orifice Diameter .... 0.012 in.
Roosa-Master No. 18767
New .... 3150-3250 psi
Used .... 2950-3050 psi
Orifice Diameter .... 0.012 in.

Opening pressure should be within specifications and should not vary more than 50 psi between nozzles. If necessary, adjust opening pressure and valve lift as follows: Loosen locknut (7 – Fig. 79 or Fig. 80) while holding pressure adjusting screw (8) from turning, then back out lift adjusting screw (9) at least two turns to insure against bottoming. Turn adjusting screw (8) until specified opening pressure is obtained. While holding pressure adjusting screw, turn lift adjusting screw (9) until it bottoms, then back out ¾-turn. Tighten locknut to a torque of 110-115 in.-lbs. on early style screw (Fig. 81) or 70-75 in.-lbs. on late style screw. Recheck opening pressure.

NOTE: When adjusting a new injector or overhauled injector with a new pressure spring, set pressure to maximum limit specified to allow for initial pressure loss as spring takes a set.

88. SPRAY PATTERN. The finely atomized nozzle spray should be evenly distributed around nozzle. Check for clogged or partially clogged orifices or for a wet spray which would indicate a sticking or improperly seating nozzle valve. If spray pattern is not satisfac-

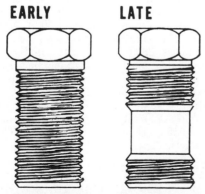

EARLY   LATE

Fig. 81 — View of early and late type pressure adjusting screws used in Roosa-Master injector. Tighten locknut on early screw to a torque of 110-115 in.-lbs., and on late models to 70-75 in.-lbs.

tory, disassemble and overhaul injector as outlined in paragraph 91.

89. SEAT LEAKAGE. Pump tester handle slowly to maintain a gage pressure of approximately 200 to 300 psi under opening pressure psi while examining nozzle tip for fuel accumulation. If nozzle is in good condition, there should be no noticeable accumulation for a period of at least 10 seconds. If a drop or undue wetness appears on nozzle tip, renew injector or overhaul as outlined in paragraph 91.

90. BACK LEAKAGE. Loosen compression nut and reposition nozzle so spray tip is slightly higher than adjusting screw end of nozzle, then maintain a gage pressure of 1500 psi. After first drop falls from adjusting screw, leakage should be at the rate of 3-10 drops in 30 seconds. If leakage is excessive, renew injector.

91. OVERHAUL. First clean outside of injector thoroughly. Place nozzle in a holding fixture and clamp fixture in a vise. NEVER tighten vise jaws on nozzle body without the fixture. Refer to Fig. 79 or Fig. 80. Loosen locknut (7) and back out pressure adjusting screw (8) containing lift adjusting screw (9). Slip nozzle body from fixture, invert body and allow spring seat (5) and spring (6) to fall from nozzle body into your hand. Early style injectors use two spring seats (5) and a ball washer (10) as shown in Fig. 79. Catch nozzle valve (4) by its stem as it slides from body. If nozzle valve will not slide from body, use special retractor (16481) or discard injector assembly.

Nozzle valve and body are a matched set and should never be intermixed. Keep parts for each injector separate and immerse in clean diesel fuel in a compartmented pan as injector is disassembled.

Clean all parts thoroughly in clean diesel fuel using a brass wire brush. Hard carbon or varnish can be loosened with a suitable non-corrosive solvent.

NOTE: Never use a steel wire brush or emery cloth on spray tip.

Clean spray tip orifices using appropriate size cleaning needle. Roosa-Master No. 18270 nozzles have 0.011 inch diameter orifices, and all other nozzles have 0.012 inch orifices.

Clean valve seat using a Valve Tip Scraper (No. 16482 or 17712) and light pressure. Use a Sac Hole Drill (No. 16476) to remove carbon from inside of tip.

Piston area of valve can be lightly polished by hand if necessary, using Roosa-Master No. 16489 lapping com-

pound. Use valve retractor to turn valve. Move valve in and out slightly while turning but do not apply down pressure while valve tip is in contact with seat.

Valve and seat are ground to a slight interference angle. Seating areas may be cleaned up if necessary using a small amount of 16489 lapping compound, very light pressure and no more than 3 to 5 turns of valve on seat. Thoroughly flush all compound from valve body after polishing.

When assembling, back out lift adjusting screw (9), and reverse disassembly procedure using Fig. 79 or Fig. 80 as a guide. Adjust opening pressure and valve lift as outlined in paragraph 87 after valve is assembled.

Models 4320-4520 (Engine S.N. 303519 and After)-4620

These models are equipped with Robert Bosch KDL (21 mm) injector nozzles. Refer to Fig. 82.

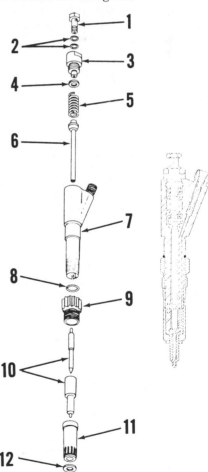

Fig. 82 — Exploded view of Robert Bosch KDL (21mm) injector used on 4320, 4520 (engine S.N. 303519 and after) and 4620 models.

1. Screw	7. Body
2. Gasket (2)	8. "O" ring
3. Plug	9. Gland nut
4. Adjusting shim	10. Nozzle
5. Spring	11. Retaining nut
6. Spindle	12. Washer

92. REMOVE AND REINSTALL. Refer to paragraph 84 for testing procedures to determine if injector nozzle shows indication of a malfunction before removing for service.

If nozzle to be removed is near alternator, disconnect battery ground strap to prevent a short circuit through tools. Wash injector and surrounding area, then disconnect leak-off line and fuel pressure line and use special tool (JDE-69 or JDE-69A) to remove gland nut (9 – Fig. 82). The gland nut will raise the nozzle out of cylinder head as it is removed.

When reinstalling, make sure injector and hole in cylinder head are clean and dry. Nozzle seat reamer (JDE-99) can be used to clean nozzle seat in head. Threads in head for gland nut (9) can be cleaned using a metric (M24 x 1.5) tap. Renew nozzle gasket. Apply anti-seize compound to bottom, inner bore and threads of gland nut (9), then tighten gland nut to 35-45 ft.-lbs. Renew leak-off line gaskets; hold fitting and tighten 12 millimeter head screw to 20 ft.-lbs. Bleed injectors as outlined in paragraph 83 and tighten pressure lines to 35 ft.-lbs. torque.

93. NOZZLE TEST. A complete job of testing and adjusting in injector requires the use of special test equipment. Only clean approved testing oil should be used in tester tank. The nozzle should be tested for opening pressure, seat leakage and spray pattern. When tested, nozzle should open with a soft chatter, and then only when lever is moved very rapidly. A bent or binding nozzle valve can prevent chatter. Spray will be broad and well atomized if injector is working properly.

Use tester to check injector as outlined in following paragraphs:

CAUTION: Fuel leaves nozzle tip with sufficient force to penetrate skin. Keep unprotected parts of body clear of nozzle spray when testing.

94. OPENING PRESSURE. Before conducting test, operate tester lever until fuel flows, then attach injector using proper adapter. Close valve to tester gage and pump tester lever a few quick strokes to be sure nozzle valve is not plugged, that all spray holes are open and that possibilities are good that injector can be returned to service without overhaul.

Open valve to tester gage and operate tester lever slowly while observing gage reading. Opening pressure should be 3100 psi; if it is not, recheck by releasing pressure and retesting. If pressure is still not correct, remove plug (3 – Fig. 82) and change shim (4) until opening pressure is correct. Use only specially

hardened shims. Shims are available in 0.002 inch steps from 0.043 to 0.059 inch. Each 0.002 inch increase in shim thickness varies pressure by about 75 psi. Opening pressure should not vary more than 50 psi between nozzles. If pressure is not correct after changing shims, disassemble injector and recondition.

NOTE: When adjusting a new injector or an overhauled injector with a new pressure spring, set pressure at 3200 to 3350 psi to allow for initial pressure loss as spring takes a set.

95. SPRAY PATTERN. The finely atomized nozzle spray should be evenly distributed around nozzle. Check for clogged or partially clogged orifices. Also check for a wet spray which would indicate a sticking or improperly seating nozzle valve. If spray pattern is not broad and even, and very rapid stroking of tester handle does not cause injector to chatter softly, disassemble and overhaul injector as outlined in paragraph 97.

96. SEAT LEAKAGE. Pump tester handle slowly to maintain a gage pressure of 2800 psi while examining nozzle tip for fuel accumulation. If nozzle is in good condition, there should be no noticeable accumulation for a period of at least 10 seconds. If a drop or undue wetness appears on nozzle tip, renew injector or overhaul as outlined in paragraph 97.

97. OVERHAUL. First clean outside of injector thoroughly and place in a soft jawed vise. Remove plug (3 – Fig. 82), then carefully withdraw shim (4), spring (5) and spindle (6). Turn nozzle body over in vise and remove nozzle retaining nut (11) with a box end wrench. DO NOT use a pipewrench. Remove nozzle valve assembly (10) and reinstall retaining nut on body (7) to protect lapped end surface. If nozzle valve cannot be removed easily, soak assembly in carburetor cleaner, acetone or other commercial solvent intended to free stuck valves. Use care to keep parts clean and free from grit by submerging in a pan of clean diesel fuel, and handle only with hands that are wet with fuel. Avoid mixing of parts with another injector, and do not allow any lapped surface to come in contact with a hard object.

Valves should be cleaned of all carbon and washed in diesel fuel. Hard carbon may be cleaned off with a brass wire brush. NEVER use a steel wire brush or emery cloth on valve or tip. Use a cleaning wire 0.003 to 0.004 inch smaller than nozzle orifices to clean nozzle tips. The number and size of orifices are etched on nozzle tip such as "4 x 0,33". The "4" in-

dicates four orifices, "O,33" indicates that each is 0.33 mm diameter. The following data applies to standard nozzles on models with KDL nozzles.

Opening Pressure
New ....3200-3350 psi
Used ....3100 psi
Orifice Diameter ....0.013 in.

A stone may be used to cut a flat surface on one side of cleaning wire to aid in removing carbon from orifices. Finish cleaning orifices by using a wire 0.001-inch smaller than hole diameter. A pin vise should be used to hold cleaning wires and wire should extend only about 1/32-inch from vise to prevent breakage. Clean seat in nozzle (10) with sac hole drill furnished with cleaning kit. When held vertically, a valve that is wet with fuel should slide down to the seat in nozzle under its own weight.

Inspect all lapped and seating surfaces for excessive wear or damage. Check spindle, spring, shims and seats. Renew any parts in question and reinstall shims only if they are smooth and flat. Edge type filter in fuel inlet passage of body (7) can be cleaned by blowing air through passage from nozzle end of body. This will provide a reverse flushing action in filter.

Assemble in reverse order of disassembly. Submerge valve and nozzle in fuel while assembling, and make sure all other parts are wet with fuel. Do not dry parts with air or towels before assembly. With injector body clamped in a soft jawed vise, tighten screw plug (3) to 36-44 ft.-lbs. and retaining nut (11) to 44-58 ft.-lbs. Retest injector as outlined in previous paragraphs. Use a new gasket when reinstalling injector in engine.

INJECTION PUMP

Roosa-Master Model CBC pump is used on 3020, 4000 and 4020 models with engine serial numbers prior to 280,001. Model CBC pump is also used on some 4520 models prior to engine serial number 303,519. Roosa-Master Model JDB pump is used on 3020, 4000 and 4020 models with engine serial numbers 280,001 and after and also some 4520 models prior to engine serial number 303,519. Model 4320 uses Roosa-Master Model JDB pump. Roosa-Master Model JDC pump is used on Model 4520 with engine serial number 303,519 and after and on all 4620 models.

Roosa-Master CBC Pump

98. REMOVE AND REINSTALL. To remove fuel injection pump, first shut off fuel supply at tank and thoroughly clean dirt from pump, lines and connections.

NOTE: Do not steam clean or pour cold water on a pump while it is warm or running, as this could cause pump to seize.

Remove injection pump timing pin screw (1–Fig. 83) and timing hole cover (2) from clutch housing. Rotate engine until drill point hole in pump drive shaft is centered in pump housing hole and engine timing marks are at TDC on No. 1 piston compression stroke. Remove injector line clamps (5), then disconnect and remove lines. Plug all openings to prevent dirt from entering. Disconnect fuel inlet and return pipes (3) at pump, speed control rod at pump arm (4), fuel shut-off solenoid wire and tachometer drive cable (8). Unbolt and withdraw pump from engine.

To reinstall pump, reverse removal procedure. Make sure No. 1 piston is at TDC on compression stroke and drill point hole is centered in pump housing hole. Renew pump mounting seal ring if necessary.

The backlash in gears can allow pump timing to be off several degrees, so it is important to recheck timing after pump is installed. Rotate crankshaft in normal direction of rotation two revolutions until engine timing marks are at TDC again. If drill point hole is not aligned in center of housing hole, loosen pump mounting nuts and rotate pump housing to align timing hole.

99. **STATIC TIMING.** To check injection pump static timing, shut off fuel and remove timing pin screw (1–Fig. 83) and timing hole cover from clutch housing. Rotate crankshaft until timing marks are at TDC on No. 1 piston compression stroke; drill point hole in pump drive shaft should be centered in pump housing hole.

If adjustment is required, loosen pump mounting cap screws and turn pump body to center drill point hole in housing hole. Tighten stud nuts, then turn crankshaft in normal direction two turns and recheck setting.

Roosa-Master JDB And JDC Pumps

100. **REMOVE AND REINSTALL.** To remove fuel injection pump, first shut off fuel supply at tank and thoroughly clean dirt from pump, lines and connections.

NOTE: Do not steam clean or pour cold water on a pump while it is warm or running, as this could cause pump to seize.

On models with JDC pump (Fig. 84), remove fuel transfer pump. On all models, remove pump timing window cover and rotate crankshaft until pump timing marks (Fig. 85) line up and engine timing marks are at TDC on No. 1 piston compression stroke. Disconnect and remove injector lines. Plug all openings to prevent entry of dirt. Disconnect

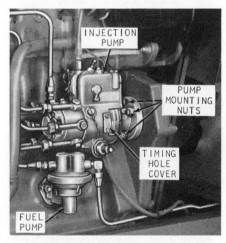

Fig. 84 – View of Roosa-Master JDC model injection pump, typical of all models with this pump.

Fig. 85 – Typical installation of Roosa-Master JDB injection pump. All models so equipped are similar.

fuel inlet and return pipes at pump, speed control linkage, fuel shut-off solenoid wire and tachometer drive cable. Remove mounting nuts and withdraw pump from engine.

NOTE: Before removing pump from engine, wire pump throttle lever in wide open position to prevent governor weights from becoming dislocated.

To reinstall pump, reverse removal procedure. Make sure No. 1 piston is at TDC on compression stroke and pump timing marks are aligned. Renew pump mounting "O" ring if necessary. Be sure seals on pump drive shaft are not damaged or turn over during installation.

The backlash in gears can allow pump timing to be off several degrees, so it is important to recheck timing after pump is installed. Rotate crankshaft in normal direction of rotation two revolutions until engine timing marks are at TDC again. If pump timing marks are not aligned, loosen pump mounting nuts and rotate pump housing to align marks.

101. **STATIC TIMING.** To check injection pump static timing, shut off fuel and remove pump timing window cover. Rotate crankshaft until engine timing marks are at TDC on No. 1 piston compression stroke and pump timing marks (Fig. 85) are aligned.

If adjustment is required, loosen pump mounting nuts and rotate pump housing to align pump timing marks. Tighten mounting nuts, then turn crankshaft in normal direction two turns and recheck timing.

All Models

102. **ADVANCE TIMING.** The injection pump has automatic speed advance which is factory set and will not normally need to be checked or reset. Minor adjustments can, however, be made without removal or disassembly of pump.

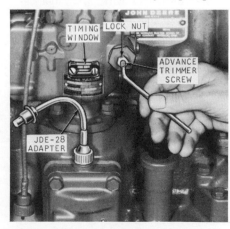

Fig. 86 – View of JDB injection pump showing timing window and advance trimmer screw, typical of all Roosa-Master pumps.

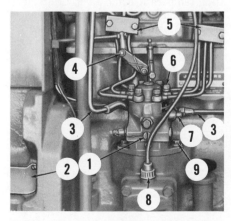

Fig. 83 – View of Roosa-Master CBC model fuel injection pump, typical of all models equipped with this pump.

1. Timing pin screw	6. Injector line screw
2. Cover	7. Shut-off solenoid
3. Fuel inlet & outlet	8. Tachometer drive cable
4. Throttle arm	9. Pump mounting cap
5. Clamps	screw

Shut off fuel and install timing window JD-270 or No. 17180 (CBC model pumps) in place of timing pin screw. On JDB and JDC model pumps, install JD-259 or No. 13366 timing window over pump timing hole (Fig. 86). Install JDE-28 adapter in tachometer drive and connect to accurate tachometer. Bleed fuel system, if necessary, then start engine and bring to operating temperature. Each mark on timing window represents 2 degrees. Timing may be changed by removing seal cap on advance trimmer screw, loosening locknut, and with engine running, turn trimmer screw in to retard or out to advance timing on CBC and JDB model pumps. Turn screw out to retard or in to advance timing on JDC model pumps. Specifications are as follows:

3020
Pump CBC431-8AL
Total Advance . . . . .7° advance
1700 rpm (No Load) . . . . .5°advance
1900 rpm (Full Load) . . .5° advance
2400 rpm (Full Load) . . .7° advance
Pump JDB431AL2401
Total Advance . . . . .5° advance
1700 rpm (No Load) . . . .4° advance
1900 rpm (Full Load) . . .4° advance
2500 rpm (Full Load) . .4½° advance
(min.)

4000 and 4020
Pump CBC633-18AL
Total Advance . . . . .8° advance
1500 rpm (No Load) . . . .5° advance
1700 rpm (Full Load) . . .5° advance
Pump CBC633-24AL
Total Advance . . . . .8° advance
1700 rpm (No Load) . . . .5° advance
1900 rpm (Full Load) . . .5° advance
Pump CBC633-28AL
Total Advance . . . . .8° advance
1700 rpm (No Load) . . . .5° advance
1900 rpm (Full Load) . . .8° advance
Pump JDB633AL2402
Total Advance . . . . .5° advance
1300 rpm (No Load) . . . .3° advance
1900 rpm (Full Load) . . .4° advance
2100 rpm (Full Load) . .4½° advance
(min.)
Pump JDB633AL2404
Total Advance . . . . .5° advance
1300 rpm (No Load) . . . .3° advance
1900 rpm (Full Load) . . .4° advance
2100 rpm (Full Load) . .4½° advance
(min.)

4320
Pump JDB633JT2400
Total Advance . . . . .6° advance
1800 rpm (No Load) . . . .5° advance
1900 rpm (Full Load) . . .5° advance

4520
Pump CBC633-38AL
Total Advance　　　　　7° advance
1700 rpm (No Load) . . . .5° advance
1900 rpm (Full Load) . . .5° advance
Pump JDB633AL2718
Total Advance . . . . .7° advance
1800 rpm (No Load) . . . .5° advance
1900 rpm (Full Load) . . .5° advance
Pump JDC625JT2398
Total Advance . . . . .6° advance
1500 rpm (No Load) . . . .5° advance
1900 rpm (Full Load) . . .5° advance

4620
Pump JDC625JT2398
Total Advance . . . . .6° advance
1500 rpm (No Load) . . . .5° advance
1900 rpm (Full Load) . . .5° advance
Full load advance must be set using a pto dynamometer. If a dynamometer is not available, the no load setting can be used, which would be the next best way to adjust pump. If proper advance cannot be obtained, pump must be either removed and adjusted on test stand, overhauled or renewed.

THROTTLE LINKAGE

Non-Diesel Models

103. **LINKAGE AND SPEED ADJUSTMENT.** Recommended idle and loaded engine speeds are as follows:

3020
Foot Pedal on Platform
(No Load) . . . . .2670-2710 rpm
Full Load . . . . .2500 rpm
Hand Lever Stop
Position . . . . .2140-2200 rpm
Pto Load . . . . .2100 rpm
Fast Idle Stop
Position . . . . .2670-2710 rpm
Full Load . . . . .2500 rpm
Slow Idle Stop Position . . . . . . .800 rpm

4000 and 4020
Foot Pedal on Platform
(No Load) . . . . .2670-2710 rpm
Full Load . . . . .2500 rpm
Hand Lever Stop
Position . . . . .2140-2200 rpm
Pto Load . . . . .1900 rpm
Fast Idle Stop
Position . . . . .2420-2460 rpm
Full Load . . . . .2200 rpm
Slow Idle Stop Position . . . . . . .800 rpm

To adjust linkage, bring engine to operating temperature and use an accurate tachometer. Depress foot control pedal and adjust jam nuts (10 – Fig. 87) on ball joint (9) to obtain specified speed. With foot pedal depressed, tighten governor speed control arm stop screw against the arm, then tighten jam nuts against governor housing. Release foot pedal and move hand control lever clockwise until it stops, then adjust turnbuckle (6) to obtain recommended speed. Pull out hand lever knob and move lever clockwise to specified fast idle stop speed, then adjust stop screw (4) until it contacts control support. Move hand lever counter-clockwise to obtain specified slow idle speed, and adjust stop screw (3) to contact bellcrank (2). With engine at slow idle speed, adjust governor arm stop screw for 1/32-inch clearance with leaf spring.

Diesel Models

104. **LINKAGE AND SPEED ADJUSTMENT.** Before making linkage adjustments, injection pump fast and slow idle stop screws must be properly adjusted. Also, engine must be at operating temperature and an accurate tachometer must be used. Recommended idle and loaded engine speeds are as follows:

Fig. 87 – View of typical throttle linkage used on 4020 non-diesel models. On diesel models, bellcrank assembly (12) is not used. Inset shows injection pump arms used on Roosa-Master CBC, JDB and JDC pumps.

1. Hand control lever
2. Hand lever bellcrank
3. Slow idle stop screw
4. Fast idle stop screw
5. Friction adjustment
6. Turnbuckle
7. Foot control pedal
8. Speed control rod
9. Ball joint
10. Jam nuts
11. To governor (non-diesel only)
12. Bellcrank assy. (non-diesel only)
13. Diesel pump arm

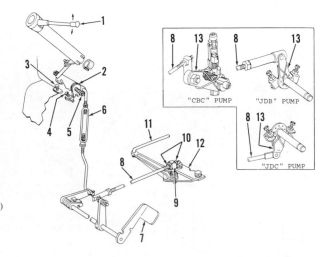

3020
Foot Pedal on Platform
  (No Load) . . . . .2650-2670 rpm
  Full Load . . . . .2500 rpm
Hand Lever Stop
  Position . . . . .2130-2170 rpm
  Pto Load . . . . .2100 rpm
Fast Idle Stop
  Position . . . . .2650-2670 rpm
  Full Load . . . . .2500 rpm
Slow Idle Stop Screw . . . . .800 rpm

All Other Models
Foot Pedal on Platform
  (No Load) . . . . .2650-2670 rpm
  Full Load . . . . .2500 rpm
Hand Lever Stop
  Position . . . . .2130-2150 rpm
  Pto Load . . . . .1900 rpm
Fast Idle Stop
  Position . . . . .2380-2420 rpm
  Full Load . . . . .2200 rpm
Slow Idle Stop Position . . . . . . .800 rpm

Disconnect throttle linkage from injection pump and check fast and slow idle speeds. Adjust injection pump stop screw settings as necessary to obtain 2650-2670 rpm fast idle and 800 rpm slow idle. Reconnect linkage to pump arm.

With foot pedal fully depressed, injection pump arm (13 – Fig. 87) should break over center against spring tension 1/16 to 3/16-inch. Adjust length of speed control rod (8) as needed. Release foot pedal and move hand lever clockwise until it stops, then adjust turnbuckle (6) to obtain recommended speed. Pull out hand lever knob and move lever clockwise until specified fast idle stop speed is reached, then adjust stop screw (4) in bellcrank until it contacts control support. Move hand lever counter-clockwise to slow idle setting. Adjust stop screw (3) until pump arm breaks over center 1/16 to 3/16-inch.

NON-DIESEL GOVERNOR

The centrifugal, flyweight type engine governor is mounted on left side of engine and is driven from a timing gear idler.

Models 3020-4000-4020

105. REMOVE AND REINSTALL. To remove governor assembly, first loosen clamps on inlet hose to carburetor, disconnect hose and move air cleaner inlet pipe out of the way. Remove linkage rods to governor, remove cap screws that hold governor to engine and lift off governor and associated parts.

When installing governor, install a new seal and place governor assembly on engine. Use sealer on cap screw threads before installing.

106. OVERHAUL. Refer to Fig. 88 for parts identification. Inspect all shafts, bushings and bearings for wear. Renew as necessary. Bearings (15) can be removed with JDE 11 puller if necessary. Governor drive shaft (11) O.D. should be 0.7495-0.7505 inch at bushing area. Bushing I.D. should be 0.7520-0.7540 inch; I.D. of thrust bearing (5) is 0.4500-0.4530 inch; Drive shaft (11) O.D. at thrust bearing is 0.4360-0.4380 inch; I.D. of weight carrier (7) is 0.4965-0.4985 inch. Shaft (11) O.D. at weight carrier is 0.4990-0.5010 inch.

When reassembling, make sure loose end of thrust ball bearing (5) goes towards governor weight arms, and lip of seal (14) is facing inward. After internal parts are assembled, place Woodruff key into drive shaft (11) and press drive gear on. End play of shaft and gear should be 0.006-0.012 inch.

COOLING SYSTEM

RADIATOR

All Models

107. To remove radiator, drain cooling system and remove muffler, side panels, grilles, screens and hood. Remove all brackets attached to radiator, disconnect hydraulic fluid line clamps and re-

move screws retaining fan shroud. Remove air cleaner hose if necessary, and disconnect radiator hoses. Remove radiator retaining cap screws and slide radiator out right or left side of tractor after fuel return line is removed. Install by reversing removal procedure.

FAN AND WATER PUMP

All Models

108. REMOVE AND REINSTALL. To remove water pump, drain cooling system and remove muffler, hood, side panels and grilles. Disconnect electrical wiring and remove alternator. Remove screws attaching fan shroud to radiator and cap screws attaching fan to pump hub, then remove fan and shroud from tractor. Remove inlet and outlet pipes from water pump. Unbolt and remove water pump. Reinstall by reversing removal procedure.

109. OVERHAUL. To disassemble water pump, refer to Fig. 89 and remove pulley and hub (if equipped) using a suitable puller which attaches to two fan screw holes; then press bearing, seal and impeller as an assembly from pump housing. Press bearing assembly out of impeller and remove ceramic sealing insert and rubber cup from impeller bore.

To reassemble, coat bearing with light oil and press into housing bore until bearing is flush with front edge of housing. Press on outer race of bearing only. Coat outside of seal with sealant that is resistant to heat and ethylene glycol antifreeze, then press seal into housing until it bottoms. Apply light coat of sealant to impeller shaft bore and install rubber cup and ceramic insert in impeller with polished side of insert facing

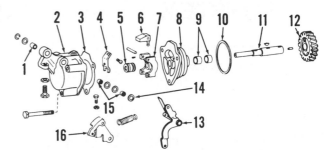

Fig. 88 — Exploded view of engine governor used on non-diesel models.

1. Bushing
2. Housing
3. Gasket
4. Governor linkage
5. Thrust bearing
6. Weight
7. Weight carrier
8. Drive shaft housing
9. Bushings
10. Seal
11. Drive shaft
12. Drive gear

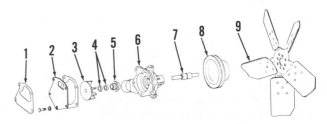

Fig. 89 — Exploded view of typical water pump used on all models.

1. Gasket
2. Rear cover
3. Impeller
4. Insert & cup
5. Seal
6. Pump housing
7. Shaft & bearing
8. Pulley
9. Fan

out. Make sure seal lip and insert face are perfectly clean and apply coat of light oil to insert.

Press impeller onto bearing shaft until clearance between impeller and machined surface of housing is 0.015-0.025 inch on 3020, 4000 and 4020 models or 0.015-0.035 inch on 4320, 4520 and 4620 models.

NOTE: Support front end of shaft on bed of press when installing impeller; and impeller end of shaft when installing fan hub or pulley.

Install fan pulley and rear cover, then reinstall water pump and associated parts by reversing removal procedure.

THERMOSTAT AND WATER MANIFOLD

All Models

110. The thermostats are contained in a thermostat housing in water manifold. Refer to Fig. 90 for view of typical manifold. All models except 3020 and early Model 4520 use double thermostats. On all models except 3020 to remove thermostats, drain system, detach upper radiator hose at front and remove housing cap screws; then remove hose and housing as a unit. On 3020 models, remove cap screws holding water outlet collector to thermostat housings, remove water by-pass pipe connectors from engine, and lay assembly to one side. If a thermostat is suspected of being faulty, check temperature range of unit suspected and test in heated water with a thermometer to be sure opening temperature is correct.

ELECTRICAL SYSTEM

ALTERNATOR AND REGULATOR

All Models

111. All models use either a Delco-Remy "DELCOTRON" alternator or a Motorola alternator. All Delco-Remy alternators include an internally mounted solid state regulator. Motorola RA series and MR series alternators are equipped with integrally mounted solid state regulator. Motorola MA series alternators use a separately mounted regulator.

CAUTION: Because certain components of the alternator can be damaged by

procedures that will not affect a D.C. generator, the following precautions MUST be observed.

a. When installing batteries or connecting a booster battery, negative post of battery must be grounded.

b. Never short across any terminal of alternator or regulator unless specifically recommended.

c. Do not attempt to polarize alternator.

d. Disconnect all battery ground straps before removing or installing any electrical unit.

e. Do not operate alternator on an open circuit and be sure all leads are properly connected before starting engine.

Specification data for alternators is as follows:

Motorola
35 Amp RA Model
Field Current at 75°F.
Amperes 1.95-2.55
Volts 10
Output at 13-15 Volts (min.)
Amperes at 1660 rpm 13
Amperes at 3000 rpm 25
55 Amp RA Model
Field Current at 75°F.
Amperes 1.85-2.25
Volts 10
Output at 13-15 Volts (min.)
Amperes at 1660 rpm 28
Amperes at 3000 rpm 45
72 Amp MA or MR Model
Field Current at 75°F.
Amperes 1.85-2.25
Volts 10
Output at 13-15 Volts (min.)
Amperes at 2288 rpm 40
Amperes at 4000 rpm 65

Delco-Remy
55 Amp Model
Field Current at 75°F.
Amperes 4.0-4.5
Volts 12

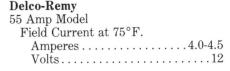
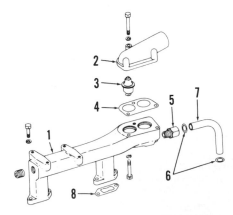

Fig. 90 — Exploded view of typical water manifold and thermostat housing.

1. Water manifold
2. Thermostat cover
3. Thermostat (2)
4. Gasket
5. Connector
6. "O" ring
7. By-pass pipe
8. Gasket (2)

Output at 13-15 Volts (min.)
Amperes at 5000 rpm 45
72 Amp Model
Field Current at 75°F.
Amperes 4.0-4.5
Volts 12
Output at 13-15 Volts (min.)
Amperes at 5000 rpm 60

STARTING MOTOR

All Models

112. A Delco-Remy or John Deere starting motor is used on all models. Refer to appropriate starter model number for specification data.

John Deere
Model 028000-3290-ND
Volts 12.0
Brush Spring Tension-Ounces 40
No Load Test:
Volts 9.0
Amperes (min.) 70
Amperes (max.) 110
Rpm (min.) 2500
Rpm (max.) 4500

Delco-Remy
Model 1107350 & 1108494
Volts 12.0
Brush Spring Tension-Ounces 35
No Load Test:
Volts 9.0
Amperes (min.) 55
Amperes (max.) 80
Rpm (min.) 3500
Rpm (max.) 6000
Model 1107578 & 1109145
Volts 12.0
Brush Spring Tension-Ounces 35
No Load Test:
Volts 9.0
Amperes (min.) 50
Amperes (max.) 80
Rpm (min.) 5500
Rpm (max.) 9000
Model 1113672 & 1113692
Volts 12.0
Brush Spring Tension-Ounces 80
No Load Test:
Volts 9.0
Amperes (min.) 130
Amperes (max.) 160
Rpm (min.) 5000
Rpm (max.) 7000

DISTRIBUTOR

Non-Diesel Models 3020-4000-4020

113. **TIMING.** Timing marks are located on flywheel and timing window on right side of clutch housing. Suggested method of timing is to use a power timing light with engine running at specified rpm. Ignition should occur at 20 degrees BTDC at 2200 rpm on 3020

gasoline models; at 25 degrees BTDC at 2100 rpm on 3020 LP-Gas models; at 20 degrees BTDC at 2000 rpm on 4000 and 4020 gasoline models; or 25 degrees BTDC at 2000 rpm on 4000 and 4020 LP-Gas models.

Recommended breaker point gap on 3020 models is 0.022 inch with cam angle 31-34 degrees. On 4000 and 4020 models prior to engine S.N. 280001, gap is 0.016 inch and cam angle 31-34 degrees. On 4000 and 4020 models after engine S.N. 280000, gap is 0.021 inch with cam angle 22-26 degrees.

114. REMOVE AND REINSTALL. Before removing distributor assembly, remove distributor cap and turn engine crankshaft until front (No. 1) piston is at TDC on compression stroke. Disconnect distributor to coil primary lead wire, remove distributor clamp screw, then withdraw distributor assembly from cylinder block.

Install distributor as follows: Check to be sure crankshaft is set with front (No. 1) piston at TDC, then install distributor with offset drive tang aligned with coupling in cylinder block. Firing order is 1-3-4-2 on 3020 models or 1-5-3-6-2-4 on 4000 and 4020 models.

On all models, check running ignition timing as outlined in paragraph 113.

115. OVERHAUL. Refer to Fig. 91 for exploded view of distributor used on all models. Distributor is driven by an offset drive tang coupled to oil pump

drive shaft. Distributor shaft diameter (new) is 0.4895-0.4900 inch and new shaft should be installed if worn to less than 0.4875 inch. Housing bushings are not renewable separately, but housing should be renewed if bushing diameter exceeds 0.494 inch. Distributor shaft end play should be limited to 0.002-0.010 inch by adding or removing shims between drive coupling and bottom of distributor housing. New drive coupling must be drilled for installation of spring pin. Coupling offset tang must be correctly positioned in relation to rotor position as shown in Fig. 92. Centrifugal advance in distributor degrees and distributor rpm should be as follows: 0-2 degrees at 225 rpm; 5-7 degrees at 350 rpm; 10-12 degrees at 825 rpm; 14-16 degrees at 1200 rpm.

SPARK PLUGS

All Models So Equipped

116. Recommended spark plugs for gasoline models are Champion D-14, AC TC-83 or Prestolite 18 4. Recommended spark plugs for LP-Gas models are Champion D-9, AC C-82 or Prestolite 18 3. Electrode gap should be 0.025 inch for gasoline models and 0.015 inch for LP-Gas models.

START-SAFETY SWITCH ADJUSTMENT

All Models

117. The start-safety switch is located high on right side of transmission housing, on all but Power Shift models, which has switch located in transmission control valve.

To adjust switch, place transmission in neutral, remove switch and place enough washers under switch to close switch when reinstalled. Remove switch

again, remove one washer and check selector in all positions. Switch should only close in neutral and park positions. On Power Shift tractors, it may be necessary to remove one additional washer. Switch will not operate properly on a Syncro-Range tractor if shifter camshaft end play exceeds 0.005 inch (Fig. 93).

ENGINE CLUTCH (SYNCRO-RANGE)

This section covers only the clutch used in tractors equipped with Syncro-Range transmissions. For models equipped with Power Shift, refer to paragraphs 142 and 143.

The clutch assembly includes two spring-loaded, dry-disc clutches mounted in tandem which operate independently of each other. Front clutch transmits power to transmission and rear clutch transmits power to pto.

The transmission clutch is manually controlled by a foot pedal. The pto clutch is hydraulically applied by actuating pto control lever.

LINKAGE ADJUSTMENT

Syncro-Range Models

118. **TRANSMISSION CLUTCH.** Clutch pedal free travel should be 1½ inches. Refer to Fig. 94. Measure free travel with transmission in park and engine running at 1900 rpm. If free travel is less than ¾-inch adjust linkage as follows:

Loosen clutch pedal adjusting screw (1) and slotted nut (2). With engine running at 1900 rpm and transmission in park, move pedal in slotted hole to obtain 1½ inches free travel. Tighten adjusting screw and slotted nut, then recheck free travel.

On tractors with Power Front Wheel Drive, clutch switch (4) may need to be adjusted after adjusting pedal free travel. Loosen switch mounting screws and adjust switch until front wheels engage just prior to rear wheels.

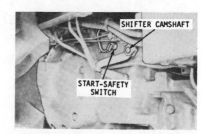

Fig. 93 — Start-safety switch adjustment should include end play check of shifter camshaft.

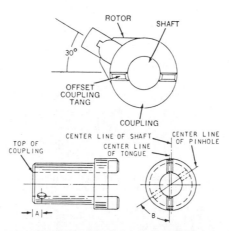

Fig. 92 — When renewing drive coupling for Delco-Remy distributor, coupling must be drilled as shown for roll pin installation.

Fig. 91 — Exploded view of Delco-Remy distributor used on all non-diesel models.

1. Drive coupling	
2. Roll pin	10. Cam
3. Spacer washer	11. Hold down plate
4. Shim	12. Base plate
5. Housing	13. Condenser
6. Spacer washer	14. Breaker points
7. Drive shaft	15. Cover
8. Weights	16. Rotor
9. Advance springs	17. Cap

119. PTO CLUTCH. To adjust pto linkage, first disconnect signal spring (2 – Fig. 95) and inner spring (3) from clutch operating rod clip. Lengthen operating rod (5) from yoke until valve spool is pushed solidly against bottom of pto valve body (4) when pto lever is moved to engaged position. Then shorten control rod one turn back into yoke.

Move control lever to neutral position and connect springs to control rod clip. Adjust position of eyebolt (1) in slotted mounting hole to provide approximately 5/16-inch clearance between signal spring hook and operating rod clip.

TRACTOR CLUTCH SPLIT

All Models

120. To detach (split) engine from clutch housing for access to engine clutch, proceed as follows:

Drain cooling system and disconnect battery cables. Remove muffler, cowl, side shields, grille screens, hood, control support covers, steps and left battery and battery support.

Discharge brake accumulator by opening right brake bleeder screw and holding brake pedal down a few minutes.

If equipped with air conditioning, remove compressor with hoses attached. Bend hoses around and place compressor inside cab.

On LP-Gas models, remove fuel withdrawal valve handles. Disconnect control rods at withdrawal valves and pull rods rearward.

On all models, disconnect hydraulic pump oil seal drain tube, tachometer cable, oil pressure switch wire and throttle rod at fuel pump (diesel models). Disconnect wiring harness, hydraulic lines, temperature indicator sender unit, ether starting aid pipe (diesel models), heater hoses (if equipped) and choke cable and throttle rod (non-diesel models). Disconnect hydraulic pump inlet pipe. If equipped with Power Front Wheel Drive, disconnect drain pipe.

Remove front end weights (if equipped). Support front and rear of tractor with suitable stands. Remove cap screws securing engine to clutch housing and roll rear half away from engine.

Fig. 95 — View of typical pto control linkage used on Syncro-Range models.

1. Eyebolt
2. Signal spring
3. Inner spring
4. Pto valve
5. Operating rod

R&R AND OVERHAUL

Syncro-Range Models 3020-4000-4020-4320

121. To remove clutch assembly after engine has been separated from clutch housing, remove the six retaining cap screws and withdraw assembly. Transmission clutch disc (14 – Fig. 96) will be freed by the removal, and may be renewed without disassembly of clutch unit. Check friction faces of flywheel and transmission clutch pressure plate (11) for heat checks, wear or scoring. Pressure surface of flywheel must be true to 0.006 inch when measured with a straightedge and feeler gage.

NOTE: If flywheel friction surface is to be machined, manufacturer cautions that distance from friction face to mounting face for clutch cover must not exceed 2.909 inches on 3020, 4000 and 4020 models or 3.004 inches on 4320 models. Exceeding these limits could make flywheel unsafe.

Disassembly and reassembly of removed unit will be facilitated by removal of flywheel or use of spare flywheel. To disassemble, place flywheel, front side down, on a bench. Place clutch assembly in operating position in flywheel and secure with three alternately spaced jack screws, jam nuts and flat washers as shown in Fig. 97. Tighten jam nuts evenly to take pressure from transmission clutch operating fingers (1); then remove jam nuts, washers and operating bars from transmission clutch operating bolts (12). Back off jack screw jam nuts evenly until spring pressure is released,

Fig. 94 — Recommended clutch pedal free travel is 1½ inches. Clutch switch is used on models equipped with Power Front Wheel Drive.

1. Adjusting screw
2. Slotted nut
3. Battery box latch
4. Clutch switch mounting screws

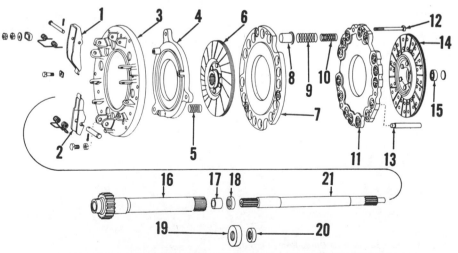

Fig. 96 — Exploded view of engine clutch assembly used on 3020, 4000 and 4020 models equipped with Syncro-Range transmission. Parts 19 and 20 are used only on tractors not equipped with pto.

1. Transmission clutch finger
2. Pto clutch finger
3. Drive plate
4. Pto clutch plate
5. Release spring
6. Pto disc
7. Pto pressure plate
8. Retainer
9. Spring (outer)
10. Spring (inner)
11. Transmission clutch pressure plate
12. Bolt
13. Flywheel pin
14. Transmission clutch disc
15. Pilot bearing
16. Pto clutch shaft
17. Bushing
18. Oil seal
19. Quill
20. Oil seal
21. Transmission clutch shaft

remove jack screws, then remove clutch components.

Inspect and renew parts as necessary. Friction surfaces of separator plate (4 – Fig. 96) and pressure plates (7 and 11) must not be scored, grooved or out-of-true more than 0.006 inch. Check springs against following specifications:

3020, 4000 and 4020 Models
Outer Spring
Approximate Free Length . . . . 3¼ in.
Lbs. Pressure at 1¾ in. 105-109
Inner Spring
Approximate Free Length . . . 2-7/8 in.
Lbs. Pressure at 1¾ in. 79-97
Pto Release Spring
Approximate Free Length . . . . 2¼ in.
Lbs. Pressure at ⅝-in. 13.5-17.5

4320 Model
Outer Spring
Approximate Free Length . . . . 2¾-in.
Lbs. Pressure at 1-3/16 in. . . . 153-187
Inner Spring
Approximate Free Length 2-23/32 in.
Lbs. Pressure at 1-29/32 in. . 51.3-62.7
Pto Clutch Spring
Approximate Free Length . . . . 2¼ in.
Lbs. Pressure at ⅝-in. 13.5-17.5
Tightening Torques
Flywheel Attaching Cap
Screws . . . . .85 ft.-lbs.
Clutch Cover Attaching
Screws . . . . .35 ft.-lbs.

To assemble clutch, position transmission clutch disc in flywheel with long end of clutch hub down. Insert the three clutch operating bolts (12 – Fig. 96) in pressure plate (11) making sure bolt heads fit in recesses of pressure plate. Install plate over guide dowel and install clutch springs on plate; then install pto front pressure plate (7). Position pto clutch disc (6) with long end of hub up, and place the three pto clutch springs on

pressure plate. Assemble drive cover, if disassembled, then position it over flywheel making sure guide dowels and operating bolts are aligned. Install three jack screws. Use clutch drive shafts to align discs, or fashion an aligning tool using dimensions shown in Fig. 98; then tighten jam nuts on jack screws. Install operating bars, washers and jam nuts on operating bolts (12 – Fig. 96). Reinstall flywheel and clutch unit using aligning tool or clutch shafts and adjust as outlined in paragraph 122.

122. CLUTCH ADJUSTMENT. Whenever clutch has been disassembled, or when flywheel or transmission clutch disc has been renewed, both sets of operating fingers must be adjusted as follows:

Using special gage No. JDE19 or JDE61, adjust transmission clutch fingers to just touch center leg of gage as shown in Fig. 99. Adjust pto clutch fingers to just touch upper gage surface as shown in Fig. 100.

NOTE: Pto clutch must be engaged while making adjustment. Use chisels, screwdrivers or other wedges under outer edge of two fingers as shown while adjustment is made on third finger.

Syncro-Range Models 4520-4620

123. To remove clutch after engine has been separated from clutch housing, remove the six retaining cap screws and remove clutch assembly from flywheel. Check friction faces of flywheel and transmission clutch pressure plate (14 – Fig. 101) for wear or scoring. Pressure surface of flywheel must be true to 0.006 inch when measured with a straightedge and feeler gage.

To disassemble clutch, remove flywheel and place clutch in flywheel, then secure with three alternately spaced jack screws, jam nuts and flat washers. Tighten nuts down against cover to relieve spring pressure, then remove pivot pins from transmission clutch fingers (4). Carefully back off jam nuts to release spring pressure, remove jack screws and separate clutch component parts.

Examine friction surfaces of separator plate (7) and pressure plates (10 and 14). Surfaces must not be scored, grooved or out-of-true more than 0.006 inch. Check clutch springs against the following specifications:

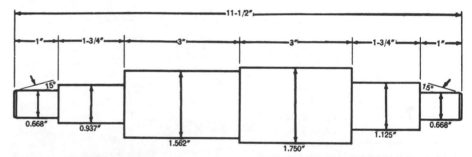

Fig. 98 – A clutch pilot shaft can be turned out of a 12-inch piece of round stock using dimensions shown. Tool is used on 3020, 4000, 4020 and 4320 Syncro-Range models.

Fig. 97 – To disassemble clutch, attach unit to a flywheel with three jack screws (J), then remove nuts and washers from bolts (12) on clutch fingers (1). Back off nuts on jack screws evenly until spring pressure is released.

Fig. 99 – Use JDE-19 or JDE-61 adjusting gage, and adjust each transmission clutch finger to just touch center leg of gage.

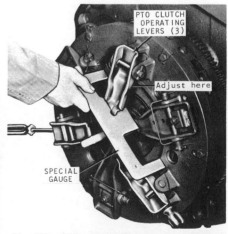

Fig. 100 – When adjusting pto clutch fingers, use chisels, screwdrivers or other wedges under outer end of two fingers to hold clutch pressure plate engaged. Adjust fingers to just touch special gage as shown.

4520 Model
Outer Spring
  Approximate Free Length ....3¾ in.
  Lbs. Pressure at 2⅜ in. 129-157
Inner Spring
  Approximate Free Length ....3½ in.
  Lbs. Pressure at 2⅛ in. 44-54
Pto Release Spring
  Approximate Free Length ....2¼ in.
  Lbs. Pressure at ⅝-in. 14-18

4620 Model
Outer Spring
  Approximate Free Length
    (min.) .... 3-5/16 in.
Inner Spring
  Approximate Free Length
    (min.) .... 3-5/32 in.
Pto Release Spring
  Approximate Free Length ....2¼ in.

To assemble clutch, position transmission clutch disc in flywheel with long end of clutch hub down. Install special aligning tool No. JDE-52 shaft with JDE-52-4 adapter through clutch disc and into pilot bearing. Reassemble clutch assembly by reversing disassembly procedure, using the three jack screws and jam nuts to compress clutch springs.

124. CLUTCH ADJUSTMENT. Whenever clutch has been disassembled, or when flywheel or transmission clutch disc has been renewed, both sets of operating fingers must be adjusted as follows:

Using special gage No. JDE-61, adjust transmission clutch fingers to just touch center leg of gage as shown in Fig. 102. Pto clutch must be engaged while adjusting pto fingers. Use chisels, screwdrivers or other wedges under outer edge of two fingers as shown in Fig. 103 while adjustment is made on third finger. Fingers should just touch gage as shown.

CLUTCH SHAFT

Syncro-Range Models

To remove either transmission or pto clutch shaft, it is necessary to split tractor between clutch housing and transmission case.

125. TRACTOR SPLIT. Open right brake bleed screw and depress pedal to discharge accumulator, and drain transmission fluid. Disconnect and remove batteries and battery boxes. Disconnect wiring harness from start-safety switch, lighting harness and dimmer switch. Disconnect clutch return spring. Remove differential lock pedal, rockshaft selector knob, platform support and platform. If equipped with Power Front Wheel Drive, remove rear drain pipe. Disconnect transmission oil temperature bulb, main hydraulic pump inlet pipe, steering return pipe, differential lock control link, pressure pipe from rockshaft housing, right and left brake pipes and brake return pipe. Place shift lever

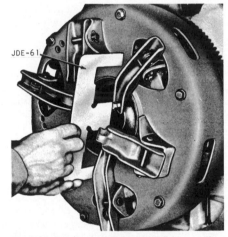

Fig. 102 — Use adjusting gage JDE-61 to adjust transmission clutch fingers on 4520 and 4620 models. Finger should just touch center leg of gage.

in tow, then pull levers outward and disconnect shift rods. Place pan under pto quill to catch trapped oil, then remove quill. Remove transmission top cover. Install suitable support stands on both ends of tractor and extend drawbar rearward and place jack under drawbar. Remove connecting cap screws and separate transmission from clutch housing. Place supports under front and back of transmission.

When reconnecting, tighten ⅝-inch retaining cap screws to a torque of 170 ft.-lbs. and ¾-inch cap screws to 300 ft.-lbs.

126. R&R AND OVERHAUL SHAFTS. After splitting tractor, clutch shafts can be removed as a unit out of clutch housing; then withdraw transmission clutch shaft from pto clutch shaft.

Inspect bushing inside pto clutch shaft. When renewing, press bushing flush with end of shaft with closed end of oil groove towards oil seal. When renewing oil seals, install seals with lips to rear. Use a suitable sleeve to protect seal when installing transmission clutch shaft into pto clutch shaft.

Install transmission and pto shafts in

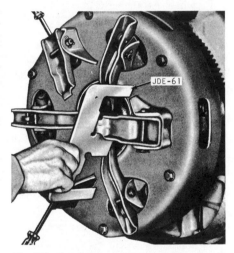

Fig. 103 — Pto clutch must be engaged when adjusting pto clutch fingers. Use screwdrivers or other wedges under two fingers as shown to engage clutch.

Fig. 104 — Some models are equipped with a notched thrust washer on pto clutch shaft. Be sure washer engages slots in clutch housing.

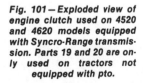

Fig. 101 — Exploded view of engine clutch used on 4520 and 4620 models equipped with Syncro-Range transmission. Parts 19 and 20 are only used on tractors not equipped with pto.

1. Adjusting screw
2. Pto clutch finger
3. Adjusting screw
4. Transmission clutch finger
5. Link
6. Clutch cover
7. Pto clutch plate
8. Pto release spring
9. Pto clutch disc
10. Pto pressure plate
11. Retainer
12. Spring (outer)
13. Spring (inner)
14. Transmission clutch pressure plate
15. Transmission clutch disc
16. Pilot bearing
17. Thrust washer
18. Clutch shaft
19. Oil seal
20. Quill
21. Oil seal
22. Bushing
23. Pto drive shaft

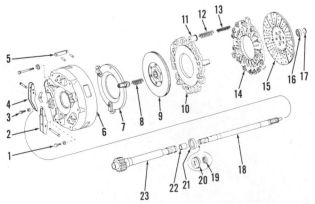

clutch housing as a unit. Some models use a notched thrust washer as shown in Fig. 104; if so equipped, be sure washer engages slots in clutch housing.

CLUTCH CONTROL LINKAGE

Syncro-Range Models

127. **OPERATING SLEEVE.** To overhaul operating sleeve assembly, clutch housing must be separated from engine as outlined in paragraph 120.

Clutch release collar (11–Fig. 105) and operating components can be removed from front of clutch housing. A counter-weighted pto clutch release finger conversion kit is available to reduce pto clutch aggressiveness. When kit is installed, a double-spring set replaces single pto clutch return spring

(16). Check return springs against following specifications:

Single Spring
Approximate Free Length . . 3-7/16 in.
Lbs. Pressure at 1-19/32 in. 20-24

Double Springs
Inner Spring
Approximate Free Length . . . 3½ in.
Lbs. Pressure at 1-19/32 in. . . . 20-24
Outer Spring
Approximate Free Length . . . 3¼ in.
Lbs. Pressure at 1-11/32 in. . . . 40-48
Inspect pto clutch operating pistons and pto brake piston. Be sure to install piston seal rings and back-up rings as shown in Fig. 106.

Transmission clutch fork (4–Fig. 105) and arm (2) are splined to operating shaft (3). Index marks on shaft, fork and

arm indicate correct position of installation.

When reassembling clutch controls, pto clutch operating sleeve travel is limited by adjusting special cap screws (15). Adjust cap screws until there is 3.91 inches between bottom of screw head and face of support (7). If pto conversion kit has been installed, distance is measured from bottom of washer (19).

128. **PTO CLUTCH VALVE.** To disassemble valve, unscrew plug (10–Fig. 107) and withdraw valve assembly from housing (17). Support valve and drive out spring pin (1), then separate component parts.

Inspect parts and renew as necessary. If counter-weighted pto release finger kit has been installed, a double-spring set (3 and 4) is used. Check springs against following specifications:

Single Spring
Approximate Free
Length . . . . .1-13/16 in.
Lbs. Pressure at 1-5/16 in. 45-55

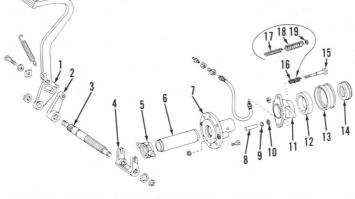

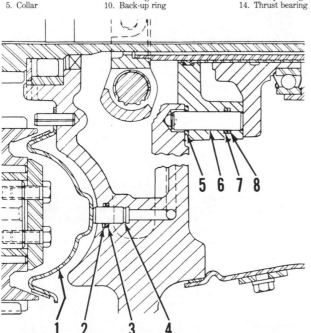

Fig. 105 — Exploded view of typical clutch controls. Double springs (17 and 18) are used when equipped with pto clutch release finger conversion kit.

1. Clutch pedal	6. Tube	11. Sleeve	15. Adjusting screw
2. Arm	7. Support	12. Ball bearing	16. Spring
3. Shaft	8. Pto operating piston	13. Pto operating collar	17. Inner spring
4. Fork	9. "O" ring	14. Thrust bearing	18. Outer spring
5. Collar	10. Back-up ring		19. Washer

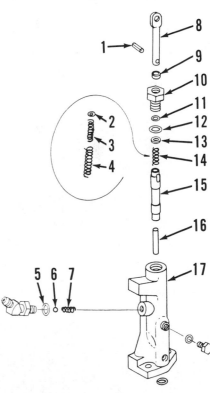

Fig. 106 — Cross-sectional view of pto clutch operating piston (two used) and pto brake piston.

1. Brake shoe
2. Back-up ring
3. "O" ring
4. Brake piston
5. Packing
6. Clutch piston (2 used)
7. "O" ring
8. Back-up ring

Fig. 107 — Exploded view of pto clutch valve. Double springs (3 and 4) are used when equipped with pto clutch release finger conversion kit.

1. Spring pin
2. Washer
3. Inner spring
4. Outer spring
5. "O" ring
6. Check ball
7. Spring
8. Operating shaft
9. Seal
10. Plug
11. "O" ring
12. "O" ring
13. Shim
14. Spring
15. Valve
16. Pin
17. Housing

Double Springs
  Inner Spring
    Approximate Free
      Length . . . . 1-7/16 in.
    Lbs. Pressure at 1-5/16 in. 21-25
  Outer Spring
    Approximate Free
      Length . . . . 1-29/32 in.
    Lbs. Pressure at 1-5/16 in. 20-24
When reassembling, install oil seal (9) in plug (10) with sealing lip up (metal side toward oil). If valve is equipped with single return spring (14), install shims (13) to obtain 0.030 inch compression of spring when spring pin (1) is against upper end of slot in valve (12). Adjustment is not necessary with double-spring set.

SYNCRO-RANGE TRANSMISSION

(For Power Shift Models, Refer to Paragraph 144.)

The "Syncro Range" transmission is a mechanically engaged transmission consisting of three transmission shafts and a single, mechanically connected, remote mounted control lever as shown in Fig. 108. The four basic gear speeds are selected by coupling one of the shaft idler gears to the splined main drive bevel pinion shaft, and can only be accomplished by disengaging engine clutch and bringing tractor to a stop. The high, low and reverse speed ranges within the four basic speeds are selected by shifting couplers on transmission drive shaft and, because of the design of the couplers, can be accomplished by disengaging engine clutch and moving control lever, without bringing tractor to a halt.

NOTE: The rotating speeds of transmission drive shaft and its idler gears are automatically equalized by synchronizing clutches. All other phases of shifting are under direct control of the operator. The fact that clashing of gears is eliminated by synchronizing clutches does not relieve him of the responsibility of using care and judgment in re-engaging the clutch after the gears have been shifted.

The idler gears and bearings on main shaft and bevel pinion shaft are pressure lubricated by a separate transmission oil pump.

Syncro-Range Models

129. **INSPECTION.** To inspect transmission gears, shafts and shifters, first drain the transmission and hydraulic fluid and remove operator's platform, then remove transmission top cover. Examine shaft gears for worn or broken teeth and shifter linkage and cam slots for wear.

CONTROL QUADRANT

This section covers disassembly and overhaul of shifter controls mounted in tractor steering support. Removal, inspection, overhaul and adjustment of shift mechanism inside the transmission housing is included with transmission gears and shafts.

Syncro-Range Models

130. **R&R AND OVERHAUL.** To overhaul control quadrant, remove cowl and raise dash enough to clear quadrant. Disconnect shifter rods (5 and 6–Fig. 109) at quadrants. While holding shaft (8) with a wrench, remove nut and cap screw securing shaft to brackets (1 and 7). Quadrant assembly may then be completely disassembled.

If lever (15) or pivot (9) are damaged and need to be renewed, proceed as follows: Clamp lower curved portion of lever in a soft-jawed vise and slip a 5/32-inch cotter pin inside roll pins (14). Grasp roll pins with a good vise-grip plier and extract with a twisting motion. When reassembling, leave at least 3/16-inch of roll pin protruding from lever. To drive pin in farther will damage pivot bushing (13).

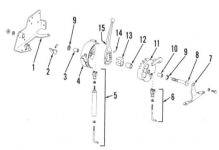

Fig. 109—Exploded view of dash mounted shifter controls used in 4520 model. Other models are similar. Notch in latch (2) fits around lower rocker of lever (15) and moves to positively lock the opposite quadrant (4 or 11) when the other is shifted.

1. Support	
2. Latch	
3. Bushing	8. Quadrant shaft
4. Speed change	9. Thrust washers
quadrant	10. Bushing
5. Speed change shift	11. Speed range quadrant
rod	12. Bushing
6. Speed range shift rod	13. Pivot
7. Outer support	14. Spring pin (2 used)
	15. Lever

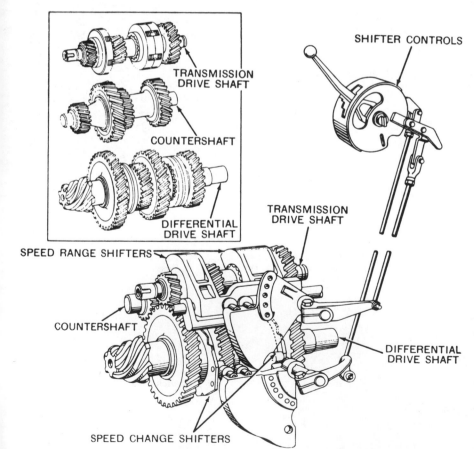

SHIFTER CONTROLS

TRANSMISSION DRIVE SHAFT

COUNTERSHAFT

DIFFERENTIAL DRIVE SHAFT

SPEED RANGE SHIFTERS

TRANSMISSION DRIVE SHAFT

COUNTERSHAFT

DIFFERENTIAL DRIVE SHAFT

SPEED CHANGE SHIFTERS

Fig. 108—Schematic view of Syncro-Range transmission components. The single shift lever moves either shifter cam to change to selected gear.

TRANSMISSION DISASSEMBLY AND REASSEMBLY

Paragraphs 131 and 132 outline general procedure for removal and installation of main transmission components. Disassembly, inspection and overhaul of removed assemblies is covered in overhaul section beginning with paragraph 134, which also outlines those adjustment procedures which are not an integral part of assembly.

Syncro-Range Models

131. **DISASSEMBLY.** Any disassembly of transmission gears and controls requires tractor to first be split between transmission and clutch housing as outlined in paragraph 125. Transmission must be disassembled in the approximate sequence outlined in following paragraphs; however, disassembly need not be completed once defective or damaged parts are removed. Transmission shafts are removed in following order:

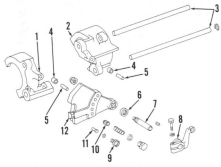

Fig. 110—Exploded view of speed range shifters used on 4520 and 4620 models. Other models are similar except only one shift rail (3) is used.

1. Reverse shifter	7. Shifter shaft
2. Low-High shifter	8. Shifter arm
3. Shift rails	9. Start-safety switch
4. Rollers	10. Shifter pawl
5. Pins	11. Safety switch follower
6. Seal	12. Shifter cam

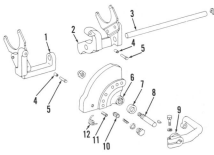

Fig. 111—Exploded view of speed change shifters used on 4520 and 4620 models. Other models are similar.

1. Rear shifter collar	7. Seal
2. Front shifter collar	8. Shifter shaft
3. Shift rail	9. Shifter arm
4. Rollers	10. Shifter pawl
5. Pins	11. Spring pin
6. Shifter cam	12. Spring

(1) Transmission drive shaft, (2) differential drive shaft and (3) countershaft.

Remove transmission top cover and rockshaft housing or transmission rear cover. Remove detent spring caps from right side of housing, raise upper shifter arm to its highest position and remove cotter pin and slotted nut from inner end of shaft. Be careful not to drop the parts in housing, as they will be difficult to remove. Withdraw shifter arm (8–Fig. 110) and shaft (7) from housing. Oil seal (6) may be renewed at this time. Slide upper shift rail (4520 and 4620 models have two rails) from housing, rotate shifters (1 and 2) upward, then lift them from housing. Shifter cam (12) may now be withdrawn.

Jack up one rear wheel of tractor and turn bevel ring gear until one of the flat surfaces of differential housing is towards transmission oil pump as shown in Fig. 112; then unbolt and remove pump, together with inlet and outlet tubes.

Working through front bearing retainer and using a brass drift, drive transmission drive shaft rearward to force rear bearing cup part of the way out of housing. Tape synchronizer clutches together to keep them from separating while shaft is being removed. Remove pto idler gear from front of transmission, remove front bearing re-

Fig. 112—To remove transmission oil pump after rockshaft housing is removed, turn differential until flat (F) is nearest pump, then remove attaching cap screws (C).

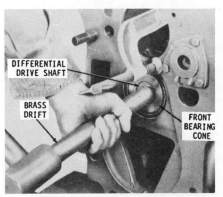

Fig. 113—Install a "C" clamp to hold front gear forward while front bearing is driven off shaft.

tainer and front bearing cup, using care not to damage or lose shims, then lift out transmission drive shaft assembly.

Remove cotter pin and nut from inner end of lower (speed change) shifter cam shaft and remove shaft and arm. Withdraw shifter rail forward out of transmission housing and lift out speed change shifter cam and shifter forks.

Remove differential assembly as outlined in paragraph 170. Remove retainer plate and pto gear from differential drive shaft, then use a puller to remove inner bearing race. Before driving pinion shaft rearward, install a "C" clamp through hole in transmission housing (Fig. 113) and tighten against front (fourth and seventh speed) gear while using a brass drift or soft hammer to drive on shaft. As shaft begins to move, slide parts forward, retighten "C" clamp and remove snap rings (Fig. 114) as they become exposed.

Continue to move shaft components and snap rings forward until rear snap ring has been unseated; then bump differential drive shaft (pinion shaft) rearward, lifting gears and associated parts out top opening as shaft is removed.

NOTE: The snap rings are the same diameter but of different thickness; thickest snap ring being in rear groove. Thus no snap ring can fall into another groove as shaft is removed or installed.

Remove countershaft front bearing retainer (10–Fig. 115) and shim pack (9). Remove snap ring (1) retaining rear bearing cup (2); then using a brass drift, drive countershaft rearward until rear bearing cup is removed. Countershaft can now be lifted out top opening of transmission case.

Overhaul transmission main components as outlined in paragraphs 134 through 141; assemble as outlined in paragraph 132.

132. **ASSEMBLY.** To assemble transmission unit, proceed as follows: Install countershaft, rear bearing cup and re-

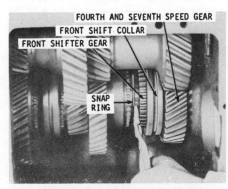

Fig. 114—Remove four snap rings as they become exposed as shaft moves rearward. Keep snap rings in order, with thickest to the rear.

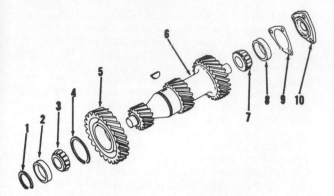

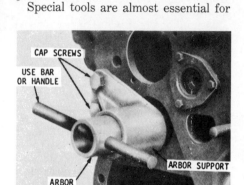

Fig. 115—Exploded view of countershaft and associated parts.

1. Snap ring
2. Bearing cup
3. Bearing cone
4. Snap ring
5. Gear
6. Countershaft
7. Bearing cone
8. Bearing cup
9. Shim
10. Bearing quill

taining snap ring, then install front bearing retainer using removed shim pack. Tighten retaining cap screws to a torque of 35 ft.-lbs.; then using a dial indicator, check countershaft end play. Adjust end play if necessary, to recommended 0.001-0.004 inch by varying thickness of shim pack (9—Fig. 115).

If bevel pinion shaft or transmission housing are being renewed; or if shaft bearings require renewal or adjustment;

refer to paragraph 141 for adjustment procedure.

Special tools are almost essential for installing bevel pinion shaft and gears. Tools needed are as follows:

JDT-2 Expander cone (3020, 4000, 4020 and 4320 models)
JDT-11 Expander Cone (4520 and 4620 models)
JDT-3 Snap Ring Retainers (4)
JDT-8 (1 & 2) Installing Arbor (4000, 4020 and 4320 models)
JDT-9 (1 & 2) Installing Arbor (3020 model)

NOTE: Special tools must be modified as shown in Fig. 116.

Expand snap rings and install expander plates as shown in Fig. 117, then lay out snap rings in order according to

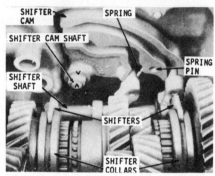

Fig. 118—Install arbor and support to hold parts while assembling differential pinion shaft gears.

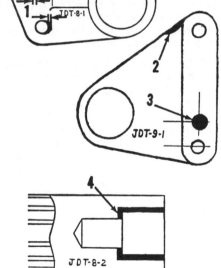

Fig. 116—Special service tools must be modified as shown: (1) Elongate holes 1/8-inch. (2) Grind relief notch 1/4-inch deep. (3) Drill 7/8-inch hole centered as follows: 1 1/4-inches above centerline of lower hole, 7/8-inch from edge. (4) Enlarge bore to 2.000-2.004 inches, 1 1/2-inches deep.

Fig. 119—Speed change shifter mechanism showing component parts. Note "V" alignment marks on shifter cam and shaft. Install shifter cam spring as shown.

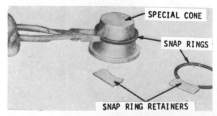

Fig. 117—Special assembly tools are needed to expand and install snap rings.

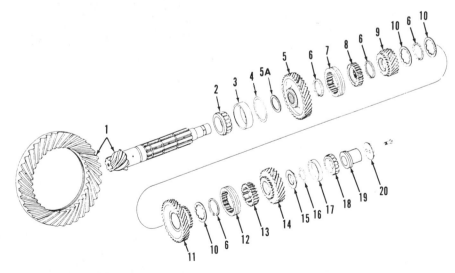

Fig. 120—Exploded view of bevel pinion shaft and associated parts.

1. Pinion shaft & gear
2. Bearing cone
3. Bearing cup
4. Shim
5. 1st-3rd gear
5A. Shim (3020 only)
6. Snap ring (4)
7. Shift collar
8. Rear shifter gear
9. 6th-8th gear
10. Thrust washer (3)
11. 2nd-5th gear
12. Shift collar
13. Front shifter gear
14. 4th-7th gear
15. Thrust washer
16. Shim
17. Bearing cup
18. Bearing cone
19. Inner bearing race
20. Retainer plate

thickness. Refer to Fig. 120 for order of reassembly of shaft components. Install arbor and support as shown in Fig. 118 (or other suitable support) to hold parts while assembling. Insert bevel pinion shaft (1 – Fig. 120) through rear bore, install largest gear (5) and thickest snap ring (6) with expander plate in place. Continue to move shaft forward, installing remaining parts in proper order. Second thickest snap ring is installed next, and so on, with thinnest at front. After fourth speed gear (14) is installed, remove expander plates and seat snap rings in their grooves; then install remainder of parts using removed (or previously determined) shim pack (16). Tighten cap screws retaining end plate (20) to a torque of 35 ft.-lbs.

Install speed change shifter forks and cam, and insert shift rail through forks. Install camshaft and arm, making sure index marks are aligned as shown in Fig. 119. Tighten locknut securely, then adjust camshaft end play to 0.002-0.005 inch if necessary, by loosening clamp nut and sliding actuating arm on shaft. Too much end play in shifter camshaft can allow detent springs to force shift cam out of proper operating position. If shift cam spring (Fig. 119) has been removed or renewed, make sure long end of spring is installed towards shifter camshaft.

To install transmission drive shaft and shifter mechanism assembly, place shaft in transmission housing and install front bearing cup, shims and retainer. Tighten cap screws securely then install rear bearing cup and transmission oil pump. Tighten pump cap screws to 20 ft.-lbs. torque. Check transmission drive shaft end play using a dial indicator and adjust to 0.004-0.006 inch by means of shims (21 – Fig. 121 or 122).

Reinstall speed range shifters and associated parts, making sure index marks are aligned and camshaft end play is within the recommended range of 0.002-0.005 inch.

133. **ADJUSTMENT.** Adjust shifter linkage rods as follows: With rods disconnected from quadrants, place both rods in uppermost position. Place shift lever in sixth speed position in quadrant, then adjust speed change rod yoke until it aligns with hole in quadrant and install retaining pin. Place shift lever in reverse position. With speed range shifter arm in top position, adjust shifter rod yoke until hole in yoke aligns with quadrant hole and install retaining pin.

OVERHAUL

The following paragraphs cover overhaul procedure of transmission main components after transmission has been disassembled as outlined in paragraph 131.

Syncro-Range Models

134. **SHIFTER CAMS AND FORKS.** Refer to Fig. 110 for an exploded view of speed range shifter mechanism and Fig. 111 for speed change shifter. Examine shifting grooves in cams (12 – Fig. 110 and 6 – Fig. 111) for wear or other damage. Parking lock spring (12 – Fig. 111) is retained to cam by spring pin (11). Spring must have sufficient tension to shift the front shift coupling into engagement.

135. **TRANSMISSION PUMP.** The removed transmission pump may be disassembled by removing cover (15 – Fig. 123). Check all surfaces, bushings and bushing bores. Clearance between pump body and gears should be 0.003-0.005 inch, and cover should be flat within 0.001 inch. All parts are available individually.

136. **COUNTERSHAFT.** Refer to Fig. 115 for an exploded view of countershaft and bearings. The shaft is a one-piece unit except for high-speed gear (5). Gear is keyed to shaft and retained by snap ring (4). Countershaft should have 0.001-0.004 inch end play in bearings when properly installed.

137. **TRANSMISSION DRIVE SHAFT.** To disassemble removed transmission drive shaft, remove snap ring (1 – Fig. 121 or 122); then remove rear bearing cone (3) using a suitable press. Lift off reverse range pinion (4) and reverse range shift collar (5), or reverse range synchronizer on models so equipped.

CAUTION: The four detent balls and springs (12) will be released when blocker is withdrawn. Use care not to lose these parts.

Remove snap ring (6) and use a press or puller to remove drive collar (7). Remove snap ring (8) and withdraw low

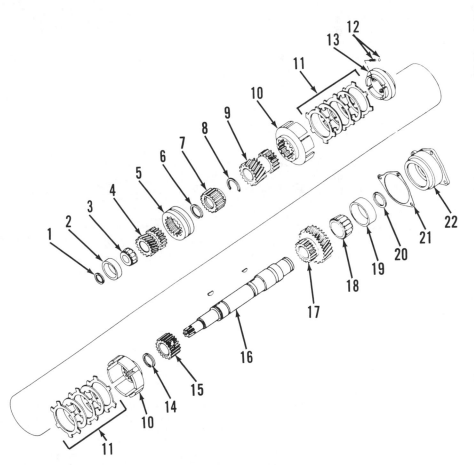

Fig. 121—Exploded view of transmission drive (input) shaft and associated parts used on models with collar shift reverse.

1. Snap ring	7. Reverse range drive collar
2. Bearing cup	8. Snap ring
3. Bearing cone	9. Low range pinion
4. Reverse range pinion	10. High-Low range drum
5. Shift collar	11. Synchronizer plates
6. Snap ring	12. Spring & ball
13. High-Low range blocker	17. High range pinion
14. Snap ring	18. Bearing cone
15. High-Low range drive collar	19. Bearing cup
16. Drive shaft	20. Snap ring
	21. Shim
	22. Bearing housing

range pinion (9); then remove high-low synchronizer observing precautions outlined in note above. High-low range drive collar (15) can be pressed from shaft after removing snap ring (14). High range pinion (17) can be removed after removing high-low range collar (15) or bearing cone (18).

138. SYNCHRONIZER CLUTCHES. The purpose of synchronizer clutches is to equalize speeds of transmission drive shaft and selected range pinion for easy shifting without stopping the tractor. The synchronizer clutches operate as follows:

The range drive collars (7 or 15 – Fig. 122) are keyed to drive shaft. Synchronizer clutch drums (5 and 10) are splined to the range pinions. The blocker rings (6 and 13) are centered in drive collar slots by detent assemblies (12). Synchronizer clutch discs (11) are connected alternately (by drive tangs) to clutch drum and blocker ring. When engine clutch is disengaged and control lever moved to change gear speeds, the first movement of shifter linkage applies contact pressure to clutch discs (11) causing blocker to try to rotate on drive collar. Drive lugs inside blocker ring ride up ramps in drive collar preventing further movement of blocker ring until shaft and gear speeds are equalized. When rotative speeds are equal, thrust force on blocker ring is relieved and synchronizer drum couples gear to shaft without clashing.

139. INSPECTION AND ASSEMBLY. Inspect transmission drive shaft for scoring or wear in areas of range pinion rotation and make sure oil passages are open and clean. Inspect drive lugs of blocker rings and friction faces of blocker rings and synchronizer drums. Check synchronizer discs for wear using a micrometer. Renew entire set if any disc measures 0.060 inch or less on 3020, 4000, 4020 and 4320 or 0.103 inch or less on 4520 and 4620 models. Check drive tangs on disc for thickening due to peening, and renew disc if badly peened.

Reassemble transmission drive shaft by reversing disassembly procedure. Special installing cone, JD-4A (3020, 4000, 4020 and 4320 models) or JDT-10 (4520 and 4620 models) is required to install detent assemblies in blocker rings and blocker assemblies on drive collar; refer to Fig. 124.

140. BEVEL PINION SHAFT. Except for bearing cups in housing and rear bearing cone on shaft, bevel pinion unit is disassembled during removal. Refer to Fig. 120 for exploded view.

All gears should have a diametral clearance of 0.004-0.006 inch on shaft.

First-third gear contains a bushing but bushing is not available for service. The bevel pinion shaft is available only as a matched set with bevel ring gear. Refer to paragraph 173 for information on renewal of ring gear. Refer to paragraph 141 for mesh position adjustment procedure if bevel gears, bearings and/or housing are renewed.

141. PINION SHAFT ADJUSTMENT. The cone point (mesh position) of main drive bevel gear and pinion is adjustable by means of shims (4 – Fig. 120). The cone point will only need to be checked if transmission housing, ring gear and pinion or rear bearing will be renewed. To make adjustment, proceed as follows:

The current cone point of housing and pinion are factory determined and assembly numbers should be etched on left upper housing flange and rear face of pinion as shown at CP – Fig. 125. If no

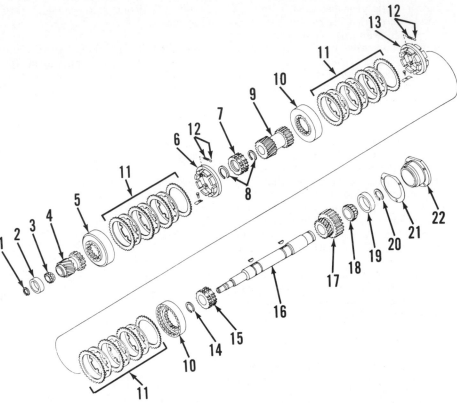

Fig. 122 – Exploded view of transmission drive (input) shaft and associated parts used on models with synchronized reverse.

1. Snap ring	7. Reverse range drive collar
2. Bearing cup	8. Snap rings
3. Bearing cone	9. Low range pinion
4. Reverse range pinion	10. High-Low range drum
5. Reverse range drum	11. Synchronizer plates
6. Reverse range blocker	

12. Spring & ball	17. High range pinion
13. High-Low range blocker	18. Bearing cone
14. Snap ring	19. Bearing cup
15. High-Low range drive collar	20. Snap ring
16. Drive shaft	21. Shim
	22. Bearing housing

Fig. 123 – Exploded view of Syncro-Range transmission oil pump and associated parts.

1. Plug w/bushing
2. "O" ring
3. Hex drive shaft
4. Outlet tube
5. Intake screen
6. Clamp
7. Plug
8. "O" ring
9. Shim
10. Spring
11. Ball
12. Manifold
13. Tube
14. Intake tube
15. Cover
16. Gears
17. Dowel pin
18. Body

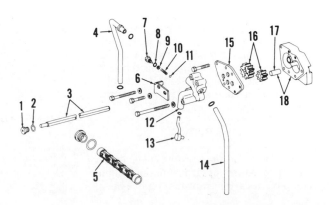

figure is stamped on transmission housing, use nominal figure listed in following procedures.

3020 MODEL. Add 1.755 to number etched on pinion gear, then subtract the total from number on transmission housing, or nominal figure 8.912. Difference will be recommended shim pack thickness.

4000, 4020 and 4320 MODELS. Add 1.755 to number etched on end of pinion gear, then subtract total from number on transmission housing, or nominal figure 10.081. Difference will be recommended shim pack thickness.

4520 and 4620 MODELS. Add 1.751 to number etched on pinion gear, then subtract total from number on transmission housing, or nominal figure 11.290. Difference will be recommended shim pack thickness.

On all models, use a punch and drive out rear bearing cup (3 – Fig. 120) and add or remove shims (4) until correct total thickness is obtained.

The bevel pinion bearings are adjusted to a preload of 0.004-0.006 inch by means of shims (16). If adjustment is required, it should be made before installing gears as follows:

First make sure cone point is correctly adjusted as previously outlined and bearing cone (2) is bottomed on pinion shaft shoulder. Install shaft, thrust washer (15), removed shim pack plus one 0.010 inch shim and front bearing. Install a suitable pipe spacer in place of the press fit inner bearing race and secure with retainer plate and cap screws. Measure shaft end play using a dial indicator as shown in Fig. 126, then remove shims equal to observed end play plus 0.005 inch. Assemble shaft and gears as outlined in paragraph 132.

ENGINE CLUTCH (POWER SHIFT)

The engine disconnect clutch used on tractors equipped with Power Shift transmission is a spring-loaded, dry-disc type clutch. Disconnect clutch is used to disengage transmission from engine for cold weather starting only, and is not to be used for starting or stopping tractor motion.

Power Shift Models

142. LINKAGE ADJUSTMENT. Place control lever in latch position and disconnect control rod (Fig. 127) from lever arm, then move control rod until release bearing contacts clutch release levers. Adjust yoke until pin hole in yoke is aligned with hole in lever arm, then unscrew yoke (lengthen rod) an additional six turns. Release latch and connect control rod to lever arm.

143. R&R AND OVERHAUL. To service clutch, tractor must be split between engine and clutch housing as outlined in paragraph 120. Unbolt and remove clutch assembly from flywheel. Check clutch disc for loose rivets, broken springs or other damage. Facings should be free of grease or oil. Thickness of a new disc is 0.427-0.447 inch.

Friction surface of flywheel must not be out-of-true more than 0.006 inch, scored or grooved. If machining is necessary, do not remove more than 0.060 inch of material. Inspect pilot bearing in flywheel and renew if loose, rough or dry.

Clutch cover and pressure plate assembly must be disassembled in a suitable press, or by using a spare flywheel and three jack screws, nuts and flat washers. To disassemble clutch cover using a flywheel, install the three jack screws in alternately spaced cover mounting holes (A – Fig. 128). Tighten nuts against clutch cover, then remove release finger cotter pins and adjusting nuts (B). Back off nuts on jack screws until spring pressure is relieved; remove jack screws, and lift out clutch cover, spring cups, springs and pressure plate.

Check pressure plate for scoring of friction surface or excessive wear in

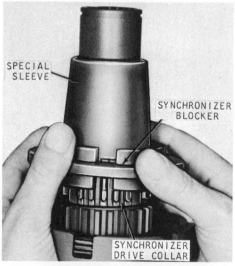

Fig. 124—Assembling synchronizer using special tool.

Fig. 125 – Cone point (mesh) adjustment of bevel ring gear and pinion is factory determined. Installation numbers (CP) are stamped on transmission housing and rear end of pinion as shown. See text for method of determining shim pack thickness.

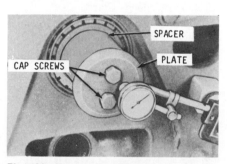

Fig. 126 – Use a pipe spacer behind retainer plate to check preload with dial indicator. Refer to paragraph 141.

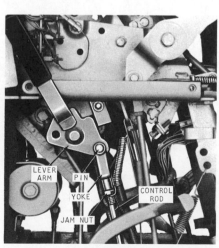

Fig. 127 – View of transmission disconnect clutch control linkage used on Power Shift models.

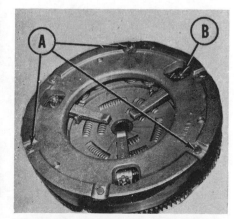

Fig. 128 – Disconnect clutch can be disassembled by installing jack screws in screw holes (A) and removing adjusting nuts (B). Refer to text.

drive pin notches. Friction surface must be true within 0.006 inch. Check clutch springs for rust, pitting or distortion. Renew as necessary.

Reinstall clutch disc with long side of hub facing flywheel. Use JDE-52 aligning tool or other suitable pilot tool to align clutch disc. Tighten clutch cover cap screws to 35 ft.-lbs. torque. On 3020, 4000 and 4020 models, adjust release fingers to height of 59/64-inch from flat surface of clutch disc. A simple adjusting gage can be made as shown in Fig. 129 to simplify adjustment. On 4520 and 4620 models, use special gage JDE-60 to adjust finger height.

POWER SHIFT TRANSMISSION

The Power Shift Transmission provides 8 forward speeds and 4 reverse speeds. Gear changes are accomplished by moving a shift lever, without stopping tractor or operating foot controlled feathering valve. Transmission contains two compound planetary pinion sets controlled by three clutches and four brakes.

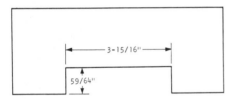

Fig. 129 — Clutch adjusting gage can be made using dimensions shown. Adjust fingers to just touch gage.

OPERATION

Power Shift Models

144. **POWER TRAIN.** The power shift transmission is a manually controlled, hydraulically actuated planetary transmission consisting essentially of a clutch pack and planetary pack as shown schematically in Fig. 130.

Hydraulic control units consist of three clutch packs (C1, C2 and C3) and four disc brakes (B1 through B4). In addition, a multiple disc clutch (pto) is housed in the clutch pack and used in the pto drive train. All units are hydraulically engaged, and mechanically disengaged when hydraulic pressure to that unit is interrupted. The power train also contains a hand operated, single disc transmission disconnect clutch (DC) mounted on engine flywheel, a foot operated inching pedal, a mechanical disconnect for towing and a park pawl.

Three hydraulic control units are engaged for each of the forward and reverse speeds. In 1st speed, Clutch 1 is engaged and power is transmitted to front planetary unit by the smaller input sun gear (C1S); Brake 1 is engaged, locking front ring gear to housing, and planet carrier walks around ring gear at its slowest speed. Clutch 3 is also engaged, locking rear planetary unit, and output shaft turns with planet carrier. Second speed differs from 1st speed only by disengaging Brake 1 and engaging Brake 2, causing planet carrier and output shaft to rotate at a slightly faster speed.

Third speed and 4th speed are identical to 1st and 2nd except that Clutch 1 is disengaged and Clutch 2 engaged, and power enters front planetary unit through larger input sun gear (C2S).

Fifth speed and 6th speed differ from 3rd and 4th speeds in rear planetary unit. Clutch 3 is disengaged and Brake 4 engaged, and output shaft turns faster than planet carrier through action of rear planet pinions and output sun gear.

In 7th and 8th speeds, both Clutch 1 and Clutch 2 are engaged, locking input planetary unit, and planet carrier turns with input shaft at engine speed. Engaging the three clutch units locks both planetary units, therefore 7th speed is a direct drive, with transmission output shaft turning with, and at same speed as the engine. Eighth speed is an overdrive, with transmission output shaft turning faster than engine speed.

Reverse speeds are obtained by engaging Brake 3, which locks output planetary ring gear to housing, and output shaft turns in reverse rotation through action of the two sets of output planetary pinions (RP & OP).

It will be noted that front planetary unit is an input unit controlled by two front clutch units in clutch pack and two front brake units in planetary pack. Two input control units must be engaged to transmit power, and five input speeds are obtained by selectively engaging input brake and clutch units.

The rear planetary unit is an output unit controlled by two rear brakes and rear clutch. One of the rear control units must be engaged to complete the power train. Two forward ranges and one

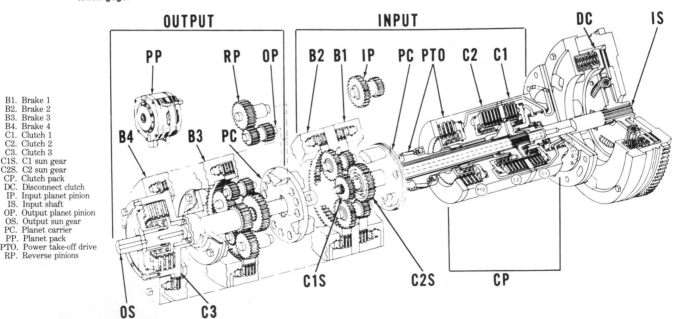

B1. Brake 1
B2. Brake 2
B3. Brake 3
B4. Brake 4
C1. Clutch 1
C2. Clutch 2
C3. Clutch 3
C1S. C1 sun gear
C2S. C2 sun gear
CP. Clutch pack
DC. Disconnect clutch
IP. Input planet pinion
IS. Input shaft
OP. Output planet pinion
OS. Output sun gear
PC. Planet carrier
PP. Planet pack
PTO. Power take-off drive
RP. Reverse pinions

Fig. 130 — Schematic view of Power Shift transmission showing primary function of units. Disconnect Clutch (DC) is for cold weather starting only, and is not to be used for starting or stopping tractor motion. The Power Take-off (pto) clutch and gear are located in, but are not a part of, transmission power train.

reverse output range are provided, depending on which rear control unit is engaged.

The accompanying table lists control units actuated to complete power flow in each shift position:

Front (Input) Control Units		Rear (Output) Control Unit
Forward Speeds		
1st	C1	B1	C3
2nd	C1	B2	C3
3rd	C2	B1	C3
4th	C2	B2	C3
5th	C2	B1	B4
6th	C2	B2	B4
7th	C1	C2	C3
8th	C1	C2	B4
Reverse Speeds		
1st	C1	B1	B3
2nd	C1	B2	B3
3rd	C2	B1	B3
4th	C2	B2	B3

145. CONTROL SYSTEM. The control valve unit consists of manually actuated speed selector and direction selector valves which operate through four hydraulically controlled shift valves to engage desired clutch and brake units. The valve arrangement prevents engagement of any two opposing control units which might cause transmission damage or lockup.

Power to operate transmission system is supplied by an internal gear hydraulic pump mounted on transmission input shaft, which also supplies charging fluid for tractor main hydraulic system. Fluid from hydraulic pump first passes through a full flow oil filter to transmission oil gallery, where pressure is regulated at approximately 140-160 psi on 3020, 4000 or 4020 models and 165-185 psi on 4520 or 4620 models for transmission control functions. Excess oil passes through regulating valve to oil cooler and main hydraulic pump.

Fluid from transmission main oil gallery is routed through inching pedal valve to Clutches 1 and 2, and through a spring-loaded accumulator to brake actuating pistons and to Clutch 3.

Refer to Fig. 131 for a schematic view of control circuits. The direction selector valve and shift valve 4 controls routing of pressure to output control units (Clutch 3, Brake 3 and Brake 4). The speed selector valve contains four pressure ports (1, 2, 3 and 4) which control movement of the four shift valves by pressurizing the closed end (opposite return spring) when port is open to pressure. Neutral position is provided by the selector valves or by depressing inching pedal.

When direction selector valve is moved to forward detent position,

system pressure is routed in Shift Valve 4. In low range positions (1st through 4th gears), speed selector valve charging port to the top of Shift Valve 4 is open to pressure, Shift Valve 4 moves downward against spring pressure and Clutch 3 is actuated. In high range positions (except 7th gear which is direct and uses all three clutch units), charging pressure is cut off to Shift Valve 4, shift valve return spring moves valve upward and Brake 4 is actuated. When direction selector valve is moved to reverse detent position, system pressure by-passes Shift Valve 4 and is routed directly to Brake 3. In neutral detent position, system pressure is cut off from all three output control units.

Shift Valves 1, 2 and 3 direct system pressure to input control units (Clutches 1 and 2 and Brakes 1 and 2). Shift Valve 1 directs pressure to Clutch 2 when hydraulically actuated and to Clutch 1

when charging port to top of Valve 1 is closed.

Shift Valve 2 routes pressure to Shift Valve 3 when hydraulically actuated, and permits simultaneous engagement of Clutches 1 and 2 when charging port to top of Valve 2 is closed.

Shift Valve 3 directs pressure to Brake 2 when hydraulically actuated and to Brake 1 when charging port to top of Valve 3 is closed.

146. LINKAGE ADJUSTMENT. To adjust control linkage, refer to Fig. 132 and proceed as follows: Disconnect park lock cable from actuating arm. Move shift lever to "Park" position; pull up on park cable until park pawl is fully engaged, and adjust cable yoke until pin can be inserted through yoke and actuating arm holes.

Disconnect direction control rod yoke (7) from bellcrank (6). Move shift lever to

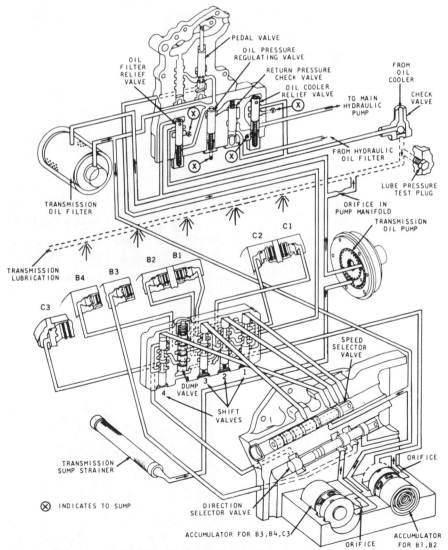

Fig. 131—Schematic view of oil control circuits, valves and accumulators in typical Power Shift transmission.

"Neutral" position. With direction control valve in center (Neutral) detent, adjust length of control rod until pin hole is aligned in bellcrank and yoke. Install pin and tighten jam nut.

With speed control rod yoke (8) disconnected from arm on lever pivot (5), move shift lever to "Neutral" position. Place speed control rod in uppermost detent position and adjust control rod yoke until connecting pin can be inserted.

To adjust pedal valve linkage, refer to Fig. 133. Disconnect operating rod (4) from valve operating arm (3). With pedal fully depressed, turn operating arm fully counter-clockwise and thread operating rod in or out of yoke until

hooked end of rod aligns with hole in operating rod. Lengthen rod additional ½-turn and reconnect.

147. ADJUSTING SPEED OF SHIFT. To adjust shifting engagement rate, stop engine and remove adjustable orifice plug at bottom of valve housing on left side of tractor (Fig. 134). Turn slotted-head adjustment screw clockwise to slow shift rate. For faster shift rate, turn adjusting screw out (counter-clockwise). Turn screw ½-turn at a time, then recheck shift speed.

NOTE: On some later models, speed of shift is set at factory and can not be adjusted.

148. PRESSURE TEST AND ADJUSTMENT. Before checking transmission operating pressure, first be sure transmission oil filter is in good condition and oil level is at top of "SAFE" mark on dipstick. Place towing disconnect lever in "TOW" position, start engine and operate at 1900 rpm until transmission oil is at operating temperature. If tractor is equipped with Power Front Wheel Drive, put main hydraulic

pump out of stroke by turning in shut-off screw. Raise and block rear of tractor so rear wheels are free to rotate.

Stop engine and install a 0-300 psi pressure gage in test port marked "CLUTCH" (Fig. 135) or "SYSTEM PRESSURE" (Fig. 136). At 1900 rpm, pressure should be 140-160 psi on 3020, 4000 and 4020 models, or 165-185 psi on 4520 and 4620 models.

If oil pressure is not as indicated, add or remove shims (17 – Fig. 145) under oil pressure regulating valve spring (12) to obtain correct oil pressure.

If pressure cannot be adjusted, other possible causes are:
1. Incorrect pedal valve linkage adjustment, refer to paragraph 146.
2. Malfunctioning regulator valve, pedal valve or oil filter relief valve. Overhaul as outlined in paragraphs 157, 158 or 159.

If adjustment of operating pressure does not cure malfunction, check pressures as outlined in following TROUBLESHOOTING paragraphs.

Fig. 134 — To adjust shifting engagement rate, remove plug indicated by arrow. Turn adjusting screw in or out as outlined in text.

Fig. 136 — On late models, clutch valve cover was changed and operating pressure is checked at "SYSTEM PRESSURE" test port. Check transmission pressures at test ports as indicated. Refer to text.

Fig. 132 — Exploded view of shift control linkage.

1. Control support	5. Control lever pivot
2. Control lever	6. Bellcrank
3. Spring	7. Direction control rod
4. Bushings	8. Speed control rod

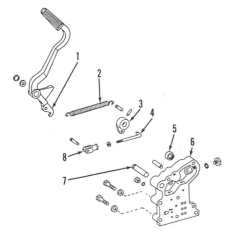

Fig. 133 — Exploded view of pedal valve assembly.

1. Pedal	5. Oil seal
2. Return spring	6. Valve housing
3. Operating arm	7. Pedal shaft
4. Operating rod	8. Yoke

Fig. 135 — On early models, transmission operating pressure is checked at "CLUTCH" test port. Remove plug and install 0-300 psi pressure gage. Refer to text.

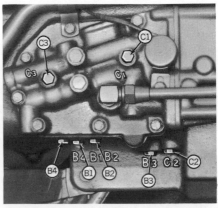

Fig. 137 — On early models, check transmission pressures at test ports as shown.

TROUBLESHOOTING

Power Shift Models

149. PRESSURE TEST. To make a complete check of transmission hydraulic system pressures, first check and adjust operating pressure as outlined in paragraph 148, then proceed as follows:

On tractors with early style valve covers (Fig. 137), install 0-300 psi pressure gages at C1, C2, B1, B2, B3, B4 and C3. On later style valve covers (Fig. 136), install pressure gages at C1, C2, B1-B2 and C3-B3-B4 test ports.

NOTE: A single gage can be used to perform tests, but the use of additional gages makes testing much quicker.

Raise and block rear of tractor so rear wheels are free to rotate. If equipped with Power Front Wheel Drive, put main hydraulic pump out of stroke. Place towing disconnect lever in "TOW" position. With engine running at 1900 rpm and operating temperature up to normal, place shift lever in each position of following table. Observe pressures at appropriate test port for each element listed.

Neutral		B1		
Neutral Fwd.		B1		C3
1st Fwd.	C1	B1		C3
2nd Fwd.	C1		B2	C3
3rd Fwd.	C2	B1		C3
4th Fwd.	C2		B2	C3
5th Fwd.	C2	B1		B4
6th Fwd.	C2		B2	B4
7th Fwd.	C1 C2	*	*	C3
8th Fwd.	C1 C2	*	*	B4
Neutral Rev.		B1	B3	
1st Rev.	C1	B1	B3	
2nd Rev.	C1		B2 B3	
3rd Rev.	C2	B1	B3	
4th Rev.	C2		B2 B3	

*7th and 8th speed check shows pressure on gage at B1, B2 pressure port but elements are not engaged since pressure is stopped at accumulator.

Excessive leakage is indicated if pressures are more than 15 psi below system pressure in B1, B2, B3, B4 or C3, or more than 25 psi below in C1 or C2.

NOTE: Because of line restriction, C1 and C2 pressure is approximately 10 psi below system pressure in normal operation.

If pressure is observed on any element that should not be engaged, check for sticking shift valves or leakage within control valve housing.

With shift lever in any forward or reverse gear and engine speed at 1900 rpm, depress inching pedal while noting C1 or C2 pressure gage reading. Gage pressure on either or both gages showing pressure should drop to zero with pedal fully depressed.

Release pedal slowly; gage pressure should rise at a smooth, even rate until approximately 80 psi is registered with pedal ½ to 1 inch from top, then move quickly to operating pressure with further pedal movement.

Failure to perform as outlined would indicate maladjustment of pedal valve linkage (see paragraph 146) or malfunction of the valve (overhaul as outlined in paragraph 157).

150. LUBRICATION PRESSURE TEST. Before overhauling or repairing Power Shift transmission, a lubrication pressure test can help isolate the problem, and prove the results after repair.

The lubrication pressure plug to be removed for test is located in clutch housing on upper right side, just to right of oil return line. It will be necessary to remove footrest panel to reach the plug for this test. Install a 0-60 psi gage that does not contain a dampener orifice. In PARK position and with oil at operating temperature, the pressure should be at least 30 psi at 2200 rpm. Excessive pressure indicates a blockage in lube circuit. If pressure is low, put main hydraulic pump out of stroke by turning in shut-off screw. If pressure comes up to normal, there is a leak in hydraulic system. Next run engine at a slow enough speed to get a reading of 5-10 psi. Place towing lever in "TOW" position. Shift transmission through positions shown in table in paragraph 149 while observing pressure gage. Pressure should drop momentarily, then come back to within 3 psi of original value. If pressure fails to return at any selected speed, depress clutch pedal to cut off pressure to C1 and C2. If pressure comes back up, a clutch unit is leaking. If it does not come up with clutch depressed, planetary pack is leaking. Test in several speeds to pinpoint leaking unit while consulting shift table.

Engage pto clutch and observe gage. If pressure fails to return to within 3 psi of original pressure, pto clutch is leaking or valve is out of adjustment. Refer to paragraph 161 for overhaul of pto clutch or paragraph 191 for pto valve adjustment.

151. BEHAVIOR PATTERNS. Erratic behavior patterns can be used to pinpoint some systems malfunctions.

ODD SHIFT PATTERN. If tractor slows down when shifted to a faster speed; speeds up when shifted to a slower speed; or fails to shift when selector lever is moved; a sticking shift valve is indicated.

SLOW SHIFT. Possible causes are: improper regulating valve adjustment; improper pedal linkage adjustment;

plugged fluid filter; malfunctioning regulating valve, pedal valve or oil filter relief valve; broken accumulator spring; sticking accumulator piston; or slipping clutch or brake unit or units.

ROUGH PEDAL ENGAGEMENT. If tractor jumps rather than starts smoothly when pedal valve is actuated, a sticking clutch pedal valve or broken pedal valve spring is indicated.

SLIPPAGE UNDER LOAD. If transmission slips, partially stalls or stalls under full load, first check adjustment of pedal valve as outlined in paragraph 146, then check transmission pressures as outlined in paragraphs 148 and 149. If trouble is not corrected, one of the clutch units; or transmission disconnect clutch is malfunctioning.

If a clutch or brake unit is suspected of slipping, it will be necessary to determine which of the three units is at fault in that speed. Refer to table in paragraph 149 to determine which three units are involved in the speed range in question. Then prove one unit at a time by choosing a speed that utilizes that particular unit, and if that speed does not slip, choose another speed that changes only one unit whenever possible. In this way the slipping unit can be isolated.

NOTE: 4th to 5th and 5th to 4th change two units at once, as do 6th to 7th and 7th to 6th.

If clutch C1 or C2 is suspected, inching pedal can be used to determine if either clutch is bad. Since the only difference between 2nd and 4th speed is the change from C1 to C2, these two speeds can be used to isolate faulty unit. (If either clutch is slipping, it will also slip in 7th and 8th speeds since both clutches are applied in both speeds.) Place shift lever in speed desired and allow brake and clutch units time to engage before releasing inching pedal.

If one or more units are found to be slipping in every gear in which unit is engaged, remove and overhaul transmission.

TRACTOR FAILS TO MOVE. If tractor fails to move when transmission is engaged, first check to see that tow disconnect is fully engaged. If tow disconnect unit is engaged, check to see that park pawl operates properly and is correctly adjusted. Park pawl is engaged by cam action and disengaged by a return spring. If spring breaks or becomes unhooked, pawl may remain engaged even though linkage operates satisfactorily. To examine or renew park pawl return spring, remove transmission housing cover as outlined in paragraph 165.

TRACTOR CREEPS IN NEUTRAL. A slight amount of drag is normal in the clutch and brake units, especially when transmission oil is cold. Excessive creep is usually caused by warped clutch or brake plates; observe the following:

If tractor creeps when inching pedal is depressed and properly adjusted, either Clutch 1 or Clutch 2 is malfunctioning. Check as follows: With engine speed at 1500 rpm and transmission fluid at operating temperature shift to 2nd speed on a flat surface. Depress the inching pedal; if tractor continues to roll forward at approximately the same speed, Clutch 1 is malfunctioning, if tractor speed increases, Clutch 2 is dragging.

Place shift lever in NEUTRAL position. Disconnect yoke from direction selector arm on transmission. This will leave output section of transmission in neutral. Shift into 1st forward with shift lever, but do not depress clutch pedal. If tractor creeps forward with throttle set at 1500 rpm and transmission oil at operating temperature, Clutch 3 or Brake 4 is dragging; if tractor creeps backward, Brake 3 is malfunctioning.

NOTE: Dragging clutch or brake units, aside from causing creep, will contribute to loss of power, heat and excessive wear. Creep is merely an indication of possibly more serious trouble which needs to be corrected for best performance, or to prevent future failure.

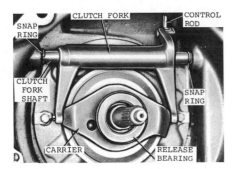

Fig. 138 — Clutch release fork and shaft must be removed before transmission pump and clutch pack can be removed.

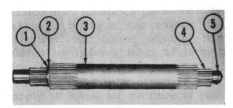

Fig. 138A — The connecting shaft assembly splines into operating clutch hubs of clutch pack and sun gears of front planetary unit.

1. Snap ring
2. Thrust washer
3. C2 (outer) clutch shaft
4. C1 (inner) clutch shaft
5. Sealing ring

REMOVE AND REINSTALL

Power Shift Models

152. **PEDAL AND REGULATING VALVES.** Pedal valve housing and regulating valve housing attach to left side of clutch housing using common gaskets and gasket plate. Housings may be removed separately, but both should be removed to renew gaskets.

To remove housings, remove left battery and battery box if necessary. Remove pedal return spring and disconnect rod from valve operating arm. Disconnect lube pipe, inlet pipe and outlet pipe from housings. Disconnect pto valve operating rod at clutch housing, then unbolt and remove housings, gasket plate and gaskets.

Overhaul pedal valve housing as outlined in paragraph 157 and regulating valve housing as in paragraph 158.

When installing, use light, clean grease to position gaskets and gasket plate, making sure gaskets are installed on proper sides of plate as shown in Fig. 144. Install regulator valve housing and retaining cap screws, then install pedal valve housing. Tighten retaining cap screws evenly and securely, and complete the assembly by reversing disassembly procedure. Adjust as outlined in paragraph 146.

153. **CONTROL AND SHIFT VALVES.** To remove control and shift valve housing, drain transmission and remove right battery and battery box if necessary. Disconnect wiring to start-safety switch and remove cotter pins which retain control cable yokes to control arms. Remove retaining cap screws, then remove valve housing, being careful not to lose detents and springs.

Overhaul removed unit as outlined in paragraph 159 or 160 and install by reversing removal procedure. Make sure inner and outer gaskets are in proper order since they are not interchangeable. Tighten retaining cap screws evenly and securely. Adjust as outlined in paragraph 146.

154. **CLUTCH PACK AND TRANSMISSION PUMP.** The transmission pump and clutch pack can be removed as a unit after draining hydraulic system and detaching clutch housing from engine as outlined in paragraph 120.

Disconnect and remove clutch release bearing, fork and shaft (Fig. 138). Then, unbolt and remove transmission pump and clutch pack assembly from clutch housing.

Clutch shafts (3 and 4 – Fig. 138A) may be removed with clutch pack, or may be withdrawn after clutch unit is out. Be sure sealing ring (5) is in good condition and shafts are in position when unit is reinstalled.

Overhaul removed clutch pack and pump as outlined in paragraphs 161 and 162.

To assist in easier installation of clutch pack, use alignment studs in the two side holes of housing and position gasket using light grease. Make sure oil passages in clutch housing and gasket are properly aligned. Insert connecting shafts with sealing ring (5 – Fig. 138A) to rear. Install clutch pack with oil passages in mounting flange aligned with those of gasket and clutch housing. Tighten cap screws to 35 ft.-lbs. torque.

155. **CLUTCH HOUSING.** The clutch housing must be removed for access to pto drive gear train or removal of planetary unit. To remove clutch housing, first split tractor between engine and clutch housing as outlined in paragraph 120 and remove clutch pack as in paragraph 154.

Remove batteries and disconnect wiring. Remove differential lock pedal pivot pin and rockshaft selector lever knob, then remove platform.

Mark and diagram location of hydraulic tubing if necessary, then remove hydraulic tubes and system control linkage as necessary.

Remove snap ring (Fig. 139) securing pto clutch gear bearing to bore and withdraw clutch gear and bearing as a unit. Remove mid pto bearing retainer.

NOTE: A slide hammer may be required to remove pto clutch gear.

Support clutch housing and steering support assembly from a hoist and remove clutch housing flange cap screws. Two upper, center screws are accessible through inside front of clutch housing as shown in Fig. 139. Pry clutch housing from its doweled position on transmission case and swing housing away from rear unit.

Use new gasket and "O" rings when reinstalling clutch housing. "O" rings

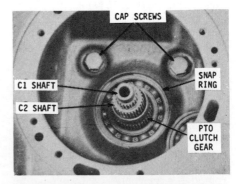

Fig. 139 — View of clutch housing with clutch pack removed. Snap ring, pto clutch gear and the two hidden cap screws must be removed before clutch housing can be separated from transmission.

may be held in position with grease. Tighten ⅝-inch cap screws to a torque of 170 ft.-lbs. and ¾-inch cap screws to 300 ft.-lbs.

156. PLANETARY PACK. To remove transmission planetary pack, first detach (split) engine from clutch housing as outlined in paragraph 120; remove clutch pack as in paragraph 154 and clutch housing as in paragraph 155. Remove rockshaft housing as in paragraph 203 and transmission top cover plate. Remove pto idler gear (3–Fig. 140) and pto clutch gear (1).

Fig. 142—Remove bearing retainer and C3 oil pipe and withdraw reduction gear shaft before attempting to remove planetary pack.

Fig. 140—Pto idler gear (3) and clutch gear (1) must be removed to remove planetary pack. Be sure thrust washers (4 and 6) and brake spring (8) are in position when reinstalling clutch housing.

1. Pto clutch gear
2. Bearings
3. Idler gear
4. Thrust washer
5. Snap ring
6. Thrust washer
7. Brake idler gear
8. Brake return spring
9. Pto drive gear
10. Snap ring
11. Pto shaft

Fig. 143—To remove planetary pack after cover is off, remove output shaft and attaching cap screws (S). Grasp with lifting tool at points indicated by arrows and lift from transmission housing.

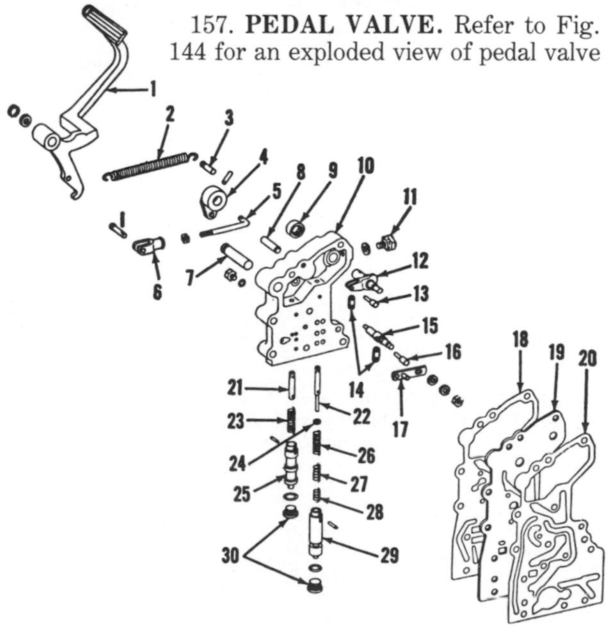

Fig. 144—Exploded view of pedal valve housing and component parts.

1. Pedal
2. Return spring
3. Anchor pin
4. Operating arm
5. Operating rod
6. Yoke
7. Pivot shaft
8. Stop pin
9. Oil seal
10. Valve housing
11. Plug
12. Operating shaft
13. Link pin
14. Link
15. Pto valve operating shaft
16. Link pin
17. Pto valve operating arm
18. Outer gasket
19. Gasket plate
20. Inner gasket
21. Pto valve rod
22. Pedal valve rod
23. Spring
24. Spring retainer
25. Pto valve
26. Upper spring
27. Center spring
28. Lower spring
29. Pedal valve
30. Plug

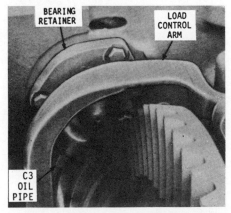

Fig. 141—Loosen cap screws to load control arm support for more wrench clearance to C3 oil pipe fitting.

Disconnect Clutch 3 pressure tube from output reduction gear rear bearing retainer and remove retainer (Fig. 141). If necessary to obtain more clearance to C3 tube fitting, loosen cap screws holding load control arm support at bottom of housing. Using a long brass drift and reaching through center of planetary pack from front, drive reduction gear rearward until bearing cup is removed from rear bore. Withdraw shaft as shown in Fig. 142.

Remove the four retaining cap screws and, using a hoist and ice tongs or similar tool, grasp planetary pack as indicated by arrows (Fig. 143) and lift unit straight upward out of transmission housing. Overhaul removed planetary unit as outlined in paragraph 163.

Before installing planetary pack, inspect or renew the four brake passage "O" rings in bottom of transmission housing. Reduction gear shaft should be installed with 0.000-0.002 inch bearing preload. To check bearing adjustment, make a trial installation of shaft, bearing cup and retainer, using one additional shim between bearing retainer and housing. Tighten retaining screws securely and measure shaft end play using a dial indicator. Remove retainer and deduct from shim pack, shims equal to measured end play plus 0.001 inch. Keep remainder of shim pack together for final installation.

Lower planetary unit straight downward being careful not to dislodge brake passage "O" rings. Tighten four retaining cap screws alternately and evenly to 55 ft.-lbs. torque, and complete assembly by reversing disassembly procedure.

OVERHAUL

Power Shift Models

157. PEDAL VALVE. Refer to Fig. 144 for an exploded view of pedal valve and associated parts. Refer to paragraph 152 for removal and installation information.

Pedal shaft (7) is renewable and is a press fit in housing. Install shaft with inner edge of snap ring groove 1-3/32 inches (3020, 4000 and 4020 models) or 1-21/32 inches (4520 or 4620 models) from machined surface of housing. If operating shaft oil seal (9) must be renewed, install seal with lip towards inner side of housing.

Valves (25 and 29) must slide smoothly in their bores and must not be scored or excessively loose. Check pto and pedal valve springs for distortion and against values which follow:

Pto Valve Spring
Free Length, Inches 1-25/32
Lbs. Test at Inches . . . 16.6-20.4 at 1⅜

Pedal Valve Spring (Lower)
Free Length, Inches . . . . ½
Lbs. Test at Inches . . . 3.3-4.1 at 11/32

Pedal Valve Spring (Center)
Free Length, Inches 1-5/16
Lbs. Test at
   Inches . . . . 15.1-18.3 at 1-3/32

Pedal Valve Spring (Upper)
Free Length, Inches 2
Lbs. Test at
   Inches . . . . 11.7-14.3 at 1-13/32

158. REGULATING VALVE. Refer to Fig. 145 for an exploded view of regulating valve and associated parts.

Refer to paragraph 152 for removal and installation information.

The three valves (11, 14 and 15) are identical, but valve springs are different and must be marked or tested for correct installation. Return oil check valve (14) is installed upside-down from other two valves.

Check valves and their bores for sticking or scoring. On late models, inner spring (5) was added and lower spring (7) was also changed. Check springs against test values which follow:

Filter By-Pass
Top Spring (Inner)
   Approximate Free Length . . . . 2.0 in.
   Lbs. Test at 1-13/32 in. 12-14
Top Spring (Outer)
   Approximate Free Length . . 2-5/16 in.
   Lbs. Test at 1-9/16 in. 19-23
Lower Spring (Early Models)
   Approximate Free Length . . . . . ¾-in.
   Lbs. Test at ½-in. 18-22
Lower Spring (Later Models)
   Approximate Free Length . . . . 7/8-in.
   Lbs. Test at ½-in. 12-15

Pressure Regulating
Approximate Free Length . 3-15/16 in.
Lbs. Test at 3-7/16 in. 58.5-71.5

Return Oil Check
Approximate Free Length . . 2-5/16 in.
Lbs. Test at ¾-in. 11-13

Cooler Relief
Approximate Free Length . 3-25/32 in.
Lbs. Test at 3-7/32 in. 33-41

159. CONTROL VALVE (EARLY MODELS). The shift valve housing is attached to inner face of control valve housing by a cover and six cap screws.

Refer to Fig. 146 for an exploded view. Refer to paragraph 154 for removal and installation procedure.

To disassemble removed unit, proceed as follows: Remove the six cap screws retaining shift valve housing (6 – Fig. 146) and lift off cover, housing, gasket plate (10) and gaskets (9 and 11). Note that the two gaskets are not interchangeable. Mark the removed gaskets "Outer" and "Inner" as they are removed to aid in correct installation of new gaskets when unit is reassembled. Lift out inner detent springs (12) and plungers (13) to prevent loss. Invert control valve housing (14) and remove outer detent retaining plugs, springs and plungers. Remove the six cap screws retaining cover (24) to housing and withdraw cover, control valves and operating linkage as a unit from control valve housing. Remove plug (3) from shift valve housing and withdraw spring (4) and dump valve spool (5). Remove the four snap rings and washers, then withdraw shift valve springs (8) and spools (7). The four shift valve spools and springs are identical.

Clean all parts in a suitable solvent and check for scoring or other damage, and for free movement of valve spools in bores. Control valve actuating mechanism need not be disassembled unless renewal of parts is indicated.

Check valve springs for damage or distortion and against values which follow:
Dump Valve Spring
Approximate Free Length . . 1-7/16 in.
Lbs. Test at 13/16-in. 7-8
Shift Valve Spring
Approximate Free Length . . . . 1¼ in.
Lbs. Test at ¾-in. 16-20
Reassemble by reversing disassembly procedure. Tighten cap screws retaining

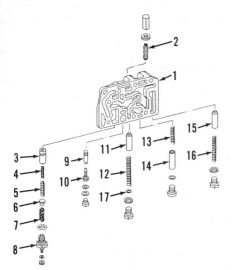

Fig. 145—Exploded view of typical Power Shift regulating valve assembly. Some later models do not use orifice (9) and adjusting screw (10).

1. Valve body	10. Adjusting screw	
2. By-pass screw	11. Pressure regulating	
3. Filter by-pass valve	valve	
4. By-pass spring (outer)	12. Spring	
5. By-pass spring (inner)	13. Spring	
6. Guide	14. Return oil check valve	
7. Spring	15. Cooler relief valve	
8. Switch	16. Spring	
9. Orifice	17. Shim	

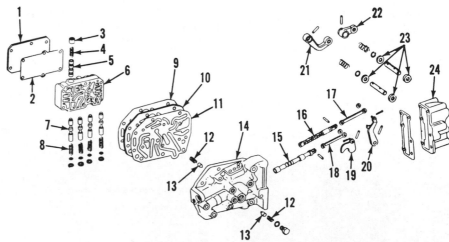

Fig. 146—Exploded view of control and shift valves used on early model Power Shift transmission.

1. Cover	7. Shift valve	13. Detent plunger	19. Operating arm
2. Gasket	8. Spring	14. Control valve housing	20. Operating arm
3. Plug	9. Gasket	15. Direction valve	21. Arm
4. Spring	10. Plate	16. Speed control valve	22. Arm
5. Dump valve	11. Gasket	17. Link	23. Oil seals
6. Shift valve housing	12. Spring	18. Link	24. Cover

shift valve housing to control valve to 20 ft.-lbs. torque.

160. CONTROL VALVE (LATE MODELS). The shift valve housing is attached to inner face of control valve housing by a cover and six cap screws. Refer to Fig. 147 for an exploded view. Refer to paragraph 153 for removal and installation procedure.

To disassemble removed unit, remove the six cap screws retaining shift valve housing (5) and lift off cover (1), housing, gasket plate (12) and gaskets (11 and 13). Note that the two gaskets are not interchangeable. Mark the removed gaskets "Outer" and "Inner" as they are removed to aid in installation of new gaskets when unit is reassembled. Lift out inner detent springs (14B and 14C), plungers (15) and balls (16) to prevent loss as housing (17) is removed. Invert control valve housing (17) and remove the eight cap screws retaining accumulator cover (26), gasket (25), pistons (24) and springs (21 and 22). Note that accumulator pistons face opposite each other, and washer (23) is installed behind outward facing piston only. Remove the six cap screws retaining cover (27) to housing and withdraw cover, control valves and operating linkage as a unit from control valve housing. Remove plug (2) from shift valve housing and withdraw spring (3) and dump valve spool (4). Remove the four retaining rings (10) and washers (9), then withdraw shift valve springs (8) and spools (7). The four shift valve spools and

springs are identical. The two replaceable accumulator charging orifices can be removed at this time.

Clean all parts in a suitable solvent and check for scoring or other damage, and for free movement of valve spools in bores. Control valve actuating mechanism need not be disassembled unless renewal of parts is indicated.

Check valve springs for damage or distortion and against following values:

Dump Valve Spring
Approximate Free Length . . 1-7/16 in.
Lbs. Test at 13/16-in. 7-8

Shift Valve Spring
Approximate Free Length . . . . 1¼ in.
Lbs. Test at ¾-in. 16-20

Accumulator Valve Outer Spring
Approximate Free Length . . 4-3/32 in.
Lbs. Test at 3¾ in. 53-64

Accumulator Valve Inner Spring
Approximate Free Length . . . . 3¾ in.
Lbs. Test at 2-7/8 in. 100-122

Reassemble by reversing disassembly procedure, using Fig. 147 as a guide. Tighten the six cap screws retaining shift valve housing to control valve evenly and alternately to a torque of 20 ft.-lbs. and tighten accumulator cover screws to 35 ft.-lbs. torque.

161. CLUTCHES. To disassemble removed clutch pack, use a holding fixture with a 2-inch hole (or drill a 2-inch hole near edge of a table or bench). Insert in-

put shaft (25 – Fig. 148) and release bearing sleeve through hole, with pto clutch pressure plate (60) up. Remove through-bolts (61), then lift off pto clutch

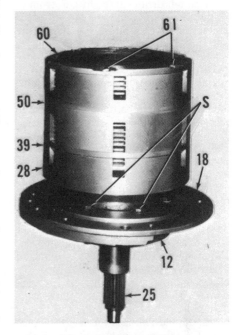

Fig. 148 — View of removed clutch assembly. The input shaft (25) and release bearing sleeve can be inserted into hole in bench or holding fixture to facilitate disassembly and reassembly. Screws (S) attach manifold plate (18) and pump housing (12) together.

12. Pump housing	39. Clutch separator plate
18. Clutch manifold	50. Rear clutch drum
25. Input shaft	60. Clutch backing plate
28. Front clutch drum	61. Through-bolts

1. Cover
2. Plug
3. Spring
4. Dump valve
5. Shift valve housing
6. "O" ring
7. Shift valve
8. Spring
9. Washer
10. Retaining ring
11. Gasket
12. Plate
13. Gasket
14B. Detent spring
14C. Detent spring
15. Detent cone
16. Detent ball
17. Control valve housing
18. Ball
19. Plug
20. Orifice
21. Spring (inner)
22. Spring (outer)
23. Special washer
24. Accumulator piston
25. Gasket
26. Piston cover
27. Linkage cover
28. Oil seal
29. Shaft
30. Felt washer
31. Spring
32. Reverse arm
33. Speed control arm
34. Operating arm
35. Link
36. Operating arm
37. Speed control valve
38. Direction valve
39. Start-safety switch

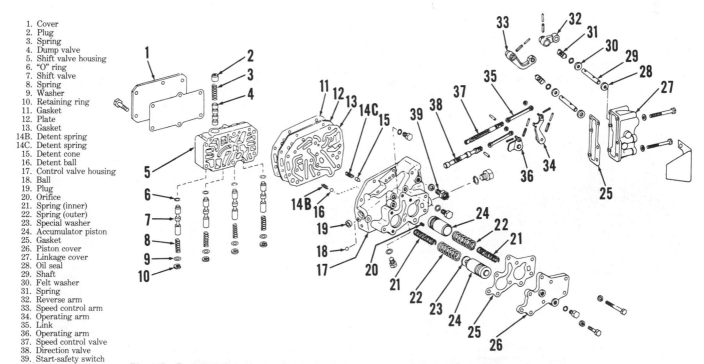

Fig. 147 — Exploded view of control valve assembly used on late model Power Shift transmission.

pressure plate, pto clutch discs and clutch hub. Pto and C2 clutch drum (50), C1 and C2 pressure plate (39) and C1 clutch drum (28) can be separated after jarring slightly.

C1 clutch drum can be lifted off input shaft and manifold assembly after removing snap ring on rear splines of input shaft (25).

Clutch plates, friction discs and other parts may be interchangeable between certain locations on some models; however, be sure to separate and identify by location all parts as they are separated to prevent improper assembly.

Coil type springs (41–Fig. 150) or Belleville washers (34) are used to return pistons when hydraulic pressure is released. A suitable fixture is necessary to compress springs for removal or installation of snap rings. Refer to Fig. 149 for typical compressing fixture. Pistons can be removed and seals renewed after snap rings and springs are removed. Refer to Fig. 151 for correct assembly of Belleville washers. Be sure to use correct compressing fixture to prevent damage to springs or other parts while assembling.

Inspect clutch discs for wear or glazing. Thickness of a new disc is 0.112-0.118 inch on 3020, 4000 and 4020 models. Discs should be replaced if less than 0.100 inch. On 4520 and 4620 models, new clutch disc thickness is 0.127-0.133 inch. Replace discs if less than 0.110 inch. Thickness of new flat clutch plate is 0.115-0.125 inch on all models. Plates with notch in one drive lug designate a wavy plate. Thickness of wavy plate is 0.085-0.095 inch.

Most clutch packs use one flat plate and remaining plates are wavy. Flat plate should be assembled next to piston. Alternate friction discs with plates, and notches in lugs of wavy plates should be staggered. Tighten throughbolts to 20 ft.-lbs. torque.

162. TRANSMISSION PUMP, MANIFOLD PLATE AND INPUT SHAFT. To overhaul transmission pump, manifold plate and input shaft, first disassemble clutch pack as outlined in paragraph 161. After C1 clutch drum (19–Fig. 150) has been removed, remove screws that hold manifold (9) to pump body (1), then slide input shaft and manifold plate assembly out of pump housing. Remove and save the two steel check balls (Fig. 152) as pump and manifold units are separated.

Remove the one remaining cap screw in rear face of pump housing (5–Fig. 150) and lift off pump body and gears.

Input shaft and bearing assembly can be removed from manifold plate after unseating and removing retaining ring (15) from rear of manifold plate. Press bearing from shaft, if renewal is required, after removing bearing snap ring (12). Chamfer of thrust washer (14) should be toward large (rear) end of shaft (16).

Examine gears and housings for scoring, wear, cracks and other damage and renew as required. Assemble by reversing disassembly procedure. Tighten cap screws in pump housing and body to a torque of 20 ft.-lbs. Reassemble clutch pack as in paragraph 161.

163. PLANETARY PACK. Refer to paragraph 156 for removal of planetary assembly. To disassemble, place unit on

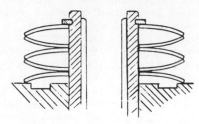

Fig. 151 — Cross-sectional schematic view showing correct method of installing five Belleville washers in clutch drum. Regardless of number used (4, 5 or 7), the inner diameter of outer washer must always fit against snap ring as shown.

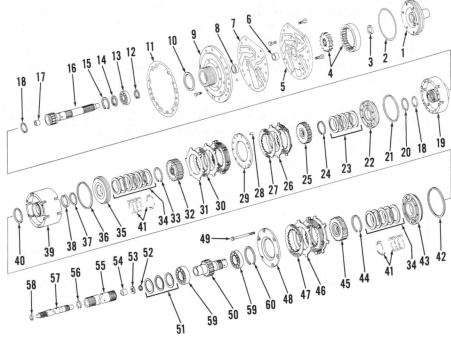

Fig. 150 — Exploded view of transmission oil pump, manifold and clutch assembly. The C1 clutch parts are 18 through 27; C2 clutch is shown at 30 through 37; pto clutch parts are 40 through 47. Coil spring and retainer (41) are used in place of Belleville washer springs (34) on some models.

1. Pump body	22. Low range clutch piston	46. Clutch plates (flat plate next to piston 43)
2. Packing ring	23. Belleville spring washers	47. Clutch friction discs (alternate with plates 46)
3. Seal	24. Snap ring	48. Clutch backing plate
4. Pump gears	25. Clutch hub	49. Through-bolts
5. Pump housing	26. Clutch plates (flat plate next to piston 22)	50. Pto drive shaft
6. Bushing	27. Clutch friction discs (alternate with plates 26)	51. Thrust bearing
7. Gasket	28. Dowel pin	52. Snap ring
8. Bushing	29. Clutch separator plate	53. Thrust washer
9. Manifold	30. Clutch friction discs (alternate with plates 31)	54. Bushing
10. Seal ring (4)	31. Clutch plates (flat plate next to piston 35)	55. High range clutch shaft (C2)
11. Gasket	32. Clutch hub	56. Snap ring
12. Snap ring	33. Snap ring	57. Low range clutch shaft (C1)
13. Ball bearing	34. Belleville spring washers	58. Seal
14. Thrust washer	35. High range clutch piston	59. Ball bearings
15. Snap ring	36. Piston outer seal	60. Snap ring
16. Input shaft	37. Piston inner seal	
17. Bushing	38. Bushing	
18. Snap ring (2)	39. Rear clutch drum	
19. Front clutch drum	40. Piston outer seal	
20. Piston inner seal	41. Coil spring and retainer (used in place of 34 on some models)	
21. Piston outer seal	42. Piston outer seal	
	43. Pto clutch piston	
	44. Snap ring	
	45. Clutch hub	

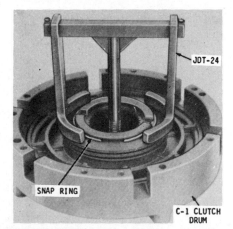

Fig. 149 — Clutch 1 drum with piston and springs installed. Refer to paragraph 161 for disassembly procedure.

bench with output end up as shown in Fig. 153. Remove cap screws (C) and lift off C3 clutch piston housing (1). Remove through-bolts (T), then lift housing (2), clutch drum, housing (3) and associated parts off until planetary assembly can be

Fig. 152 — When installing assembled pump on gasket and manifold, be sure steel check balls are in place as shown.

removed. Refer to paragraph 164 for disassembly of planetary assembly.

Check for warped, worn or scored brake discs and plates. Thickness of both brake discs and plates when new is

Fig. 153 — View of removed planetary pack.

C. Cap screw
P. Pressure ports
T. Through-bolts
1. C3 piston housing
2. B4 piston housing
3. B3 piston housing
4. B2 piston housing
5. B1-B2 pressure plate
6. B1 piston housing
7. Piston return springs

0.117-0.123 inch on 3020, 4000 and 4020 models. On 4520 and 4620 models, thickness of a new brake disc is 0.127-0.133 inch and thickness of a new brake plate is 0.117-0.123 inch.

To disassemble planetary clutch piston (Fig. 154), a suitable fixture, such as the tool shown in Fig. 149, is necessary to compress return springs for removal of snap ring. Renew seal rings on pistons whenever a piston is removed from housing (Fig. 155).

NOTE: Parts from various tractor models and from different locations within same tractor model may be similar, but not identical. DO NOT attempt to install similar, but incorrect parts at any location within this transmission.

Bushing (32 – Fig. 154) should be pressed into housing (34) using a suitable piloted driver. Bushing should be flush with piston side of housing. Be sure to align cut-out in bushing (32) with similar cut-out in housing.

Bushings (46 and 49) should be

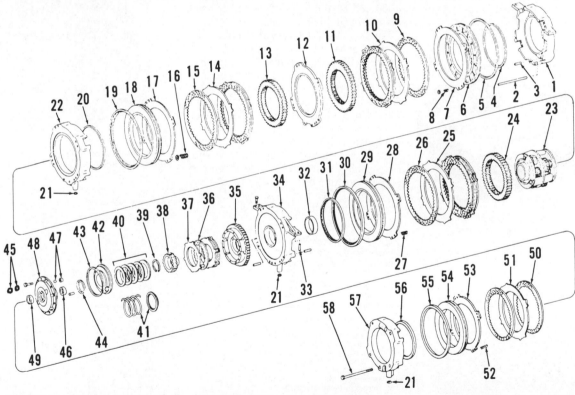

Fig. 154 — Exploded view of the planetary pack. The B1 brake includes parts (4 through 11); B2 brake is (13 through 20); B3 brake is (24 through 31); C3 clutch is parts (36 through 44); B4 brake includes parts (50 through 55). Parts shown are typical and differences may be noted.

1. First planet piston housing
2. Pin (4 used)
3. Dowel pins
4. Piston inner seal
5. Piston outer seal
6. Brake piston
7. Brake piston return plate
8. Springs
9. Friction discs
10. Separator plates
11. First planet ring gear
12. Brake facing plate
13. Second planet ring gear
14. Separator plate
15. Friction discs
16. Springs
17. Brake piston return plate
18. Brake piston
19. Piston outer seal
20. Piston inner seal
21. "O" rings
22. Second planet piston housing
23. Planetary carrier assembly (Refer to Fig. 157)
24. Third planet ring gear
25. Separator plate
26. Friction discs
27. Springs
28. Brake piston return plate
29. Brake piston
30. Piston outer seal
31. Piston inner seal
32. Bushing
33. Dowel pins
34. Third planet piston housing
35. Planetary clutch drum
36. Friction discs (alternate with 37)
37. Clutch plates
38. Planetary clutch hub
39. Snap ring
40. Belleville spring washers
41. Coil spring and cup (used in place of 40 on some models)
42. Planetary clutch piston
43. Piston outer seal
44. Piston inner seal
45. Ball and plug
46. Bushing
47. Ball and plug
48. Clutch piston planetary housing
49. Bushing
50. Friction discs
51. Separator plate
52. Springs
53. Brake piston return plate
54. Brake piston
55. Piston outer seal
56. Piston inner seal
57. Fourth planet brake piston housing
58. Cap screws

pressed into bores of clutch housing (48) using appropriate size pilots. Large chamfer in inside diameter of rear bushing (49) should be towards outside (rear) of clutch housing (48). Press both bushings to bottom of bores.

NOTE: Coat all parts with John Deere Hy-Gard transmission and hydraulic oil or equivalent while assembling.

To assemble planetary pack, place B1 piston housing (Fig. 156) closed end down on a bench. Install four guide studs in threaded holes and alignment dowels in remaining holes as shown.

The piston return plates (7, 14, 28 and 53 – Fig. 154), which are installed next to pistons, have drilled holes in the four extended lugs to serve as seats for brake return springs.

NOTE: Be sure to align oil holes in brake facing plate with oil holes in B1 housing when assembling. If not correctly assembled, B1 brake will not receive pressure to apply brake.

Refer to paragraph 164 for assembly of planetary assembly. Remainder of assembly procedure should be accom-

plished by reversing disassembly procedure and observing Fig. 153 and Fig. 154. Alternate separator plates and lined friction brake discs at all locations. All oil ports (P – Fig. 153) must be aligned as shown. Remove aligning studs, install through-bolts (T) and tighten to 35 ft.-lbs. torque. Tighten screws (C) to 20 ft.-lbs. torque.

Before reinstalling planetary brake pack, apply 50-80 psi air pressure to each of the oil passage ports (P – Fig. 153) in turn; listen for air leaks and note action of brake plates. If leaks are noted or if brake return springs do not compress, recheck assembly procedure and correct trouble before reassembling tractor. Be sure "O" rings (21 – Fig. 154) in bottom of transmission case do not move when installing planetary pack. Tighten cap screws to 55 ft.-lbs. torque.

164. **PLANETARY ASSEMBLY.** The planetary assembly shown in Fig. 157 is located in planetary brake pack as shown at 23 – Fig. 154. Two different types of planetary units are used: Type "A" (Fig. 157) is typical of units on 3020, 4000 and 4020 models; Type "B" is used on 4520 and 4620 models.

On Type "A" planetary assemblies disassemble as follows: Place unit on a bench with rear retainer (12) up. Remove three cap screws holding retainer (12) to carrier (1). Pry cover plate (11) from dowels; remove three planet pinions (10), being careful not to lose roller bearings. All rear planet pinions are equipped with **TWO** rows of 31 bearings and one spacer each. Turn planet assembly upside down and remove three cap screws holding retainer (25) to carrier. Pry cover (24) from carrier. Remove ring gear and pinions (17), being careful

not to lose roller bearings.

When assembling Type "A" planetary unit, refer to Fig. 158. Install the three planet pinion shafts (18) in plate (24) and locate using the steel balls (15 – Fig. 157). Position retainer (25) under cover plate to hold shafts in place. Position C2-B1 sun gear (22) and B1 ring gear (11 – Fig. 154) on plate (24 – Fig. 158). Locate timing marks on ring gear at center of planet pinion shafts. Install three planet pinions (17) with bearing rollers, thrust washers and spacers and with timing marks aligned with marks on C1-B2 sun gear (21). Position thrust washer (20 – Fig. 157) on sun gear, then position carrier (1) over planet assembly being sure that dowel pins (19) properly engage holes in plate. Carefully turn assembly over and install the three cap screws through retainer (25) and plate (24). Tighten screws to 35 ft.-lbs. and

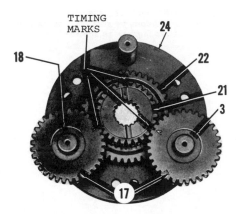

Fig. 158 – View of Type "A" planetary assembly showing C1 and C2 sun gears and related timing marks.

3. Thrust washer	21. Second planet sun gear
17. First & second planet pinions	22. First planet sun gear
18. Pinion shaft	24. Pinion carrier front cover

Fig. 155 – View of B1 piston housing with piston withdrawn showing seal rings.

1. Housing	5. Piston ring
4. Sealing ring	6. Piston

Fig. 157 – Exploded view of two different planetary types used. Type "A" is used on 3020, 4000 and 4020 models; Type "B" is used on 4520 and 4620 models. Parts shown are typical and differences may be noted.

1. Planet pinion & carrier housing
2. Dowel pins
3. Thrust washers (18 used)
4. Needle rollers
5. Spacer (6 used)
6. Third planet pinion (3 used)
7. Shafts (6 used)
8. Third planet sun gear
9. Fourth planet sun gear
10. Fourth planet pinion (3 used)
11. Pinion carrier rear cover (Type "A")
12. Pinion shaft rear retainer
13. Thrust washer (Type "A")
14. Gasket (Type "B")
15. Steel balls (9 used)
16. Spacer (3 used)
17. First & second planet pinion (3 used)
18. Pinion shaft (3 used)
19. Thrust washer (Type "A")
20. Thrust washer (Type "A")
21. Second planet (C1-B2) sun gear
22. First planet (C2-B1) sun gear

23. Bushing
24. Pinion carrier front cover (Type "A")
25. Retainer
26. Needle thrust bearing (Type "B")

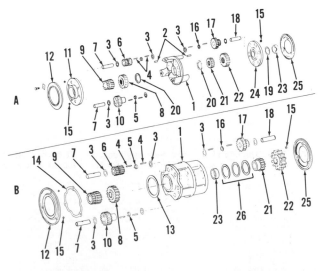

Fig. 156 – Install guide screws and dowels in place of through-bolts.

bend lock plates around heads. Install six pinion shafts (7) and steel balls (15) in cover plate (11). Place retainer (12) under cover plate to hold shafts in plate. Install planet pinions (6 and 10) and B3 sun gear (8) as shown in Fig. 159. On models so equipped, be sure thrust washer is correctly centered on B3 sun gear. On all models, position planet carrier assembly (1–Fig. 157) on pinion shafts correctly engaging hollow dowels (2). Turn assembly over carefully, install cap screws through lock plates, retainer (12) and plate (11). Tighten cap screws to 35 ft.-lbs. torque, then bend lock plates up around heads of cap screws.

Models 4520 and 4620 planetary assembly is Type "B" shown in Fig. 157. Each planet pinion shaft is retained in carrier housing by a steel ball (15) and retainer plates (12 and 25). Shaft is a slip fit in carrier housing bores. Each planet pinion contains two rows of 20 loose needle rollers (4) separated by a spacer for a total of 40 bearings to each pinion.

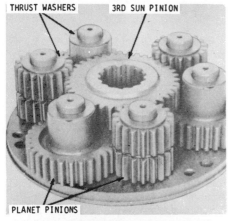

Fig. 159—Assemble planet pinion shafts, steel balls and planet pinions in rear cover plate as shown before installing sun pinion gears on Type "A" unit.

Fig. 160—Numbers on pinion gears must align with corresponding number stamped in housing on "B" type planetary units.

Withdraw pinions carefully to prevent loss of bearings. All planet pinion bearing rollers are interchangeable, but should be kept in sets. Front and rear bearing spacers (5 and 16) are of different thickness, the thicker spacers being used in rear (output) planetary unit.

Inspect bushing (23) in planetary carrier. Installed diameter of new bushing is 2.505-2.507 inches. When renewing bushing, notched end of bushing goes towards bottom of bore. Press in flush to 0.020 inch below flush.

When assembling, note first and second planet pinion (17) is index-marked "1", "2" and "3" and corresponding marks appear on planet carrier adjacent to pinion shaft bore (Fig. 160). When assembling planet carrier, place carrier on a bench front down, install first sun gear (22–Fig. 157) with cupped out side down and install second sun gear (21) with gear side down. Special loading tools (JDT-22-1 and JDT-22-2) can be used to more easily load roller bearings in pinion gears. When inserting loaded pinion gears, corresponding index marks on pinion and carrier MUST be aligned as shown in Fig. 160. No indexing is required on other planetary unit, and third sun gear (8–Fig. 157) can be installed with either side up.

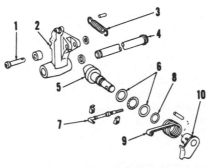

Fig. 161—Exploded view of typical transmission park lock pawl and associated parts.

1. Pin
2. Lock pawl
3. Spring
4. Pivot shaft
5. Lock cam
6. Thrust bearing
7. Operating cable
8. "O" ring
9. Spring arm
10. Arm hub

Fig. 162—Exploded view of typical tow disconnect control linkage.

1. Latch
2. Shift lever
3. Spring
4. Stop plate
5. Retainer
6. L.H. shaft
7. "O" ring
8. Shift yoke
9. Adjusting screw
10. Aluminum washer
11. R.H. shaft
12. Shoe (2)

REDUCTION GEARS, TOW DISCONNECT AND PARK PAWL

All Power Shift Models

165. **TRANSMISSION TOP COVER AND PARK PAWL.** If park pawl remains engaged even though linkage is properly adjusted and operates satisfactorily, the park pawl return spring may be unhooked or broken. The unit can be inspected and spring renewed after removing transmission top cover. Proceed as follows:

Remove operator's platform, steering support side panels and interfering hydraulic tubes and linkage. Unbolt and remove transmission top cover. Remove damaged or broken spring and install a new spring. Malfunction can also occur if engagement spring arm (9–Fig. 161) is damaged or broken or if camshaft (5) binds. Pivot shaft (4) on some models is retained by a vertical dowel in transmission case. Adjust arm on cam to give 0.005-0.010 inch end play in cam.

Fig. 163—To adjust tow disconnect on early models, first loosen locknut (N). Pry output gear (G) rearward and turn adjusting screw (A) until clearance between collar (C) and gear is 0.010-0.050 inch. (P) and (S) are attaching points for park pawl spring.

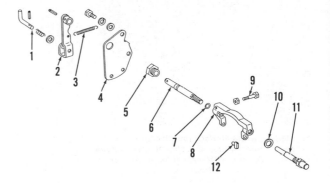

166. TOW DISCONNECT. To remove tow disconnect mechanism, first remove planetary output shaft as outlined in paragraph 156. Unscrew retaining nut (5 – Fig. 162) and shaft (11), then lift out disconnect fork (8), collar (4 – Fig. 165), reduction gear (3) and bearing (2). Front bearing cup can be drifted rearward out of housing after gear has been removed.

Reinstall by reversing removal procedure. Adjust reduction gear shaft bearings as outlined in paragraph 156. After tow disconnect parts have been installed, adjust as follows: On early models, pry reduction gear (G – Fig. 163) rearward. While holding back on disconnect lever, loosen locknut (N) and turn adjusting screw (A), if necessary, until clearance (B) between face of gear and collar (C) is 0.010-0.050 inch. Tighten locknut and recheck adjustment. On late models, shift disconnect lever into full forward position with spring unhooked (Fig. 164). With stop plate cap screws loose, align mark on stop plate with front of lever as shown and tighten cap screws. Reattach spring and shift lever to detent (engaged) position.

167. IDLER GEAR AND SHAFT. Idler gear (17 – Fig. 165) can be removed after removing planetary unit as outlined in paragraph 156 and tow disconnect and reduction gear as in paragraph 166.

Remove cotter pin, nut and washer from rear of idler shaft and snap ring (14) from front bearing; then drift or pull shaft forward out of housing and gear. Front end of shaft contains a 5/8-11 tapped hole to facilitate removal. Remove gear (17) and spacer (18) through top opening as shaft is removed. Assemble by reversing disassembly procedure; tighten shaft nut to 180 ft-lbs. torque and install cotter pin.

168. BEVEL PINION SHAFT. Bevel pinion shaft and bearings can be removed for service after removing planetary unit as outlined in paragraph 156 and differential assembly as in paragraph 170. Remove idler shaft and gear by following procedures outlined in paragraph 167.

Remove nut (24 – Fig. 165) from front end of bevel pinion shaft and drift shaft rearward until front bearing cone, shims (26) and spacer (27) can be removed. Withdraw shaft and rear bearing cone from rear while lifting gear (28) and spacer (29) out top opening.

The bevel pinion shaft is available only as a matched set with bevel ring gear. If rear bearing cup is renewed, keep cone point adjusting shim pack (30) intact and reinstall same pack or shims of equal thickness. If gears and/or housing are renewed, check and adjust cone point as outlined in paragraph 169.

NOTE: If main drive bevel pinion and/or transmission housing are renewed, cone point (mesh position) of gears must be checked and adjusted BEFORE adjusting bearing preload.

When reinstalling bevel pinion shaft and bearings, make a trial installation using removed shim pack (26) plus one additional 0.010 inch shim. Tighten shaft nut, then measure shaft end play using a dial indicator. Remove shims equal to measured end play plus 0.005 inch, to obtain recommended 0.004-0.006 inch bearing preload. A preferred method of measuring bearing preload is by wrapping string around pinion shaft just in front of pinion and measuring rolling torque of shaft with a spring scale. The spring scale should indicate 2½-7½ pounds of pull required to rotate shaft one revolution per second with nut (24) tightened to 400 ft.-lbs. torque.

After tightening nut, stake it in at least two locations to maintain setting.

169. CONE POINT SETTING. The cone point, mesh position of main drive bevel gear and pinion, is adjusted by means of shims (30 – Fig. 165). The cone point will only need to be checked if transmission housing or ring gear and pinion assembly are renewed.

The correct cone point of housing and pinion are factory determined and assembly numbers are etched on rear face of pinion and top center of left-hand side of transmission housing. To determine correct thickness of shim pack, proceed as follows:

3020 MODELS. Add 1.442 to figure etched on pinion shaft. Subtract the total from figure stamped on transmission case. If no figure is stamped on transmission use nominal figure 11.290. Difference will be correct shim pack.

4000 AND 4020 MODELS. Add 1.755 to figure etched on pinion shaft. Subtract total from number on transmission case, or use nominal figure 10.081. Difference will be correct shim pack.

4520 AND 4620 MODELS. Add 1.751 to figure on pinion shaft. Subtract total from number on transmission, or use nominal number 11.290. Difference will be correct shim pack.

Adjust pinion shaft bearings to 0.004-0.006 inch preload after cone point is correctly set. Refer to paragraph 168.

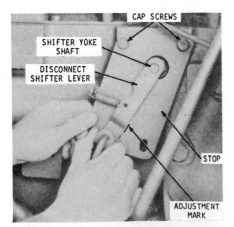

Fig. 164 — On late models, tow disconnect is correctly adjusted when adjustment mark aligns with front of lever with lever in full forward position.

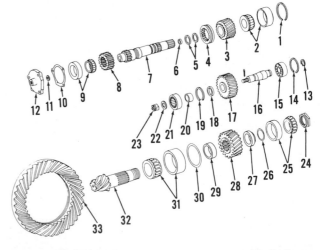

Fig. 165 — Exploded view of output gear reduction unit, main drive bevel pinion, ring gear and associated parts.

1. Snap ring
2. Bearing
3. Reduction gear
4. Shift collar
5. Seal rings
6. Oil seal
7. Reduction shaft
8. Drive sleeve
9. Rear bearing
10. Shim pack
11. Seal ring
12. Retainer
13. Snap ring
14. Snap ring
15. Bearing
16. Idler shaft
17. Idler gear
18. Spacer
19. Snap ring
20. Inner race
21. Bearing
22. Washer
23. Nut
24. Nut
25. Bearing
26. Shim pack
27. Spacer
28. Gear
29. Spacer
30. Shim pack
31. Bearing
32. Pinion shaft
33. Ring gear

DIFFERENTIAL AND MAIN DRIVE BEVEL RING GEAR

All Models

170. **REMOVE AND REINSTALL.** To remove differential assembly, first drain transmission and hydraulic fluid. Remove rockshaft housing as outlined in paragraph 203. Unbolt load control arm support from bottom of housing. Load control arm comes out with differential. On Syncro-Range models, remove transmission oil pump inlet and outlet lines.

Block up tractor and remove both final drive units as outlined in paragraph 178. Remove brake backing plates and brake discs, then withdraw both differential output shafts.

If two of the ring gear to housing cap screws are recessed, rotate ring gear until the two screws are horizontal, which will provide more clearance for removal.

Place a chain around differential housing as close to bevel ring gear as possible, attach a hoist and lift differential enough to relieve the weight on carrier bearings. Remove both bearing quills using care not to lose, damage or intermix shims located under bearing quill flanges. Differential assembly (and load control arm) may now be removed.

Overhaul removed unit as outlined in paragraph 171.

When installing, place an additional 0.010 inch shim on bearing quill on ring gear side, tighten retaining cap screws and measure differential bearing end play using a dial indicator. Preload carrier bearings by removing shims equal in thickness to measured end play plus 0.002-0.005 inch. Shims are available in thicknesses of 0.003, 0.005 and 0.010 inch.

After correct carrier bearing preload is obtained, attach a dial indicator, zero indicator button on one bevel ring gear tooth and check backlash between bevel ring gear and pinion, in at least two places 180 degrees apart. Proper backlash is 0.008-0.015 inch. Moving one 0.005 inch shim from one bearing quill to the other will change backlash by about 0.010 inch.

When bearing preload and backlash are established, tighten differential bearing quill cap screws to a torque of 85 ft.-lbs. and bend up lock plates. Tighten rockshaft housing bolts to 85 ft.-lbs. torque. Complete tractor assembly by reversing disassembly procedure.

171. **DIFFERENTIAL OVERHAUL.** To overhaul removed differential assembly, refer to Fig. 166 or 167. Differential gears and differential lock parts can be inspected after removing cover (6). Differential pinion shaft (18) is retained by special cap screw. Thrust washers are not used on differential pinions, but are used on axle gears. Refer to paragraph 172 for service on differential lock clutch and to paragraph 173 for bevel ring gear. Tighten cover cap screws to 55 ft.-lbs. torque if equipped with differential lock, and 85 ft.-lbs. without lock.

172. **DIFFERENTIAL LOCK CLUTCH.** The multiple disc differential clutch can be overhauled after removing unit as outlined in paragraph 170 and removing cover (6—Fig. 167). The three internally splined clutch drive discs (8) are 0.112-0.118 inch thick. Discs should be renewed when thickness is less than 0.100 inch. Externally splined clutch plates (9) are 0.115-0.125 inch thick. The three piston return springs ride on guide dowels contained in cover (6) as shown in Fig. 168.

Remove piston (11—Fig. 167) with air pressure or by grasping two opposing strengthening ribs with pliers. Renew sealing "O" rings whenever piston is removed. Examine sealing surface in bore of bearing quill which houses sealing rings (19), and renew quill if sealing area is damaged. Renew cast iron sealing rings (19) if broken, scored or badly worn.

173. **BEVEL RING GEAR.** The main drive bevel ring gear and pinion are available only as a matched set. Always replace both gears. To remove ring gear, remove retaining cap screws and use a heavy drift, hammer or suitable press. Heat new gear to 300° F. in an oven and position gear. Tighten retaining cap screws to 85 ft.-lbs. torque on all models except 4520 and 4620, which should be tightened to 170 ft.-lbs. Renew pinion shaft as outlined in paragraph 131 for Syncro-Range models or paragraph 168 for Power Shift models.

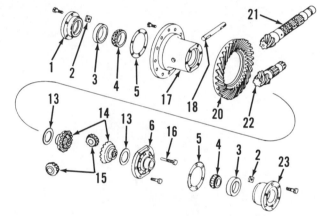

Fig. 166—Exploded view of differential assembly used on 4020 models not equipped with differential lock. Other models are similar in major details.

1. Bearing quill L.H.
2. Square nut
3. Bearing cone
4. Bearing cup
5. Shim pack
6. Cover
13. Thrust washer
14. Axle gears
15. Pinion gears
16. Special bolt
17. Housing
18. Pinion shaft
20. Ring gear
21. Pinion (Syncro-Range)
22. Pinion (Power Shift)
23. Bearing quill R.H.

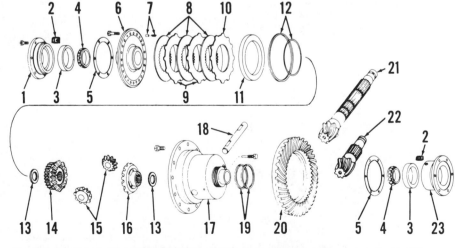

Fig. 167—Exploded view of differential assembly used on 4020 models equipped with differential lock. Other models are similar in major details.

1. L.H. bearing quill	7. Spring and pin (3)	13. Thrust washer	19. Sealing rings (2)
2. Square nut (2)	8. Drive disc (3)	14. Gear (diff. lock)	20. Bevel ring gear
3. Bearing cup	9. Clutch plate (2)	15. Differential pinions	21. Pinion (Syncro-Range)
4. Bearing cone	10. Backing plate	16. Axle gear	22. Pinion (Power Shift)
5. Shim pack	11. Piston	17. Differential housing	23. R.H. bearing quill
6. Housing cover	12. "O" rings	18. Pinion shaft	

DIFFERENTIAL LOCK

Tractors may be optionally equipped with a hydraulically actuated differential lock which may be engaged to insure full power delivery to both rear wheels when traction is a problem. The differential lock consists of a foot operated control and regulating valve and a multiple disc clutch located in differential housing.

Models So Equipped

174. **OPERATION.** When pedal is depressed, pressurized fluid from hydraulic system is directed to clutch piston (11 – Fig. 167) locking axle gear (14) to differential case. Elimination of differential as a working part causes both differential output shafts and main drive bevel gear to turn together as a unit. Available power is thus transmitted to both rear wheels equally, despite variations in traction.

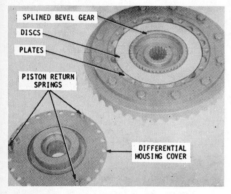

Fig. 168 – Differential lock clutch partially disassembled. Piston return springs ride on guide dowels in cover and enter blank splines in clutch plate lugs.

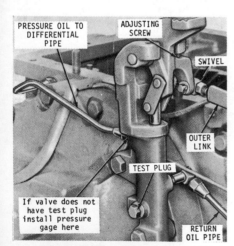

Fig. 169 – Differential lock control valve and associated parts used on early models. Late models are similar.

To release differential lock, slightly depress either brake pedal which releases system pressure by acting through linkage.

175. **ADJUSTMENT.** Differential lock valve and linkage should be adjusted for operating pressure and release linkage length. Proceed as follows:

To check operating pressure, refer to Fig. 169 and install a 0-1000 psi gage at test port shown. With engine running at rated speed, depress lock actuating pedal. Pressure should be 420-480 psi; if not, remove plunger (8 – Fig. 170 or 171) and add or remove shims (10). One shim should change pressure 25-30 psi.

To adjust linkage, place lock pedal in fully engaged position with engine running. On early models, turn swivel adjusting nuts (3 – Fig. 170) until link assembly (1) is just snug between brake pedals and lock arm. On late models, adjust nuts (3 – Fig. 171) until release link (1) is snug between brake pedals and lock pedal. On all models, differential lock must release with a small movement of either brake pedal, but release link must not be tight enough to cause lock to release from tractor vibration. When adjustment is correct, tighten adjusting nuts securely.

OVERHAUL

Models So Equipped

176. **CONTROL VALVE.** To remove differential lock control valve, bleed off hydraulic system pressure by actuating brakes with engine not running. Disconnect control linkage and oil pipes at valve. Unbolt and lift off valve and spacers.

Refer to Fig. 170 or 171 and disassemble valve. Examine parts for wear or scoring and springs for distortion. Keep shim pack (10) intact for use as a starting point when readjusting operating pressure. Renew "O" rings whenever valve is disassembled.

Assemble by reversing disassembly procedure. Tighten control valve housing cap screws to 170 ft.-lbs. torque.

177. **CLUTCH.** Refer to paragraph 172 for overhaul procedure on differential lock clutch unit.

REAR AXLE AND FINAL DRIVE

Some tractors are available in high clearance (Hi-Crop) models equipped

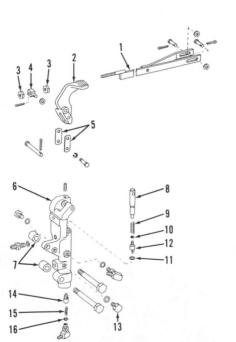

Fig. 170 – Exploded view of differential lock control valve and linkage used on early models.

1. Release link	9. Spring
2. Pedal	10. Shim pack
3. Nuts	11. "O" ring
4. Swivel	12. Flow control valve
5. Links	13. Test port plug
6. Valve body	14. Check valve
7. Spacers	15. Spring
8. Plunger	16. "O" ring

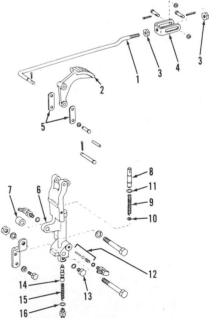

Fig. 171 – Exploded view of differential lock control valve and linkage used on late models.

1. Release link	9. Spring
2. Pedal	10. Shim pack
3. Nuts	11. "O" ring
4. Yoke	12. Check valve
5. Links	13. Test port plug
6. Valve body	14. Control valve
7. Spacer	15. Spring
8. Plunger	16. "O" ring

with drop housings containing a final reduction bull gear and pinion. Standard equipment on all models is a planetary reduction final drive gear located at inner ends of rear axle housings.

REMOVE AND REINSTALL

All Models

178. To remove either final drive as a unit, first drain transmission and hydraulic fluid, suitably support rear of tractor and remove rear wheel or wheels. Remove cab if so equipped.

On standard models, remove fenders and light wiring if so equipped. On Hi-Crop models, remove 3-point hitch lift links and draft links or entire drawbar assembly. On right final drive, remove differential lock pressure pipes and control valve (models so equipped). On left final drive, remove rockshaft return pipe (Syncro-Range models).

Support final drive assembly with a jack or a hoist, remove attaching cap screws and swing unit away from transmission housing.

On Hi-Crop models, remove the six stud nuts securing drop housing to shaft housing; remove the two retainer plugs and thread jack screws into retainer plug holes. Tighten jack screws evenly to force housings apart.

When reinstalling, be sure sun gear (25 – Fig. 172) on standard models stays all the way in, to prevent brake disc from falling behind teeth on gear. Tighten retaining cap screws to transmission case to a torque of 170 ft.-lbs. on standard models or 150 ft.-lbs. on Hi-Crop models. Complete installation by reversing removal procedure. On Hi-Crop models, tighten outer gear housing (11 – Fig. 174) to shaft housing (8) to a torque of 275 ft.-lbs.

OVERHAUL

All Models Except Hi-Crop

179. To disassemble removed final drive unit, remove lock plate (24 – Fig. 172) and cap screw (23), then withdraw planet carrier (21) and associated parts.

Planet pinion shaft (17) is retained in carrier by snap ring (16). To remove, expand snap ring and, working around carrier, tap all three shafts out while snap ring is expanded. Withdraw parts being careful not to lose any of the loose bearing rollers (20) in each planet pinion. Examine shaft, bearing rollers and gear bore for wear, scoring or other damage and renew as indicated.

NOTE: Bearing rollers should be renewed as a set.

After planet carrier has been removed, axle shaft (1) can be removed

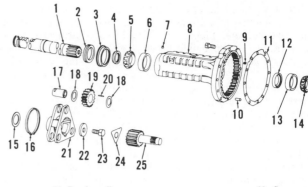

Fig. 172—Exploded view of planetary type final drive assembly used on all models except Hi-Crop.

1. Axle
2. Oil seal
3. Retainer
4. Spacer
5. Bearing cone
6. Bearing cup
7. Lube plug
8. Axle housing
9. Dowel
10. Dowel
11. Gasket
12. Oil seal
13. Bearing cup
14. Bearing cone
15. Thrust washer
16. Retaining ring
17. Pinion shaft
18. Bearing washer
19. Planet pinion
20. Bearing roller
21. Planet carrier
22. Washer
23. Cap screw
24. Lock plate
25. Sun gear

from inner bearing and housing by pressing on inner end of axle shaft. Remove bearing cone (5) and spacer (4) if they are damaged or worn. When assembling, heat spacer and cone to approximately 300 degrees F. and install on shaft making sure they are fully seated. Heat inner bearing cone (14) to 300 degrees F. Have planet carrier ready to install. Insert bearing and in-

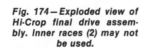

Fig. 173—Use a torque wrench calibrated in inch-pounds and special tool to measure rolling torque when adjusting rear axle bearings. Refer to text.

stall planet carrier, washer (22) and cap screw (23).

Adjustment of bearings (5, 6, 13 and 14) should be made before inner bearing cools. First measure rolling torque of axle, with end play in axle bearings, using special tool JDG-4 (3020, 4000 and 4020) or JDG-6 (4320, 4520 and 4620) or other suitable tool (Fig. 173). Then tighten special screw (23 – Fig. 172) until rolling torque is 20-70 in.-lbs. greater than initial measurement when adjusting new bearings, or 10-35 in.-lbs. greater when adjusting used bearings.

Hi-Crop Models

180. **OUTER HOUSING AND GEARS.** If only outer housing, gears, shafts, bearings or oil seals are being overhauled, complete final drive assembly will not need to be removed. Suitably support tractor and remove wheel and tire unit. Remove draft link or disconnect drawbar from gear housing.

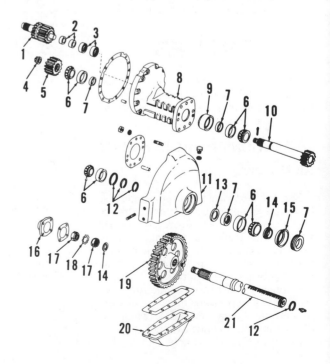

Fig. 174—Exploded view of Hi-Crop final drive assembly. Inner races (2) may not be used.

1. Differential gear
2. Inner races
3. Needle bearings
4. Nut
5. Drive shaft gear
6. Bearing & cup
7. Oil seal
8. Drive shaft housing
9. Quill
10. Drive shaft
11. Gear housing
12. Snap rings
13. Washer
14. Spacer
15. Oil seal cup
16. Bearing cover
17. Bearing nut
18. Lockwasher
19. Bull gear
20. Cover
21. Axle shaft

Remove the six stud nuts securing gear housing to shaft housing; remove the two retainer plugs and thread jack screws into retainer plug holes. Tighten jack screws evenly to force housings apart.

To disassemble removed gear housing, first remove bull gear cover and wheel axle inner bearing cover (16–Fig. 174), then remove inner bearing nuts (17) with a spanner wrench. Unseat snap ring on inner side of bull gear; install spacers between insides of gear housing to prevent damage to housing, then press or drive out rear axle shaft. Axle shaft is equipped with two oil seals, an inner seal which is pressed into housing and two-piece outer seal in housing and on shaft.

When assembling, heat bearing cones to a temperature of 300 degrees F. to facilitate installation. Install and tighten inner axle shaft nut (17) to torque of 200 ft.-lbs. to seat bearing races. Rotate axle several times, then loosen nut (17) and install dial indicator as shown in Fig. 175; retighten nut until housing wall is deflected 0.002 inch in area of shaft bearing. Install lockwasher and outer nut on shaft to lock setting. Tighten drop housing retaining stud nuts to 275 ft.-lbs. torque. Final drive gear housing uses SAE 90 multi-purpose gear lubricant.

181. SHAFT HOUSING AND INNER GEARS. To overhaul drive shaft housing assembly (8–Fig. 174), first remove cotter pin and slotted nut from inner end of drive shaft (10), then remove shaft using a brass drift. When reinstalling, heat bearing cones to a temperature of 300 degrees F. for easy installation. Tighten inner nut (4) to provide end play of 0.004 inch for shaft bearings.

BRAKES

The hydraulically actuated power disc brakes use the main hydraulic system as the power source. Discs are located on differential output shafts.

OPERATION AND ADJUSTMENT

All Models

182. Power to the wet type, single disc brakes is supplied by system hydraulic pump through foot operated control valves when engine is running. A nitrogen filled accumulator provides standby hydraulic pressure in sufficient volume to apply brakes several times after main hydraulic system ceases to operate. Control valve also contains master cylinders to permit manual operation when hydraulic pressure is not available.

Refer to Fig. 176. Parts 1 through 7

are duplicated for right and left brakes. In the first ¾-inch of pedal travel, operating rod (5) moves operating rod guide (6) which mechanically opens an equalizing valve pin and ball (2 and 8–Fig. 177). This insures equal pressure to both brakes when both pedals are depressed. Further movement of operating rod guide closes brake valve plunger (15) to close escape passage. Brake valve (13) is also unseated which allows pressure oil to fill the cavity under manual brake piston (16) and continue on to unseat check valve disc (24) which allows oil to reach brake cylinders. If pressure in valve should become too high, plunger (15) and brake valve (13) will move up as soon as pedal is released slightly, which allows escape passage to open and inlet passage to close, stabilizing pressure.

In case of pressure failure, manual braking is accomplished as follows: Pressure on operating rod and guide

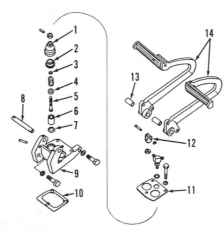

NOTE: Indicator MUST be mounted as shown.

Indicate Here

INNER BEARING LOCK NUT

Mount Indicator Base Here

Fig. 175—When installing rear axle bearings on Hi-Crop models, tighten inner bearing locknut until side of housing deflects 0.002 inch. Use dial indicator as shown to measure deflection.

closes plunger (15) and opens brake valve (13). Since oil is trapped under manual brake piston (16) and pressure line from pump is closed by a check ball and spring, pressure exerted by manual piston can open check valve seat (24) and apply brakes. As pedal is released, reservoir check ball is pulled off its seat, allowing oil from brake valve reservoir to be pulled into manual brake piston cavity, and continued pumping of the pedal can build pressure in the brake lines.

The only adjustment provided is at operating rod and yoke (5 and 12–Fig. 176). This adjustment is used to equalize brake pedal height. To adjust pedals, loosen jam nut and turn yoke in or out until pedals are equal in height.

For service on hydraulic pump, refer to paragraph 200. Refer to appropriate following paragraphs for service on brake valve, actuating cylinders, brake discs and accumulator.

BLEEDING

All Models

183. When brake system has been disconnected or disassembled, bleed system as follows:

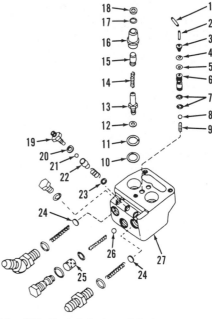

Fig. 177—Exploded view of brake valve with cover removed.

1. Pin	15. Plunger
2. Equalizing valve shaft	16. Manual brake piston
3. Sleeve	17. "O" ring
4. Packing washer	18. Washer
5. Guide	19. Connector
6. Guide	20. "O" ring
7. "O" ring	21. Ball
8. Ball	22. Inlet guide
9. Spring	23. Washer
10. "O" ring	24. Check valve disc
11. Back-up ring	25. Filter
12. "O" ring	26. Ball
13. Inlet valve	27. Valve housing
14. Spring	

Fig. 176—Exploded view of brake valve cover, pedals and associated parts. Parts 1 through 7 are duplicated for right and left brakes.

1. Boot	8. Pedal shaft
2. Retainer	9. Bracket
3. "O" ring	10. Gasket
4. Stop	11. Retainer
5. Rod	12. Yoke
6. Guide	13. Bushing (2)
7. "O" ring	14. Pedals

Pump brakes until accumulator is discharged. Start engine, loosen locknut on bleed screws, located on both sides of transmission housing, just above axle housings. Back out bleed screws two full turns, then tighten locknuts to prevent external oil leak. Fully depress brake pedal and hold down for 1-2 minutes. With pedal depressed, tighten bleed screws, then release pedal.

To test brake system, stop engine and depress each pedal once. A solid pedal feel should be obtained on first application. If pedal is spongy, repeat bleeding

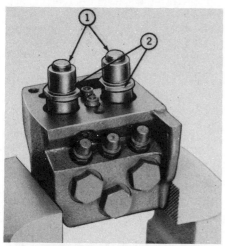

Fig. 178 – Manual brake pistons and brake valve plungers can be lifted out after cover is removed.

1. Brake valve plungers   2. Manual brake pistons

Fig. 179 – Remove guides and "O" rings from cover bore. Be sure guides are free in bores.

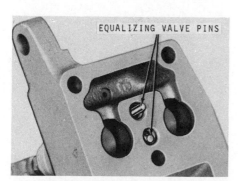

Fig. 180 – Equalizing valve pins can be withdrawn from underside of cover.

procedure and, if trouble is not corrected, overhaul system components.

BRAKE VALVE

All Models

184. To remove brake valve, first bleed down pressure in accumulator, then disconnect pressure and discharge lines at valve housing. Remove pedals, then unbolt and remove control valve.

Use Figs. 176 and 177 as a guide when disassembling brake valve. Manual brake pistons (1 – Fig. 178), brake valve plungers (2) and springs can be lifted out. Use a deep socket to remove inlet

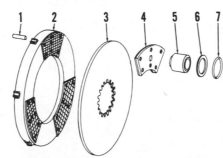

Fig. 181 – Exploded view of wet type hydraulic disc brake operating parts located on differential output shafts (final drive sun gears).

1. Dowel
2. Backing plate
3. Brake disc
4. Pressure pad
5. Piston
6. Back-up ring
7. "O" ring

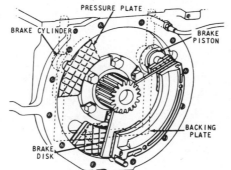

Fig. 182 – Schematic view of assembled brake unit.

valve nipples. Clean parts thoroughly and check against values which follow:
Brake Manual Piston
  Bore . . . . 0.9365-0.9375 in.
  Manual Piston O.D. . . . . . 0.933-0.935 in.
  Plunger Bore in Piston . . 0.561-0.563 in.
  Plunger O.D. . . . . 0.5595-0.5605 in.

Check for free movement of operating valve guide (Fig. 179) and equalizing valve pins (Fig. 180). Binding can cause equalizing valve to fail to close and both brakes will be applied if opposite pedal is pushed. Renew "O" rings whenever unit is disassembled.

Use Figs. 176, 177, 178 and 179 as a guide when reassembling. Tighten inlet valve nipples (13 – Fig. 177) to a torque of 40 ft.-lbs. When control valve has been installed, check for equal pedal height and bleed system as outlined in paragraph 183.

DISCS AND PADS

All Models

185. To remove brake discs or operating pistons, first remove final drive unit as outlined in paragraph 178. Remove output shaft, backing plate and brake disc. The three stationary pads are riveted to backing plate (2 – Fig. 181). The three actuating pads (4) are pressed on operating pistons (5) which can be withdrawn from transmission housing bores after brake disc is removed. Facings are available in sets of three and should only be renewed as a set.

Operating pistons (5) should have 0.0025-0.0065 inch diametral clearance in cylinder bores. Refer to Fig. 182 for a schematic assembled view of brake unit.

BRAKE ACCUMULATOR

All Models

186. **R&R AND OVERHAUL.** To remove brake accumulator, bleed fluid pressure from brake system. Disconnect pressure lines for accumulator connections, then remove retaining clip (20 – Fig. 183) and remove accumulator.

Fig. 183 – Exploded view of nitrogen filled brake accumulator.

1. Plug
2. "O" ring
3. Retaining ring
4. Gas end cap
5. Back-up ring
6. Cylinder
7. Packing
8. Washer
9. Valve
10. Spring
11. Spring guide
12. Packing (U-cup)
13. Piston
14. Retaining ring
15. Guide
16. Steel ball
17. Connector
18. Plug
19. Hydraulic cap
20. Retaining clip

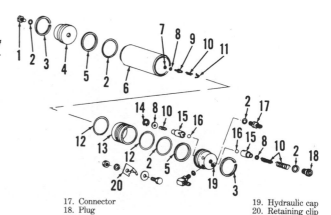

CAUTION: Bleed accumulator before attempting to disassemble. Gas side of accumulator piston is charged to 475-525 psi with NITROGEN gas.

Accumulator is discharged by removing protective plug (1) and depressing charging valve (9). With pressure removed, push cap (19) into cylinder, then unseat and remove snap ring (3). Remove gas end cap (4) by removing the other snap ring (3).

Check all parts for wear or damage, and assemble by reversing disassembly procedure. Recharge cylinder using approved charging equipment and DRY NITROGEN ONLY, to a pressure of 500 psi. Remove charging equipment and check by immersing charged accumulator in water. When it has been determined that there are no leaks, reinstall plug (1) and install accumulator by reversing removal procedure. Bleed brakes as outlined in paragraph 183. After engine has been run again and accumulator recharged, brakes should still have at least five power applications after engine is shut off.

POWER TAKE-OFF

OPERATION

All Models

187. The power take-off is driven by a hydraulically applied, independently controlled, flywheel operated, single disc clutch on Syncro-Range models. Power Shift models use a multiple disc clutch, which is part of the transmission. See paragraph 121 and 123 for clutch overhaul on Syncro-Range models and paragraph 161 for clutch pack overhaul on Power Shift models.

On all tractors the pto clutch is hydraulically engaged. The control valve on Power Shift models is contained in transmission pedal valve housing and service procedures are contained in paragraph 157. On Syncro-Range models, a separate valve is used; refer to paragraph 128 for service procedures.

R&R AND OVERHAUL

Syncro-Range Models

188. **MODELS 3020-4000-4020-4320.** To renew pto front bearing (24 – Fig. 184) or oil seal (25), first drain transmission and hydraulic system fluid. Remove front pto guard, then unbolt and withdraw quill (27). Bearing is retained in quill by a snap ring (23). Drive bearing rearward out of quill using two punch holes provided in quill. On some models, inner bearing race (22) must be removed from pto shaft. Install new oil seal with lip facing bearing.

To renew pto drive gears or transmission pto shaft, detach transmission from clutch housing as outlined in paragraph 125. After tractor is split, pto drive shaft (3 – Fig. 185), clutch shaft (2) and countershaft idler gear (1) can be re-

moved if renewal is required. Hold washers in contact with countershaft gear to retain loose roller bearings as gear is withdrawn. Remove snap ring (Fig. 186) and pto drive gear from transmission pto shaft. Remove cap screws, retaining plate and idler gear from front of transmission housing, being careful to catch loose needle rollers as gear is withdrawn. Use a suitable puller to remove idler gear inner bearing race (12 – Fig. 184). Heat new race to 300 degrees F. when reassembling.

To remove pto output shaft and reduction gears (dual speed pto), remove snap ring (1 – Fig. 187) and withdraw output stub shaft (2 or 5). Unbolt and remove rear bearing quill (7), then lift out loose 540 rpm drive gear (13). Compress snap ring (Fig. 188) and unseat from its groove; then tap transmission pto shaft (15 – Fig. 187) rearward and withdraw shaft. Remove idler shaft while removing reduction idler gear as shown in Fig. 189. Be careful not to lose needle rollers.

On single speed pto models, transmission pto shaft and rear bearing quill will be removed as a unit, after removing pto

Fig. 185 – Rear face of clutch housing showing pto drive gear (3), clutch shaft (2), countershaft idler gear (1) and pto brake shoe (B).

Fig. 186 – Front face of transmission housing on Syncro-Range 3020 tractor. Snap ring retains pto drive gear on pto shaft.

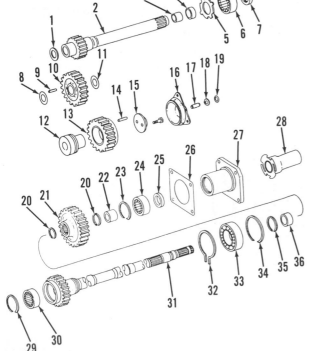

Fig. 184 – Exploded view of pto drive gears used on 4020 model equipped with Syncro-Range transmission. Other models are similar.

1. Thrust washer
2. Pto drive shaft
3. Bushing
4. Oil seal
5. Thrust washer (early models)
6. Bearing
7. Oil seal
8. Thrust washer
9. Bearing roller
10. Idler gear (countershaft)
11. Thrust washer
12. Inner bearing race
13. Idler gear (diff. shaft)
14. Bearing roller
15. Plate
16. Pto brake shoe
17. Brake piston
18. Back-up ring
19. "O" ring
20. Snap ring
21. Pto drive gear
22. Inner bearing race (late models)
23. Snap ring
24. Bearing
25. Oil seal
26. Gasket
27. Front bearing quill
28. Guard
29. Snap ring
30. Bearing
31. Pto shaft
32. Snap ring
33. Bearing
34. Snap ring
35. Snap ring
36. Bushing

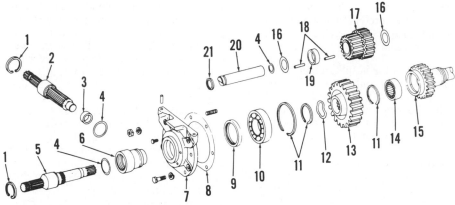

Fig. 187—Exploded view of reduction gears and associated parts used on models with dual speed pto.

1. Snap ring
2. 540 rpm shaft
3. Bearing race
4. "O" ring
5. 1000 rpm shaft
6. Pto pilot

7. Bearing quill
8. Gasket
9. Oil seal
10. Ball bearing
11. Snap ring

12. Spring washer
13. 540 rpm gear
14. Roller bearing
15. Pto shaft
16. Thrust washer

17. Countershaft gear
18. Roller bearing
19. Spacer
20. Idler shaft
21. Snap ring

drive gear from front of shaft and quill retaining cap screws.

Examine gears and bearings for wear, scoring or other damage and renew as required.

When assembling, use grease to hold bearing rollers and thrust washers in position. Reassemble by reversing disassembly procedure.

189. **MODELS 4520-4620.** To service pto shaft (34–Fig. 190) only, it is not necessary to separate tractor. To remove pto shaft, first remove front bearing quill (18). Remove snap ring (Fig. 191) and pull inner bearing race from shaft; two tapped holes are pro-

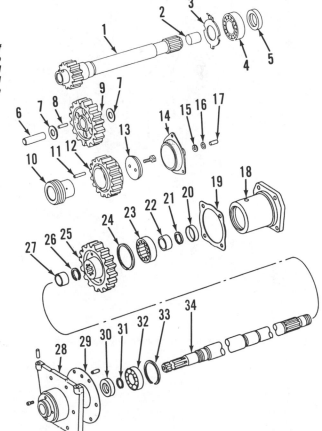

Fig. 191—Front face of transmission housing on Syncro-Range 4620 model showing pto drive gear and idler gear. Drive gear has two tapped holes for installation of a puller when removing bearing race.

vided in pto drive gear for installation of a puller. Puncture oil seal (30–Fig. 190) in rear quill; remove seal and snap ring (31), then unbolt and remove rear quill. Pto shaft can now be driven out forward.

To gain access to pto drive gears, tractor must be split between transmission and clutch housing as outlined in paragraph 125. After splitting tractor, trans-

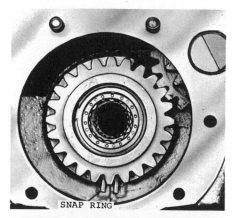

Fig. 188—To remove transmission pto shaft after removing front drive gear and output bearing quill, compress and unseat snap ring and tap shaft rearward out of housing.

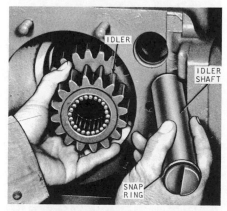

Fig. 189—When removing reduction idler gear and shaft, be careful not to lose bearing rollers.

Fig. 190—Exploded view of single speed (1000 rpm) pto drive gears and associated parts used on 4620 model equipped with Syncro-Range transmission.

1. Pto drive shaft
2. Bushing
3. Thrust washer
4. Bearing
5. Oil seal
6. Idler shaft
7. Thrust washer
8. Bearing roller
9. Upper idler gear
10. Inner bearing race
11. Bearing roller
12. Lower idler gear
13. Plate
14. Pto brake shoe
15. Back-up ring
16. "O" ring
17. Brake piston
18. Front bearing quill
19. Gasket
20. Plug
21. Snap ring
22. Inner bearing race
23. Bearing
24. Snap ring
25. Pto drive gear
26. Snap ring
27. Bushing
28. Rear bearing quill
29. Gasket
30. Oil seal
31. Snap ring
32. Bearing
33. Snap ring
34. Pto shaft

mission and pto clutch shafts can be removed and overhauled as outlined in paragraph 126. When removing idler gear, hold thrust washers in contact with gear to retain loose needle rollers as gear is withdrawn. Pto brake can also be renewed at this time. Remove cap screws, retaining plate and idler gear from front of transmission. Use a suitable puller to remove idler gear inner bearing race. Heat new race to 300 degrees F. when reassembling.

Examine gears and bearings for wear, scoring or other damage and renew as required.

When assembling, use grease to hold bearing rollers in position. Install oil seals with lip facing oil. Reassemble by reversing disassembly procedure.

Power Shift Models

190. **MODELS 3020-4000-4020-4520-4620.** To service pto drive gears, clutch housing must be separated from transmission as outlined in paragraph 155. If pto clutch is to be serviced, refer to paragraph 161 for clutch pack overhaul procedure.

After tractor is split, upper and lower idler gears (9 and 14–Fig. 192 or 194) can be removed. Be sure to catch loose roller bearings as gears are removed. If idler shafts (Fig. 193) are being renewed, use care to prevent damaging transmission case bores when removing shafts. To remove lower idler shaft, set screw must be removed (on Models 4520 and 4620 planetary pack must first be removed). Hole in end of lower idler shaft may be tapped for pulling after removing expansion plug. Be sure to reinstall ball (6–Fig. 192 or 194) in upper idler quill (5) which prevents idler shaft from turning in case.

On Models 3020, 4000 and 4020, pto drive gear (22–Fig. 192) is retained on pto shaft (35) with snap ring. On Models 4520 and 4620, inner bearing race (29–

Fig. 194) must be pulled to remove pto drive gear (22); two tapped holes are provided in drive gear for installation of puller. On single speed pto models, transmission pto shaft and rear bearing quill can be removed as a unit, after removing pto drive gear and quill retaining cap screws.

On dual speed models, remove reduction gear train and pto drive shaft as follows: Remove snap ring (1–Fig. 187) and stub shaft. Unbolt and remove rear bearing quill (7), then lift out loose 540 rpm drive gear (13). Compress snap ring (Fig. 188) and unseat from its groove; then tap transmission pto shaft rearward and withdraw shaft. Remove reduction idler shaft (Fig. 189) while removing idler gear. Be careful not to lose needle rollers.

Examine gears and bearings for wear, scoring or other damage and renew as required.

When assembling, use clean grease to hold bearing rollers and thrust washers in position. Reassemble by reversing disassembly procedure. Be sure "O" rings are in place on transmission case when reinstalling clutch housing.

LINKAGE ADJUSTMENT

Adjustment of pto clutch linkage on Syncro-Range models is outlined in paragraph 119.

Power Shift Models

191. To adjust operating linkage on Power Shift models, refer to Fig. 195 and move control lever overcenter to engaged position. Lengthen control rod from yoke until pto valve is bottomed in bore, then turn rod back into yoke one turn.

Move control lever to brake position;

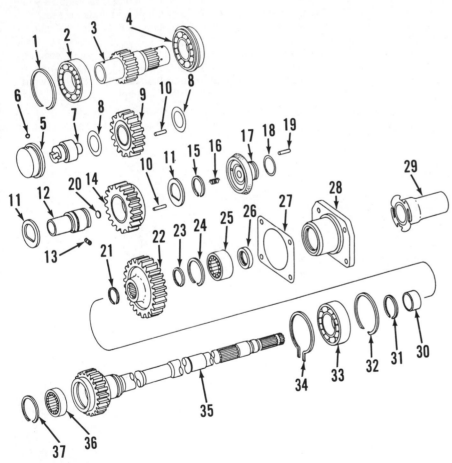

Fig. 192 — Exploded view of pto drive gears and related components used on Power Shift 4020 tractor.

1. Snap ring	11. Thrust washer	20. Expansion plug	29. Guard
2. Bearing	12. Lower idler shaft	21. Snap ring	30. Bushing
3. Pto drive shaft	13. Set screw	22. Pto drive gear	31. Snap ring
4. Bearing	14. Lower idler gear	23. Snap ring	32. Snap ring
5. Upper idler shaft quill	15. Snap ring	24. Snap ring	33. Bearing
6. Ball	16. Spring	25. Bearing	34. Snap ring
7. Upper idler shaft	17. Pto brake shoe	26. Oil seal	35. Pto shaft
8. Thrust washer	18. "O" ring	27. Gasket	36. Bearing
9. Upper idler gear	19. Dowel pin	28. Front bearing quill	37. Snap ring
10. Bearing roller			

Fig. 193 — When removing lower idler shaft, be sure to first remove set screw. On 4520 and 4620 models, planetary pack must be removed to remove set screw.

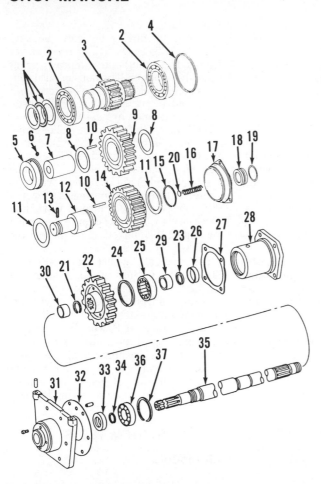

Fig. 194 — Exploded view of single speed (1000 rpm) pto drive gears and related parts used on Power Shift 4620 tractor.

1. Thrust bearing assy.
2. Bearing
3. Pto drive shaft
4. Snap ring
5. Upper idler shaft quill
6. Ball
7. Upper idler shaft
8. Thrust washer
9. Upper idler gear
10. Bearing roller
11. Thrust washer
12. Lower idler shaft
13. Set screw
14. Lower idler gear
15. Snap ring
16. Spring
17. Pto brake shoe
18. Brake piston
19. "O" ring
20. Expansion plug
21. Snap ring
22. Pto drive gear
23. Snap ring
24. Snap ring
25. Bearing
26. Plug
27. Gasket
28. Front bearing quill
29. Inner bearing race
30. Bushing
31. Rear bearing quill
32. Gasket
33. Oil seal
34. Snap ring
35. Pto shaft
36. Bearing
37. Snap ring

With engine speed at 1900 rpm and pto engaged, clutch pressure should be 140-160 psi on 3020, 4000 and 4020 models, or 165-185 psi on 4520 and 4620 models. Install gage in pto brake test port and check pressure with control lever in brake position and engine running at 1900 rpm. Brake pressure should be 140-160 psi on 3020, 4000 and 4020 models, or 165-185 psi on 4520 and 4620 models. Brake pressure should drop immediately to zero when control lever is moved to engaged position.

HYDRAULIC SYSTEM

The closed center hydraulic system provides standby pressure to all tractor hydraulic components, with a maximum available flow of 18 gpm on 3020 models, or 22 gpm on all other models without Power Front Wheel Drive. On models equipped with Power Front Wheel Drive, a larger capacity pump is used and available flow is 23 gpm on 3020 models, 27 gpm on 4020 models or 30 gpm on 4320, 4520 and 4620 models. A standby pressure of 2200-2300 psi is maintained when hydraulic functions are not in use. System operating pressure is 2000 psi.

adjust clip on operating rod to provide 1/8-inch clearance between signal spring lower hook and clip.

PRESSURE TEST

Power Shift Models

192. To check pto clutch pressure, install 0-300 psi pressure gage in test port marked "pto" shown in Fig. 196 or 197.

OPERATION

All Models

193. The main hydraulic system pump is mounted in front of radiator and coupled to engine crankshaft. This variable displacement, radial piston pump provides only fluid necessary to main-

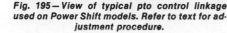

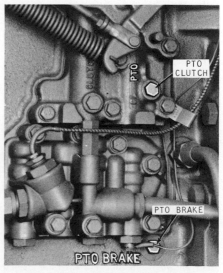

Fig. 195 — View of typical pto control linkage used on Power Shift models. Refer to text for adjustment procedure.

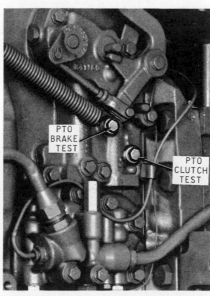

Fig. 196 — Location of pto pressure test ports on early models. Install 0-300 psi pressure gage to check pressure.

Fig. 197 — Location of pto pressure test ports on late model Syncro-Range tractors. Refer to text.

tain system pressure. When there are no demands on the system, pistons are held away from pump camshaft by fluid pressure and no flow is present. When pressure is lowered in the supply system by hydraulic demand or by leakage, stroke control valve in the pump meters fluid from camshaft reservoir, permitting pistons to operate and supply flow necessary to maintain system pressure.

The transmission pump provides pressure lubrication for transmission gears and shafts, and has a small priority valve in the pump manifold, which closes off lubricating pressure until clutch and main pump receive a preset pressure. On Power Shift models, pump also supplies operating fluid for transmission operation. On all models, excess fluid flow from transmission pump passes through the full flow system filter to inlet side of main hydraulic system pump. If no fluid is demanded by main pump, the fluid passes through the oil cooler then back to reservoir in transmission housing.

The oil cooler is mounted in front of tractor radiator, and on air conditioned models is an integral part of air conditioning condenser.

Return oil from different functions is routed through a second filter on Power Shift models.

RESERVOIR AND FILTERS

All Models

194. The hydraulic system reservoir is the transmission housing and the same fluid provides lubrication for transmission gears, differential and final drive units. The manufacturer recommends that only John Deere Hy-Gard transmission and hydraulic oil or its equivalent be used in the system. Approximate reservoir capacity in U.S. Gallons is as follows:

Power Shift Models
```
3020 ........................11
4000 ........................14
4020 ........................14
```

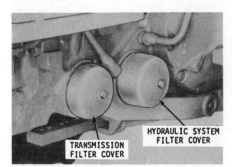

Fig. 198—Typical Power Shift dual filter arrangement showing hydraulic system filter behind transmission filter. Other models have only one filter.

```
4520 ........................16
4620 ........................16
```

Syncro-Range Models
```
3020 .........................8
4000 ........................10
4020 ........................10
4320 ........................14
4520 ........................18
4620 ........................18
```

Approximately 4½ gallons must be added to above capacities on all tractors equipped with Power Front Wheel Drive.

To check fluid level, stop tractor on level ground and check to make sure fluid level is in "SAFE" range on dipstick.

The oil filter element (or elements) may be renewed without draining fluid reservoir, by removing filter cover and extracting element. Filters are located on left side of transmission housing as shown in Fig. 198.

All filters are provided with a by-pass valve which opens to allow oil to flow when cold or with filter plugged. On

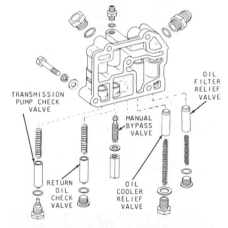

Fig. 199—Exploded view of oil filter relief valve assembly used on Syncro-Range tractors. Corresponding valves used on Power Shift models are shown in Fig. 145.

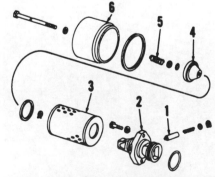

Fig. 200 — Rear hydraulic filter and by-pass valve used on Power Shift models.

```
1. By-pass valve            4. Retainer
2. Base housing             5. Spring
3. Filter element           6. Cover
```

Syncro-Range models, by-pass valve is located in oil filter relief valve housing (Fig. 199). To service filter relief valve, remove front plug and withdraw spring and valve plunger. The housing also contains an oil cooler relief valve, return oil check valve, transmission pump check valve and manual by-pass valve.

On Power Shift models, relief valve for front (transmission) filter is located in Power Shift Regulating Valve housing and is shown at (3–Fig. 145). The rear (hydraulic) filter by-pass valve is located in filter base housing as shown in Fig. 200. To renew or inspect valve, it is necessary to drain system and remove rear filter; then unbolt and remove filter housing.

The manual by-pass valve (2–Fig. 145) for Power Shift models, or Fig. 199 for other models, applies less restriction to return oil from single acting cylinder, but allows a greater quantity of oil to by-pass oil cooler on Power Shift models. Open the valve if necessary, when using a front loader or similar implement, but make sure valve is closed for normal tractor operation. By-pass valve can be turned in or out after removing hex cap nut.

SYSTEM TESTS

All Models

195. Efficient operation of tractor hydraulic units requires that each component operates properly. A logical procedure for testing system is therefore needed. The indicated tests include Transmission Pump Flow Test, System Pressure Tests and Leakage Tests, as outlined in the following three paragraphs. Unless indicated repair of hydraulic units is obvious because of breakage, these tests should be performed before attempting to repair the individual units.

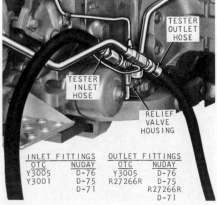

INLET FITTINGS		OUTLET FITTINGS	
OTC	NUDAY	OTC	NUDAY
Y3005	D-76	Y3005	D-76
Y3001	D-75	R27266R	D-75
	D-71		R27266R
			D-71

Fig. 201—Connect flow meter as shown for transmission pump test on Syncro-Range models.

196. TRANSMISSION PUMP FLOW TEST. A quick test of transmission pump operation can be performed by removing fluid filter (front filter on Power Shift models) and turning engine over with starter. A generous flow of fluid will be pumped into filter housing if pump is operating satisfactorily.

To more thoroughly test pump condition, connect a flow meter into main hydraulic pump supply line on left side of clutch housing (see Fig. 201 or 202). With unit at operating temperature, engine at 1900 rpm and flow unrestricted, output should be 7.5 gpm for 3020 Syncro-Range models, or 9 gpm for all other Syncro-Range models. On Power Shift models, output should be 10 gpm for 3020 models, or 12 gpm for all other models.

Slowly close test unit pressure valve while observing flow, which should remain relatively constant to 80 psi. At approximately 100 psi, relief valve should start to open and flow decrease; if it does not, overhaul Oil Cooler Relief Valve (Fig. 199) on Syncro-Range models or (15—Fig. 145) on Power Shift models.

197. MAIN PUMP PRESSURE AND FLOW. To check main pump pressure and flow, bleed off hydraulic pressure by

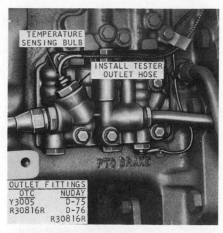

Fig. 204—Connect outlet from tester in temperature sensing bulb outlet.

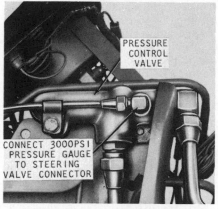

Fig. 205—To check function operating pressure, install 0-3000 psi pressure gage at priority valve outlet to steering.

opening right brake bleeder screw and holding pedal down a few moments. Disconnect pipe leading from main hydraulic pump to pressure (priority) control valve (Fig. 203); connect inlet side of flow meter to pipe and cap control valve inlet. Remove temperature sensing bulb and install outlet line from flow meter as shown in Fig. 204. Start and run engine at 1900 rpm.

NOTE: If flow meter is not inline type, install outlet hose in transmission filler tube. Run engine for short periods only while testing with this type meter.

Close test unit control valve until a pressure of 2000 psi is registered; then check fluid flow, which should be 18 gpm on 3020 models, or 22 gpm on all other models without Power Front Wheel Drive. On models equipped with Power Front Wheel Drive, flow should be 23 gpm on 3020 models, 27 gpm on 4000 and 4020 models or 30 gpm on 4320, 4520 and 4620 models.

Slowly close flow meter control valve until flow stops; pressure should be 2200-2300 psi. Adjust standby pressure

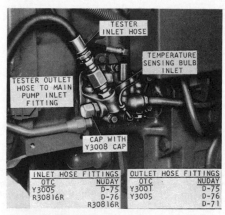

Fig. 202—Connect flow meter as shown to test transmission pump on Power Shift models.

INLET HOSE FITTINGS		OUTLET HOSE FITTINGS	
OTC	NUDAY	OTC	NUDAY
Y3005	D-75	Y3001	D-75
R30816R	D-76	Y3005	D-76
	R30816R		D-71

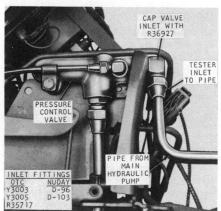

INLET FITTINGS	
OTC	NUDAY
Y3003	D-96
Y3005	D-103
R35717	

Fig. 203—To check main hydraulic pump flow, connect flow meter inlet to pipe from main pump. Cap control valve inlet.

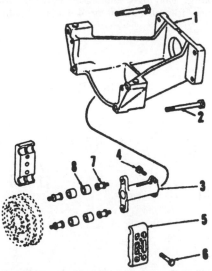

Fig. 206—Exploded view of main hydraulic pump support and drive coupling.

1. Support		5. Coupler half	
2. Cap screw		6. Cap screw	
3. Drive shaft		7. Coupler stud	
4. Cap screw		8. Bushing	

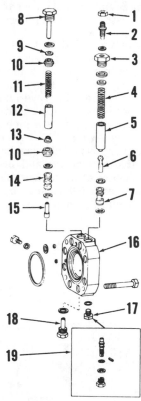

Fig. 207—Exploded view of stroke control valve housing showing component parts. Item 19 might not be used.

1. Nut		11. Spring
2. Adjusting screw		12. Filter
3. Bushing		13. Guide
4. Spring		14. Valve sleeve
5. Spring guide		15. Outlet valve
6. Stroke control valve		16. Housing
7. Valve sleeve		17. Plug
8. Plug		18. Plug
9. Shim washer		19. Lockout valve
10. Packing		

if necessary, by turning stroke control adjusting screw (2 – Fig. 207) clockwise to increase standby pressure or counterclockwise to decrease pressure.

198. PRESSURE CONTROL (PRIORITY) VALVE CHECK. Be sure hydraulic pressure is bled off. Remove flow meter hose and reconnect pipe to pressure control (priority) valve. Refer to Fig. 205 and install a 3000 psi pressure gage in priority valve outlet. Install a jumper hose in a breakaway coupler so hydraulic fluid can flow through coupler and back to reservoir, and turn metering valve arm to "Fast" position.

Start engine and operate at 800 rpm (slow idle), then move selective control valve lever to pressurize breakaway coupler. Gage reading will be minimum pressure which is maintained by priority valve to insure that steering and brakes will always be pressurized, even if other functions receive no pressure. If pressure is not within recommended range of 1650-1700 psi, refer to Fig. 13 and disassemble and inspect priority valve. BE SURE hydraulic pressure is bled off before removing any oil lines. Shims control pressure at which control valve starts to restrict oil flow to functions.

199. LEAKAGE TEST. To check for leakage at any of the system valves or components, move all valves to neutral and run engine for a few minutes at 1900 rpm. Check all of the hydraulic unit return pipes individually for heating. If temperature of any return pipe is appreciably higher than the rest of the lines, that valve is probably leaking. Disconnect that return line and measure flow for a period of one minute. Leakage should not exceed 0.8 ounce on rockshaft, 1.3 ounces on selective control

valve, or 5.0 ounces on pressure control valve. If leakage is excessive, overhaul indicated valve as outlined in appropriate sections of this manual.

MAIN HYDRAULIC PUMP

All Models

200. REMOVE AND REINSTALL. When external leaks or failure to build or maintain pressure indicates a faulty hydraulic pump, remove unit for service as follows: Relieve hydraulic pressure. The oil cooler, oil reservoir tank, and radiator must be drained and then removed. Disconnect main supply line, oil cooler line, pressure line and oil seal drain tube from pump. Remove drive shaft coupler halves and loosen clamp screws in pump half of coupler. Suitably support pump, remove screws securing pump to support and remove pump.

Install by reversing removal procedure. Tighten pump mounting bolts to a torque of 85 ft.-lbs. Other applicable torques are as follows: (Numbers in bolt description refer to identification symbols in Fig. 206).

Pump Drive Clamp
Screws (4) . . . . 35 ft.-lbs.
Drive Coupler Cap
Screws (6) . . . . 30 ft.-lbs.
Coupler Cap Screw
Locknuts . . . . 30 ft.-lbs.
Drive Coupler Studs (7) . . . . . . 35 ft.-lbs.
Use "Loctite" when installing coupler studs (7).

201. OVERHAUL. Before disassembling pump, check pump shaft end play using a dial indicator, and record measurement for convenience in reassembling. End play should be 0.001-0.003 inch for either (3.0 or 4.0 cubic inch)

pump. End play is adjusted by adding or removing shims (20 – Fig. 208) which are available in thickness of 0.006 and 0.010 inch. Bearing wear, or wrong number of adjusting shims can cause excessive end play.

NOTE: All pistons, springs, valves, seats and inlet valve plugs must be returned to bore from which removed. Identify all parts for proper reassembly.

To disassemble pump, remove four cap screws retaining stroke control valve housing (16 – Fig. 207) to front pump, and remove housing. Withdraw discharge valve plugs, guides, springs and valves (10 – Fig. 208). Remove all piston plugs (1), springs (2) and pistons (3), then carefully withdraw pump shaft (15) together with bearing cones, thrust washers (14), cam race (17) and loose needle rollers (16). Thrust washers (14) are 0.1235-0.1265 inch thick for 3.0 cubic inch pump (models without Power Front Wheel Drive); 0.0422-0.0452 inch thick for 4.0 cubic inch pump used on models with Power Front Wheel Drive.

Remove plugs retaining inlet valve assembies (5), and check valve lift using a dial indicator. Lift should be 0.060-0.080 inch. If lift exceeds 0.080 inch, spring retainers are probably worn and valves should be renewed. Also check for apparent excessive looseness of valve stem in guide. Do not remove inlet valve assembly unless renewal is indicated or discharge valve seat (9) must be renewed. To remove inlet valve, use a small pin punch and drive valve out, working through discharge valve seat (9). Discharge valve seat can be driven out after inlet valve is removed. Press in new discharge valve seat using JDH-39 seat driver or other suitable driver until seat is 0.870 inch from finished face of housing. Press in new inlet valve until end of guide is 0.535 inch from finished face of housing.

Check pistons and piston bores for scoring or pitting. Scored pistons or bores could cause pistons to stick. The manufacturer recommends that the eight piston springs (2) test within 1.5 lbs. of each other and within range of 34-40 lbs. at 1.62 inches for 3.0 cubic inch pump; 47-53 lbs. at 1.78 inches for 4.0 cubic inch pump. Springs should be renewed as a set. Be sure color code is the same on new springs.

Install seal (6) only deep enough to allow snap ring to enter groove, to avoid blocking relief valve hole in body.

Valves located in stroke control valve housing control pump output as follows: The closed hydraulic system has no discharge except through operating valves or components. Peak pressure is thus maintained for instant use. Pumping action is halted when line pressure reaches

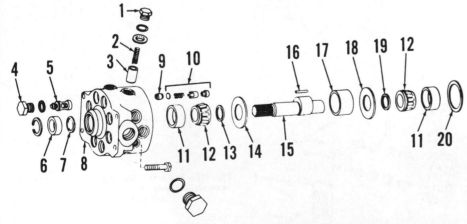

Fig. 208 – Exploded view of main hydraulic pump body, shaft and associated parts. Two additional spacers are used between bearing cones and thrust washers on 4.0 cubic inch pump.

1. Plug	6. Oil seal
2. Spring	7. Packing
3. Piston	8. Body
4. Plug	9. Seat
5. Inlet valve	10. Discharge valve

11. Bearing cup	16. Roller bearing
12. Bearing cone	17. Race
13. Spacer	18. Thrust washer
14. Thrust washer	19. Spacer
15. Pump shaft	20. Shim

a given point by pressurizing camshaft reservoir of pump housing, thereby holding pistons outward in their bores.

The cutoff point of pump is controlled by pressure of spring (4–Fig. 207) and can be adjusted by turning adjusting screw (2). When pressure reaches standby setting, valve (6) opens and meters required amount of fluid at reduced pressure into crankcase section of pump. Crankcase outlet valve (15) is held closed by hydraulic pressure and blocks outlet passage. When pressure drops as a result of system demands, crankcase outlet valve is opened by pressure of spring (11) and a temporary hydraulic balance on both ends of valve, dumping pressurized crankcase fluid and pumping action resumes. Stroke control valve spring (4) should test 125-155 lbs. pressure when compressed to 3.3 inches, and crankcase outlet spring (11) should test 45-55 lbs. at 2.2 inches.

Cut-off pressure is regulated by setting of adjusting screw (2) and adjustment procedure is given in paragraph 197. Cut-in pressure is determined by thickness of shim pack (9) and/or pressure of spring (11). A special tool (JDH-19) is available to determine shim pack; refer to Fig. 209 and proceed as follows:

Assemble outlet valve units (8 through 15–Fig. 207), using existing shim pack (9). Install special tool (JDH-19) in place of plug (18), using one ⅛-inch thick washer as shown in Fig. 209. If adjusting shim pack thickness is correct, scribe line on tool plunger should align

with edge of tool plug bore as shown; if it is not correct, remove top plug (8–Fig. 207) and add or remove shim washers (9) as required. Shims (9) are available in 0.030 inch thickness. If special tool is not available use shim washers of same thickness as those removed, then add shims to raise cut-in pressure, or remove shims to lower pressure.

When installing stroke control housing, add or remove shims (20–Fig. 208) as necessary to obtain specified pump shaft end play of 0.001-0.003 inch. Always use new "O" rings, packings and seals. Oil all parts liberally with clean hydraulic system oil. Tighten stroke control valve housing retaining cap screws to 85 ft.-lbs. and tighten piston cap plugs to 100 ft.-lbs. torque. Adjust standby pressure as outlined in paragraph 197 after tractor is reassembled.

PRESSURE CONTROL (PRIORITY) VALVE

All Models

202. The Pressure Control (Priority) valve is mounted on right side of steering column support. Refer to paragraph 14 in STEERING section for data on valve unit.

ROCKSHAFT HOUSING AND COMPONENTS

Models So Equipped

203. **REMOVE AND REINSTALL.** To remove rockshaft housing, first remove seat and operator's platform. Disconnect three-point lift links on tractors so equipped. Disconnect and remove hydraulic control rods, interfering wiring and hydraulic lines. Remove attaching bolts and lift housing from tractor using a hoist.

When installing unit, place load selector lever in "L" (load) position and make sure linkage roller is to rear of cam follower as housing is lowered. Do not bend draft linkage. Complete assembly by reversing removal procedure and tighten retaining cap screws to a torque of 85 ft.-lbs.

204. **CONTROL VALVE HOUSING OVERHAUL.** To remove control valve housing, disconnect operating rods, then unbolt and lift off cover.

NOTE: If tractor is equipped with dual or triple selective control valves, housings may be unbolted and laid aside before valve housing is removed. Be sure rockshaft valve cover remains with valve.

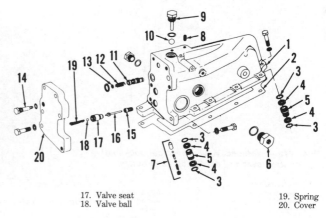

Fig. 210—Exploded view of rockshaft control valve housing, valves and associated parts.

1. Housing
2. Gasket
3. "O" ring
4. Back-up ring
5. Inlet pipe
6. Adjusting plug
7. Thermal relief valve
8. Pipe plug
9. Plug
10. Check ball
11. Flow control valve
12. Spring
13. Shim
14. Adjusting plug
15. Guide
16. Metering shaft
17. Valve seat
18. Valve ball
19. Spring
20. Cover

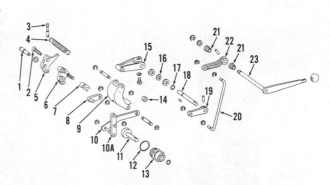

Fig. 211—Exploded view of typical rockshaft control linkage. Refer to Fig. 212 for rockshaft control lever and depth adjusting screw.

1. Valve cam shaft
2. Adjusting screw
3. Pin
4. Spring
5. Control valve adjusting cam
6. Control valve adjusting link
7. Adjusting link screw nut
8. Link
9. Load selector control link
10. Operating link
10A. Valve operating link
11. Operating shaft
12. "O" ring
13. Operating lever shaft quill
14. Cam roller
15. Load selector control arm
16. Spring washer
17. "O" ring
18. Shaft
19. Load selector operating arm
20. Operating rod
21. Bushing
22. Operating arm
23. Load selector lever

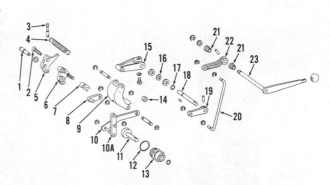

Fig. 209—Special adjusting tool (JDH19) can be used to determine stroke control valve setting.

1/8" OR .125"

SCRIBE LINE

JDH-19 CRANK-CASE OUTLET VALVE ADJUSTING TOOL

ADJUSTING WASHERS

Remove rear cover (20–Fig. 210) (or selective valve mounting cover on models so equipped), and withdraw operating valves (15 through 19) and flow control valve (11 through 13). Remove thermal relief valve assembly (7) and plug (6).

Remove load selector arm (19–Fig. 211), load selector control shaft (18), selector control arm (15) and control link (9) with roller (14). Remove valve camshaft (1), and avoid losing washer as camshaft is removed. Disconnect spring (4) from control valve adjusting cam (5), remove retainer ring holding linkage to control valve adjusting link (6) and remove linkage.

Check all linkage, springs, valves and housing for wear, scoring, or other damage and renew any parts in question. Valves can be lapped to seats, if necessary, using fine lapping compound. Inspect thermal relief valve assembly and spring, which should have 8 to 10 pounds pressure at a compressed length of 15/32-inch.

When reassembling, install thermal relief valve, spring and shims as required. Install valve operating link, link with pin and spring (Fig. 213). Assemble control valve adjusting cam, adjusting link, adjusting screw and screw nut. Install in housing and attach to link pin (Fig. 214). With control valve adjusting cam in position, install valve camshaft and washer and hook spring to adjusting cam as shown in Fig. 215. Install operating lever shaft quill (13–Fig. 211) and lower operating arm (1–Fig. 212) with operating shaft (11–Fig. 211). Be sure pin on operating link (10A) is in hole of operating shaft (11). Assemble load selector control link with roller and control arm in housing and attach valve operating link to operating link (Fig. 216). Install lower operating arm (19–Fig. 211) and shaft (18) and secure to control arm (15). Reassemble valve assembly in reverse order of disassembly, and make sure roller of load selector control link goes on backside of cam follower. Tighten retaining cap screws to torque of 85 ft.-lbs.

After housing is installed, turn adjusting screw (2–Fig. 211) counter-clockwise until bottomed, then clockwise ½-turn. This will adjust rockshaft only enough to allow operation. Refer to paragraph 205 for adjustment.

205. CONTROL VALVE ADJUSTMENT. While making any of the following adjustments, no load should be on hitch. Load selector lever should be in "D" (depth) position.

To adjust rockshaft valve clearance, disconnect linkage from rockshaft control lever arm and clamp Vise Grip to lever arm as shown in Fig. 217. Measure 10 inches from center of control lever shaft and place reference mark on Vise Grip.

CAUTION: Avoid lift arm and lift link while making this adjustment, as they may move rapidly.

Start engine and measure movement of reference mark required to change direction of rockshaft. Movement should be 3/16 to ⅜-inch. If distance is greater than ⅜-inch, insert screwdriver blade through opening in side of housing and turn adjusting screw clockwise to decrease free movement.

Raise rockshaft approximately half way and shut off engine. If rockshaft

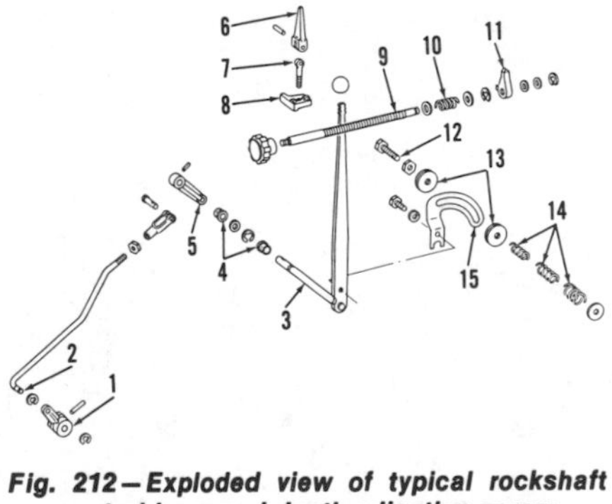

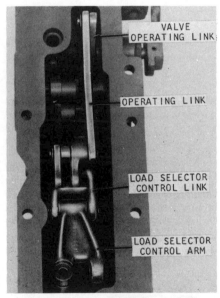

Fig. 216—Linkage assembled and installed in housing.

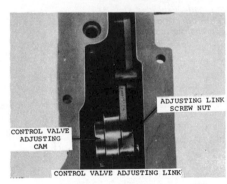

Fig. 212—Exploded view of typical rockshaft control lever and depth adjusting screw.

1. Operating arm
2. Operating rod
3. Control lever
4. Bushings
5. Upper operating arm
6. Cam
7. Special screw
8. Height stop
9. Adjusting screw
10. Spring
11. Lever stop
12. Friction screw
13. Washers
14. Springs
15. Friction plate

Fig. 214—Valve adjusting cam and adjusting link installed.

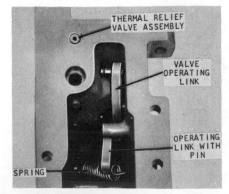

Fig. 213—Assembling linkage in valve housing.

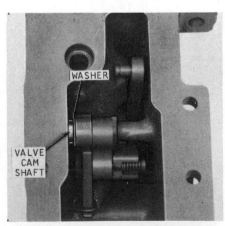

Fig. 215—Connect spring to adjusting cam.

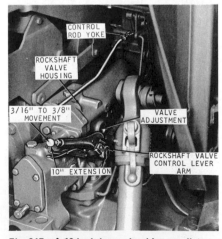

Fig. 217—A 10-inch lever should move distance shown to change direction of rockshaft. Valve adjustment is through plug hole in housing.

starts to drop, insufficient valve clearance exists or valves are leaking. Turning adjusting screw counter-clockwise increases clearance.

206. FRICTION DEVICE ADJUSTMENT. Rockshaft control lever friction adjusting screw is located on side on console (Fig. 218). With valve operating rod disconnected at lower operating arm, control lever should require a 7-8 lbs. pull to make lever move. Loosen jam nut and adjust friction screw to obtain desired resistance.

Negative Stop Screw Equipped Models

207. CONTROL LINKAGE ADJUSTMENT. Start engine and place selector lever in depth (D–Fig. 218) position. Using control valve lever arm (Fig. 217), completely raise rockshaft and hold arm in that position. Move rockshaft console lever to rear of slot. Align control rod with hole in control valve lever arm, then shorten rod one turn and attach to lever arm. When console lever is 1/16-inch from rear of slot, rockshaft should be completely raised.

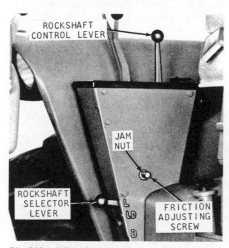

Fig. 218 — Adjust friction screw to obtain desired resistance on rockshaft control lever.

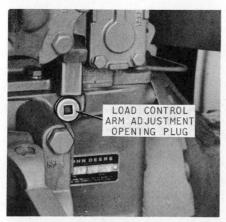

Fig. 219 — To adjust load control arm extension screw, remove plug from rear of housing.

208. LOAD CONTROL ARM EXTENSION ADJUSTMENT. Start engine and place selector lever in load (L) position. Remove adjustment opening plug (Fig. 219) and adjust load control arm (4–Fig. 224) so rockshaft is completely raised when rear edge of console lever is at "O" position. Rockshaft should begin to drop when rear edge of console lever is no farther than "1" position.

209. NEGATIVE STOP SCREW ADJUSTMENT. Turn negative stop screw (Fig. 219A) in draft link support until it is against transmission case, then turn screw counter-clockwise 1/8-turn and lock with jam nut.

Reverse Signal Lockout Equipped Models

210. CONTROL LINKAGE ADJUSTMENT. Remove load control adjustment opening plug (Fig. 219), and turn adjusting screw fully clockwise. Loosen operating shaft quill set screw (Fig. 220), and position operating shaft bushing timing mark as shown. Place load selector lever in depth (D–Fig. 218) position and start engine. Adjust operating rod so rockshaft is fully raised when front edge of console lever is 1/4-inch ahead of "O" position. Place selector

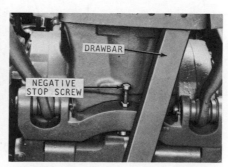

Fig. 219A — Negative stop screw (models so equipped) is located in draft link support.

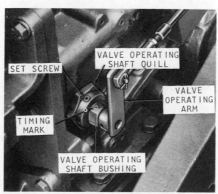

Fig. 220 — On models equipped with reverse signal locknut, position valve operating shaft bushing timing mark at 12 o'clock position when adjusting linkage.

lever in load (L) position and front edge of console lever at "O" position. With quill set screw loose, rotate operating shaft bushing clockwise until rockshaft begins to raise, then tighten set screw.

211. LOAD CONTROL ARM EXTENSION ADJUSTMENT. Place load selector lever in load (L–Fig. 218) position. Move console lever to lower, then raise, stopping control lever with front edge 1/4-inch ahead of "O". Remove adjustment opening plug (Fig. 219) and adjust extension screw (4–Fig. 224) counter-clockwise until rockshaft begins to raise. Rockshaft must lower when console lever front edge is no farther than "1½" position on console guide.

Models So Equipped

212. ROCKSHAFT HOUSING OVERHAUL. Rockshaft piston can be removed for inspection or renewal without removal of rockshaft housing. Re-

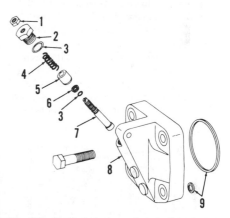

Fig. 221 — Exploded view of typical rockshaft cylinder cover and associated parts.

1. Nut
2. Bushing
3. "O" ring
4. Spring
5. Throttle valve
6. Back-up ring
7. Valve shaft
8. Cylinder cover
9. Packing

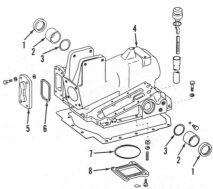

Fig. 222 — Exploded view of typical rockshaft housing and associated parts.

1. Oil seal
2. Bushing
3. "O" ring
4. Housing
5. Cover
6. Packing
7. "O" ring
8. Bottom cover

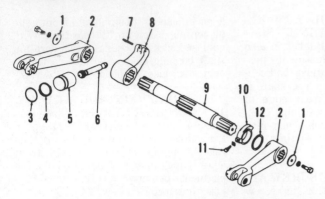

Fig. 223 — Exploded view of typical rockshaft, piston, lift arms and associated parts.

1. Washer
2. Lift arms
3. "O" ring
4. Back-up ring
5. Piston
6. Piston rod
7. Spring pin
8. Crank arm
9. Rockshaft
10. Servo cam
11. Set screw
12. Spacer washer

move cylinder end cover (8 – Fig. 221) and force piston out by pushing down on rockshaft arms with short, jerky motions.

NOTE: Be careful not to damage open end of cylinder with connecting rod or ram (crank) arm. Tighten cylinder cover retaining cap screws to a torque of 170 ft.-lbs. when reinstalling.

To disassemble rockshaft, remove housing and lower cover (8 – Fig. 222). Remove set screw (11 – Fig. 223) and lift arm (2), then slide rockshaft (9) out side of housing, removing crank arm (8) and servo cam (10) as shaft is withdrawn.

When installing rockshaft bushings, use suitable drivers and make sure oil holes are aligned. Add as many spacers (12) as will fit in housing, to eliminate

end play. Install servo cam (10) with ramp up as shown and make sure set screw (11) enters locating hole in rockshaft (9). Splines on crank arm (8) and shaft (9) are indexed for proper alignment during assembly. Tighten lift arm cap screws to 300 ft.-lbs.; strike arm with hammer and retighten.

LOAD CONTROL ARM AND SHAFT

Models So Equipped

213. **OPERATION.** When load selector lever is moved to load (L) position, operating depth of three-point hitch is

controlled by draft of attached implement acting in conjunction with position of control lever.

The amount of draft is transmitted by lower links to the drawbar support, then to the control valve by load control shaft and arm. The spring steel load control shaft is anchored in each side of transmission housing, and drawbar or draft link support frame affixed to outer ends of shaft. Positive or negative draft causes center of load control shaft to deflect a predetermined amount according to load encountered. The center arc of the flexing shaft moves the straddle mounted lower end of load control arm while upper (follower) end transmits required signal to the control valve.

214. **R&R AND OVERHAUL.** Load control arm (8 – Fig. 224) is removed in conjunction with removing differential assembly as outlined in paragraph 170.

To remove load control shaft or draft link support, drain transmission and hydraulic fluid and remove snap rings (Fig. 225 or 226) and retainers (6). Suitably support draft link support, then with a brass drift and working from right side of tractor, bump load control shaft to the left and out of housing. Bushings (7) in draft link support and (8) in trans-

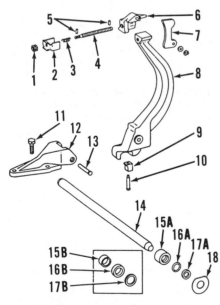

Fig. 224 — Typical load control arm and associated parts used on 4020 model. Other models are similar.

1. Nut
2. Lock plate
3. Spring
4. Adjusting screw
5. Spring pins
6. Extension
7. Follower
8. Arm
9. Follower block
10. Pin
11. Cap screw
12. Support
13. Pin
14. Load control shaft
15A. Bushing (late)
15B. Bushing (early)
16A. "O" ring (late)
16B. Oil seal (early)
17A. Sealing ring (late)
17B. Washer (early)
18. Washer

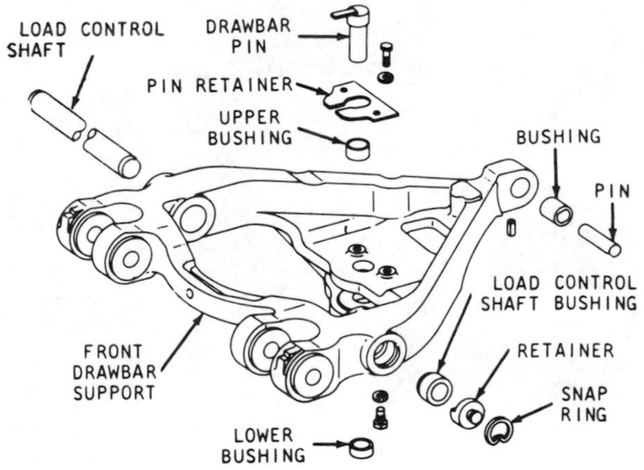

Fig. 225 — Exploded view of draft link and front drawbar support used on 3020, 4000, 4020 and 4320 models.

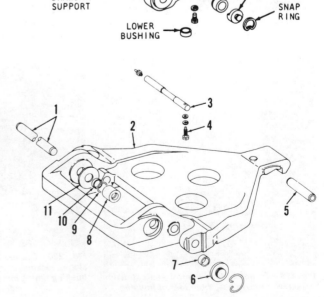

Fig. 226 — Draft link support and load control shaft used on 4520 and 4620 models.

1. Load control shaft
2. Support
3. Roller shaft
4. Locating screw
5. Roller
6. Retainer
7. Outer bushing
8. Inner bushing
9. "O" ring
10. Sealing ring
11. Washer (selective)

mission housing can be renewed at this time as can oil seals (10). Use a suitable puller when removing bushings and suitable driver when installing new bushings. Renew transmission housing bushings only if necessary.

Chill new bushings before installation. Inside diameter of bushing is tapered to provide a small bearing area for shaft (1); install bushings (8) in transmission housing with small I.D. to inside and bushings (7) in support with small I.D. to outside, away from transmission.

On some early models, a lip type oil seal is used in transmission housing. On later models an "O" ring (Fig. 227) and sealing ring are used inside control shaft bushing for oil control. When renewing sealing ring, heat ring in 160 degrees F. water to soften, then collapse as shown. Lubricate with hydraulic fluid, then install "O" ring and seal ring in bushing groove and reshape sealing ring to fit groove. On all models, carefully reinstall load control shaft using as many selective washers (11 – Fig. 226) as necessary to provide minimum clearance between transmission case and support. Install retainers and snap rings.

Adjust as outlined in paragraph 208 or 211.

SELECTIVE (REMOTE) CONTROL VALVES

Models So Equipped

215. **OPERATION.** Tractors are optionally equipped with one, two or three selective (remote) control valves for operation of remote cylinders.

As with all other units of the hydraulic system, pressure is always present at valves but no flow exists until valve is moved. Refer to Fig. 228 for an exploded view of valve mechanism. Each breakaway coupler is equipped with two return valves and two pressure valves, arranged so one of each is opened when

control lever is moved off center in either direction. Detent piston (16) is actuated by pressure differential across metering valve (26) and released by pressure equalization when flow stops at end of piston stroke. Flow control valve (25) maintains an even flow with varying pressure loads.

216. **OVERHAUL.** Refer to Fig. 228 for an exploded view of selective control valve. Clamp unit in a vise and unbolt and remove cover (23) and associated parts carefully as shown in Fig. 229. Identify parts as required for later assembly, then remove valves, springs and guides.

Rotate valve body in vise so rocker assembly is up. Rocker arm can be disassembled by driving out spring pin, and removing control arm and shaft. Remove screws holding cam and remove rocker (5 – Fig. 228). Notice how parts are assembled, to aid in reassembly. Inspect all bores, valves and valve seats (Fig. 230). Seats are non-renewable, but can be reconditioned by using NJD 150 Valve Seat Repair Kit (use exactly as directed).

Inspect valves and housing for wear, scoring or other damage. Be sure stop pin (24 – Fig. 228) exposed length from cover (23) is 15/16-inch. A worn pin can cause jerky operation. Check valve springs against the following values: Pressure valve springs should compress to 1.25 inches at 36-44 lbs. pressure. Return valve springs should compress to 1.25 inches at 19-23 lbs. pressure. Flow control valve spring should compress to

2.15 inches at 40-50 lbs. pressure.

Assemble by reversing disassembly procedure. If actuating cam was disassembled, reassemble as follows: On early models (Fig. 231), install cams with tapered end toward detent piston bore side of valve housing. Install cam with

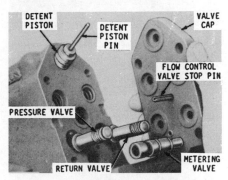

Fig. 229 – Clamp housing in a vise to remove valve cover.

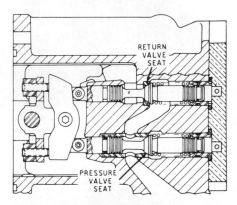

Fig. 230 – Valve seats can be reconditioned using special tool NJD150.

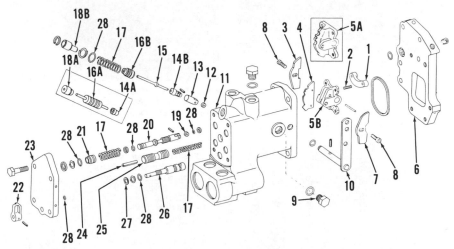

Fig. 228 – Exploded view of selective (remote) control valve showing component parts.

1. Keeper (2)	9. Plug	15. Detent pin	20. Poppet valve (4)
2. Adjusting screw (4)	10. Rocker shaft	16A. Detent piston (early)	21. Valve guide (4)
3. Regular cam	11. Housing	16B. Detent piston (late)	22. Lever
4. Detent cam	12. Detent roller	17. Spring	23. Cover
5A. Rocker (early)	13. Detent follower	18A. Outer detent guide (early)	24. Stop pin
5B. Rocker (late)	14A. Inner detent guide (early)	18B. Outer detent guide (late)	25. Flow control valve
6. Cover	14B. Inner detent guide (late)	19. Cam roller (4)	26. Metering valve
7. Float cam			27. Thrust washer
8. Cam clamp screws			28. "O" ring

Fig. 227 – When renewing load control shaft sealing ring, soften ring in 160°F. water and collapse ring as shown to ease installation.

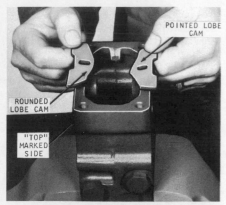

Fig. 231—On early models, install cams as shown. Cam with pointed lobe is installed on numbered side of housing.

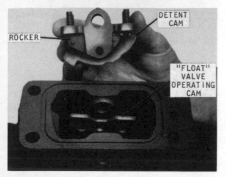

Fig. 232—On late models, float cam is shorter than regular cam and is installed on numbered side of housing.

Fig. 233—Front view of early model housing showing rocker assembly, adjustment screws and associated parts.

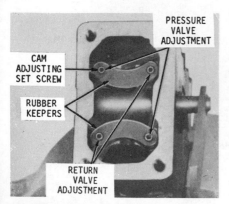

Fig. 234—Front view of late model housing showing rocker assembly, adjustment screws and rubber keepers.

pointed lobe on numbered side of housing. On late models (Fig. 232), float cam is shorter than regular cam and is installed on numbered side of housing. Install cams with pointed end toward detent piston bore. On all models, install detent cam (4-Fig. 228) in rocker (5) with two pins, then install rocker assembly and rocker shaft (10).

Adjusting valve requires use of special adjusting cover (JDH-15C) and a dial indicator. Remove the two adjusting plugs (9-Fig. 228) and loosen the two cam locking screws (8). Back out the four adjusting screws (2) at least two turns.

Back out all adjusting screws on special adjusting cover, then install cover in place of valve cover (23) and valve guides (21). Be sure cam is riding properly on valve rollers. Carefully FINGER TIGHTEN four adjusting screws contacting operating valves until valves are seated; then while holding operating lever in center position, FINGER TIGHTEN detent locking screw until detent roller is seated in neutral detent on detent cam. With operating lever in neutral position, refer to Fig. 233 or 234 and turn in two diagonally opposite pressure valve cam adjusting screws until screws, cams and follower rollers are in contact, then back

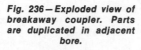

Fig. 235—Use a dial indicator to measure rocker arm movement when adjusting cams. Adjusting cover (JDH-15C) is installed on bottom of valve.

out screws ¼-turn. Turn in diagonally opposite return valve cam adjusting screws until they contact cams, then back out screws ⅛-turn and snug jam nuts (Fig. 233) or install rubber keepers (Fig. 234). Move cams into contact with adjusting screws and tighten clamp screws (8-Fig. 228).

To double-check adjustment, install a dial indicator 3 inches from center of shaft on operating arm as shown in Fig. 235. Zero dial indicator while locked in neutral detent, then back out detent locking screw on adjusting cover. Back out the two adjusting cover screws which contact operating valves on lever side and measure rocker movement, which should be 0.024-0.048 inch toward return valve or 0.084-0.108 inch toward pressure valve as shown. Valves contacting cam opposite lever side are being checked. Tighten the two adjusting cover screws on lever side and loosen other two screws, then check adjustment of valves on lever side. Readjust as necessary for correct rocker movement. This procedure will allow return valves to open before pressure valves, when selective control valve is used.

217. BREAKAWAY COUPLER. Drive a punch into expansion plugs (17-Fig. 236) and pry out of housing. Remove retainer rings and springs. Operating levers can then be removed. Drive receptacle assembly from housing. Check steel balls, springs, and all parts for wear and replace as necessary. Renew "O" rings and back-up washers. Reassemble in reverse order of disassembly.

POWER WEIGHT TRANSFER VALVE

Models So Equipped

218. OPERATION. The power weight transfer hitch uses a special coupler, a power weight transfer control valve, pressure gage, a special rockshaft piston cover, a double acting remote cyl-

Fig. 236—Exploded view of breakaway coupler. Parts are duplicated in adjacent bore.

1. Snap ring
2. Snap ring
3. Back-up ring
4. "O" ring
5. Ball
6. Back-up ring
7. Receptacle
8. Ball
9. Spring
10. Plug
11. Snap ring
12. Sleeve
13. Operating lever
14. Cam
15. Lockwasher
16. Washer
17. Expansion plug

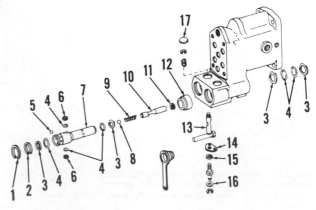

inder (used as a retracting cylinder only) and a transfer link. The remote cylinder takes the place of center link on a three point hitch, and becomes a telescoping center link.

When using power weight transfer hitch, only a drawn implement should be used and operation is in rockshaft "L" position. Excessive load causes load control arm to direct pressure oil through control valve to remote cylinder. As cylinder retracts, it tilts coupler forward at the top, which pulls on transfer link to

implement being used, causing weight of implement to bear down on draft links. This has the effect of using the rear wheels as a pivot point to unload front wheels and give added traction to the rear when needed, without having to add ballast to rear wheels. The console mounted pressure gage shows when weight is being transferred.

Refer to Fig. 237 for an exploded view of valve unit and to Fig. 238 for hose routing.

The valve is primarily a switch valve which diverts fluid from rockshaft cylin-

der to control cylinder, and rockshaft lever is used for control lever.

219. **OVERHAUL.** Refer to Fig. 237 for an exploded view of valve unit. Diverting and relief valves can be removed after removing port plugs (9 and 14). One seat for diverting valve (7) is on upper surface of plug (9); the other seating surface is in bore of body (6). Diverting valve moves upward by pressure of spring (8) and inward flow of oil when knob (1) is backed out for rockshaft operation; diverter valve seals against

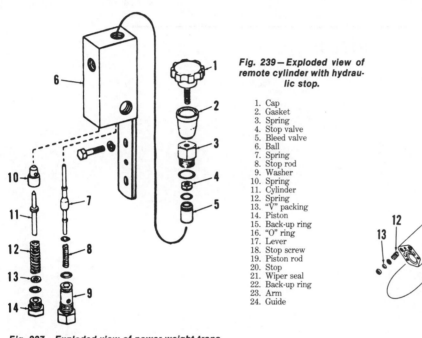

Fig. 237 — Exploded view of power weight transfer valve.

1. Valve screw	8. Spring
2. Boot	9. Plug
3. Bushing	10. Relief valve seat
4. Nut	11. Relief valve
5. Shaft	12. Spring
6. Housing	13. Shim
7. Diverting valve	14. Plug

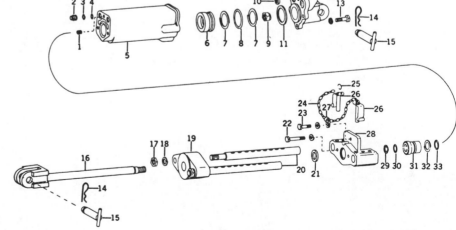

Fig. 239 — Exploded view of remote cylinder with hydraulic stop.

1. Cap
2. Gasket
3. Spring
4. Stop valve
5. Bleed valve
6. Ball
7. Spring
8. Stop rod
9. Washer
10. Spring
11. Cylinder
12. Spring
13. "V" packing
14. Piston
15. Back-up ring
16. "O" ring
17. Lever
18. Stop screw
19. Piston rod
20. Stop
21. Wiper seal
22. Back-up ring
23. Arm
24. Guide

Fig. 240 — Exploded view of remote cylinder with mechanical stop. Piston (6) may be equipped with "O" ring on inside also.

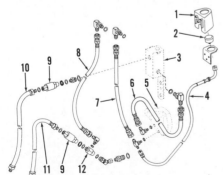

Fig. 238 — Hose routing diagram for power weight transfer valve. Hoses (7 and 8) connect to left side of control valve (3).

1. Cover	8. L.R. port hose
2. Pressure gage	9. Coupler
3. Control valve	10. Hose to head end of cylinder
4. Gage hose	11. Hose to rod end of cylinder
5. Rockshaft piston cover	12. Special connecter
6. Right port hose	
7. L.F. port hose	

1. Plug	10. Gasket (2)	18. Lockwasher	26. Stop pin
2. Plug	11. Gasket	19. Piston rod stop	27. Clamp
3. Back-up ring	12. Cap	20. Stop rod (2)	28. Stop rod guide
4. "O" ring	13. Cap screw	21. Oil seal	29. Back-up ring
5. Cylinder	14. Lock pin	22. Cap screw	30. "O" ring
6. Piston	15. Attaching pin	23. Cap screw	31. Piston rod guide
7. Back-up ring (2)	16. Piston rod	24. Chain	32. Back-up ring
8. "O" ring	17. Jam nut (2)	25. Link	33. "O" ring
9. Nut			

upper seat to close off passage to remote cylinder. Turning control knob (1) clockwise mechanically moves diverter valve into contact with seat on plug (9), closing return passage to rockshaft cylinder and opening passage to remote cylinder. Relief valve spring (12) should test 180-220 lbs. when compressed to a height of 1⅝ inches. Shims (13) may be added if necessary to increase release pressure of relief valve. Renew any parts which are worn, broken or damaged.

REMOTE CYLINDER
All Models

220. Refer to Fig. 239 for exploded view of double acting, hydraulic stop remote cylinder. To disassemble, remove end cap (1), stop rod spring (3) and valves (4 and 5), using care not to lose ball (6), if cylinder is equipped with override provision. Fully retract cylinder and remove nut from piston end of piston rod (19). To remove stop rod and springs, drive groove pin from stop rod

arm (23).

Install new wiper seal (21) in guide (24) with lip to outside, assemble rod end of cylinder, rod packing and piston rings, and have piston fully inserted in cylinder before installing rod nut. Tighten nut securely. Be sure piston rod stop (20) is located so stop lever (17) is opposite stop rod arm (23). Tighten cap screws securing piston rod guide (24) to a torque of 35 ft.-lbs. and cap screws retaining piston cap (1) to 85 ft.-lbs.

Fig. 240 shows parts identification on cylinders with mechanical stop.

NOTES

JOHN DEERE

Series ■ 6030

Previously contained in I & T Shop Service Manual No. JD-38

SHOP MANUAL

JOHN DEERE

SERIES 6030
Tractor serial number located on rear of transmission case. Engine serial number located on front right side of engine block.

INDEX (By Starting Paragraph)

CONDENSED SERVICE DATA

GENERAL
Engine Make . . . . Own
No. of Cylinders 6
Bore, Inches 4¾
Stroke, Inches 5
Displacement, Cu. In. 531
Compression Ratio 15.4:1
Induction . . . . *NA. & Turbocharged
Cylinder sleeves . . . . Wet
Forward Speeds 8
TUNE-UP
Firing Order 1-5-3-6-2-4
Compression Pressure
  Cranking Speed 325-395
Valve Tappet Gap
  Intake 0.018
  Exhaust 0.028
Injection Pump Timing . . . . 27° BTDC
Injector Opening Pressure . . . . 3100 psi
Engine Rated Speeds
  Slow Idle . . . . 800 rpm
  Working Range . . . . 1500-2100 rpm
  Maximum Transport Speed . . . . . . 2300 rpm

TUNE-UP Cont.
Horsepower @ 2100 rpm
  PTO *(N.A.) 145
  PTO (Turbo) 175
**SIZES—CAPACITIES—
CLEARANCES**
Cooling System (Quarts) *N.A.33-Turbo. 40
Crankcase Oil (Quarts with filter) . . . *N.A.20-Turbo. 26
Transmission & Hydraulic System
  (Gallons) (Refill) 16
Fuel Tank (Gallons) *N.A.68-Turbo. 73
Crankshaft Journal Diameter 3.748-3.749
Crankpin Diameter 3.498-3.499
Piston Pin Diameter 1.8739-1.8745
Crankshaft End Play 0.004-0.010
Main Bearing Clearance 0.0019-0.0049
Connecting Rod Bearing Clearance . . 0.0024-0.0054
TIGHTENING TORQUES—Ft.-Lbs.
Cylinder Head 175-185
Main Bearings 205-215
Connecting Rods 155-165

*Naturally aspirated

Printed in U.S.A.

FRONT SYSTEM

All models are equipped with either a wide fixed axle, or a wide adjustable axle. Two tie rods are attached to a bell crank mounted on a center axle unit. The bell crank is actuated by hydrostatic steering and two single acting hydraulic cylinders.

AXLE AND SUPPORT

Fixed Axle Models

1. HOUSING AND PIVOT BRACKET. The front axle attaches to tractor frame by pivot bracket (1-Fig. 1). Pivot pin bushing (5) installed I.D. should be 2.004-2.018 inches, with 2.011 nominal. Tighten pivot bracket to frame cap screws to 300 ft. lbs. torque, and pivot pin to housing cap screws to 85 ft. lbs.

2. SPINDLES AND BUSHINGS. Knuckle pin bushings (8-Fig. 1) should have an installed I.D. of 1.237-1.241 inches. The bushing hole must line up with grease fitting hole. Thrust washers (10) control the knuckle end play, which should be 0.005-0.045 inch.

Adjustable Axle Models.

3. HOUSING AND PIVOT BRACKET. The pivot pin bushings (5-Fig. 2) should have an installed I.D. of 2.004-2.018 inches, with 2.011 in. nominal. Tighten pivot bracket to frame cap screws to 300 ft. lbs. torque, and pivot pin to housing cap screws to 85 ft. lbs.

4. SPINDLE AND BUSHINGS. The O.D. of knee bushing (7-Fig. 2) should be 2.189-2.190 inches. Check bore I.D. before installing bushing to make sure that it is not worn. Bore should measure 2.185 inches. Align grease fitting hole as bushing is installed, and installed I.D. of bushing should be 2.001-2.004 inches. End play of knuckle (10) is adjusted by thrust washers (9) and should be 0.005-0.045 inch. Spring pins prevent thrust washers from turning.

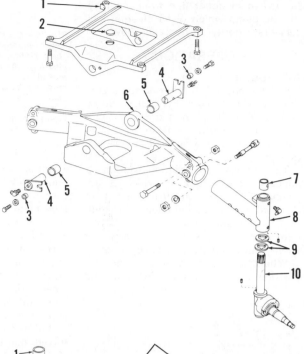

Fig. 2—Exploded view of adjustable front axle and pivot bracket assembly.

1. Pivot bracket
2. Expansion plug
3. Hollow dowel
4. Pivot pin
5. Bushing
6. Axle housing
7. Knee bushing
8. Knee
9. Thrust washers
10. Knuckle

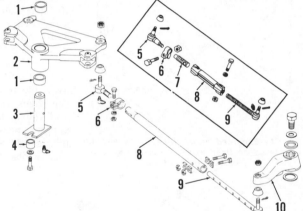

Fig. 3—Exploded view of bellcrank and tie rods used on adjustable axles. Inset shows parts used on tractors to serial no. 034597.

1. Bushing
2. Bellcrank
3. Pivot pin
4. Hollow dowel
5. Inner tie rod end
6. Clamp
7. Inner tie rod
8. Tie rod
9. Outer tie rod end
10. Steering arm

TIE RODS AND TOE IN

All Models

5. Tractors with adjustable axles were equipped with threaded tie rods and ends (inset Fig. 3) up to serial number 034597. After that serial number, outer tie rod ends were provided with holes and cap screws for easier adjustment. Toe in (all models) should be 1/8-3/8 inch measured at wheel center height, and both tie rods must be equal length so tractor will steer as far to the right as to the left. Tighten clamps in the downward position to 35 ft. lbs. torque.

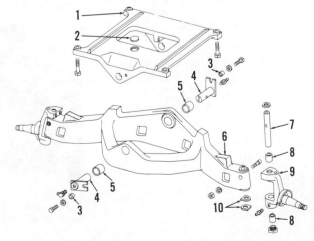

Fig. 1—Exploded view of fixed front axle and pivot bracket assembly.

1. Pivot bracket
2. Expansion plug
3. Hollow dowel
4. Pivot pin
5. Bushing
6. Axle housing
7. Knuckle pin
8. Bushing
9. Knuckle
10. Thrust washers

POWER STEERING SYSTEM

All models are equipped with a full power steering system. No mechanical linkage exists between steering wheel and steering cylinders; however, steering can be manually accomplished by hydraulic pressure when tractor hydraulic unit is inoperative. Power is supplied by the same hydraulic pump which powers the lift and brake systems. A pressure control (priority) valve is located in outlet line from main hydraulic pump which gives steering system first priority on hydraulic flow.

OPERATION

All Models

6. The power steering system consists of the tractor hydraulic supply system described in paragraph 116, plus the steering control unit and steering cylinders described in this section.

The control unit (Fig. 5) contains a double acting piston (3) which is approximately equal in displacement to the two operating cylinders. In addition, it contains two pressure valves, two return valves and an unloading valve which actuate the power assist.

When the control unit is in neutral position, there is no fluid flow but fluid

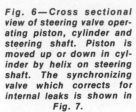

Fig. 6—Cross sectional view of steering valve operating piston, cylinder and steering shaft. Piston is moved up or down in cylinder by helix on steering shaft. The synchronizing valve which corrects for internal leaks is shown in Fig. 7.

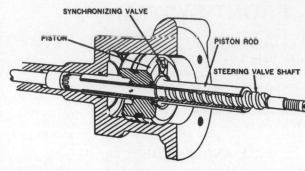

at pump pressure is available at inlet line (8). When the steering wheel is turned for a right or left turn, first movement of steering wheel reacts on the operating collar (5) and one operating lever (6 or 7) to open one pressure and one return valve leading to either end of metering cylinder (2). With continued turning of steering wheel, the pressurized fluid entering one side of metering cylinder moves actuating piston (3) toward opposite end of cylinder and the trapped fluid is forced out line to operating cylinder. Return fluid from the opposite operating cylinder passes through the return metering valve and unloading valve (9) back to the reservoir.

When the pressure required for steering effort exceeds the pressure available from hydraulic system pump, the check valve in inlet line (8) closes and the metering piston (3) manually supplies pressure to the operating cylinders and return fluid flows to back side of piston (3).

Refer also to Figs. 6 and 7 for operating principles regarding the synchronizing valve and associated parts.

BLEEDING

All Models

7. To bleed the steering system, first remove the cowling and attach a small transparent hose from bleed screw (Fig. 8) on left side of steering valve housing. Bleed hose should be long

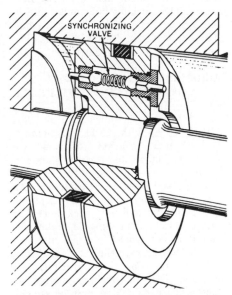

Fig. 7—Steering valve piston must be synchronized with steering cylinders for full turning action. Synchronization is automatically accomplished. When control valve piston reaches end of its stroke, the extended rod unseats the ball check valve, allowing pressurized fluid to flow through piston until cylinders complete their travel.

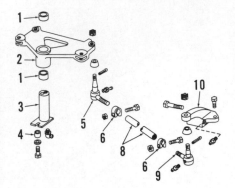

Fig. 4—Exploded view of bellcrank and tie rods used on fixed axles. Refer to Fig. 3 for parts identification.

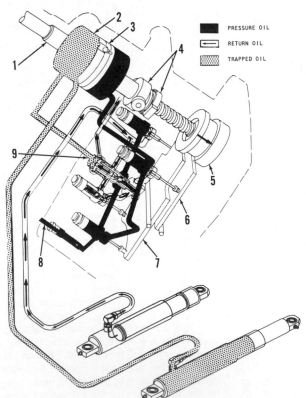

PRESSURE OIL

RETURN OIL

TRAPPED OIL

Fig. 5—Schematic view of power steering gear showing operating parts. Refer to paragraph 6 for operation.

1. Steering shaft
2. Slave cylinder
3. Piston
4. Operating nut
5. Operating collar
6. Operating lever
7. Operating lever
8. Pressure line
9. Unloading valve

Fig. 8—Cowling removed to show steering valve bleed screw. Refer to paragraph 7 for procedure.

enough to reach transmission filler opening or alternately, to a clean container.

Have tractor sitting on front wheels. Start engine and run at slow idle speed. Turn wheels to full left turn position then open bleed valve. With bleed valve open, **slowly** turn steering wheel to full right without moving front wheels. Close bleed valve and allow wheels to turn to full right. Turn wheels to full left with bleed valve closed; and repeat the operation as necessary until air-free fluid flows from bleed hose.

PRESSURE CONTROL (PRIORITY) VALVE

All Models

8. OPERATION. The Pressure Control (Priority) valve cuts off hydraulic flow to main hydraulic system whenever system pressure drops below 1600-1700 psi, thus, giving priority to steering and brake units on all tractors. Refer to Fig. 9 for a sectional view of valve and to Fig. 10 for an exploded view.

9. R&R AND OVERHAUL. To remove the priority valve, first remove cowl. Disconnect inlet, bleed and outlet lines and unbolt and remove valve assembly.

Examine valve and housing bore for scoring or other damage and housing for cracks. Spring (5-Fig. 10) should have a free length of approximately 4-5/8 inches and test 45-55 lbs. when compressed to a height of 3½ inches. Shims (4) may be added or removed to adjust valve operating pressure. Maximum leakage through bleed line should not exceed 100 cc per minute at standby pressure of 2250 psi.

STEERING CONTROL UNIT

All Models

10. REMOVE AND REINSTALL. To remove the steering control unit, first remove steering wheel using a suitable puller; then remove cowling, hand

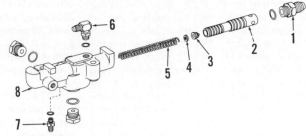

Fig. 10—Exploded view of steering priority valve showing component parts.

1. Inlet
2. Valve spool
3. Orifice
4. Shims
5. Spring
6. Steering outlet
7. Systems outlet
8. Body

throttle, hub, and instrument panel. Remove selective control valve levers, control support, speed control rod, tube and arm.

Disconnect steering fluid lines, being sure to cap all exposed connections, then unbolt and remove the complete steering unit.

When installing, bleed steering system as outlined in paragraph 7 and adjust throttle linkage as in paragraph 65. Tighten steering wheel nut to a torque of 50 ft.-lbs. Tighten 3/8" cap screws to 35 ft. lbs. and ½" cap screws to 85 ft. lbs.

11. OVERHAUL. To disassemble the removed steering control units, first remove lower cover (1—Fig. 11), cotter pin and nut (7); then unbolt and re-

move control valve housing (9) and associated parts from cylinder housing (22) and cover (17). Remove operating collar (8) with housing (9), and be careful not to drop steering check valve stop and spring (Fig. 12), as housing is removed.

Remove operating shaft nut (13—Fig. 11) and slide spring (14) and piston rod collar (15) from shaft. Turn steering wheel shaft to the left to force off cylinder cover (17). Steering shaft (21) is retained in steering column by snap ring (24). Remove shaft if service on oil seal and bushing (23), shaft or housing is indicated.

When reassembling, make sure synchronizing valves in piston are at one o'clock position and rod holes horizontal

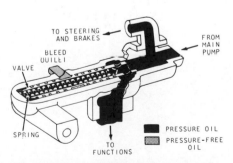

Fig. 9—Cross sectional view of steering priority valve showing fluid flow. Pressurized fluid from main hydraulic pump enters valve. Steering line is always open to pump pressure; hydraulic systems line to functions is closed off when system pressure drops below 1650-1700 psi.

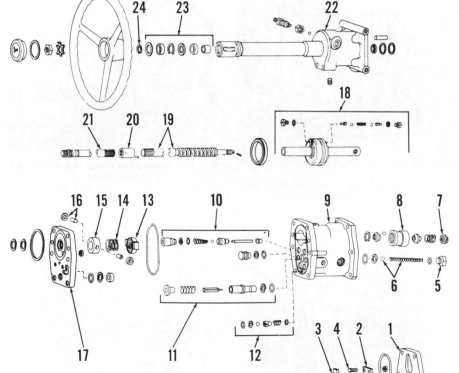

Fig. 11—Exploded view of steering valve assembly showing component parts. No mechanical linkage exists between steering valve and steering bellcrank.

1. Cover
2. Operating lever
3. Operating lever
4. Adjusting screw
5. Plug
6. Check valve
7. Nut
8. Operating Collar
9. Housing
10. Unloading valve
11. Check valve
12. Ball check
13. Nut
14. Spring
15. Collar
16. Guide roller
17. Cylinder cover
18. Piston
19. Steering shaft
20. Coupling
21. Steering shaft
22. Cylinder housing
23. Seal and bushing
24. Snap ring.

as shown in Fig. 13. Tighten spring loaded nut (13—Fig. 11) until a gap of approximately 5/16 inch exists between nut and collar (15). This tension provides the friction which gives a feeling of stability to the steering effort. End plugs for operating levers (2 & 3) should be adjusted to provide not more than 0.003 end play, but so levers will not bind. Tighten jam nuts securely when adjustment is correct. Tighten nut (7) to a torque of 5 ft.-lbs., loosen to nearest castellation and install cotter pin. Collar (8) must turn freely by hand but must have no end play.

With steering control unit completely assembled except for lower cover, temporarily install steering wheel and mount unit in a vise as shown in Fig. 14. Install special positioning clamps JDH-3C as shown. Use the side marked "3000-4000-5000". This will hold the valve operating collar 0.030 inch from inside edge of housing face (neutral position). Turn steering wheel fully to right and hold in position by attaching a weight to rim of steering wheel. Loosen the jam nuts and four adjusting

Fig. 12—Remove steering check valve stop and spring as unit is separated.

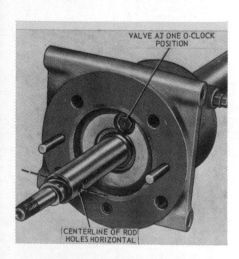

Fig. 13—Assemble piston with synchronizing valve at one o'clock and rod holes horizontal as shown.

screws (4—Fig. 11); then using a dial indicator, adjust upper left and lower right adjusting screws to a clearance of 0.0025-0.0035. With the first screws adjusted, turn remaining two screws to a clearance 0.001-0.002 less than first clearance.

NOTE: Adjustment can be made more accurately if lever is pulled outward slightly using a stiff wire hook, as adjustment is being made.

Tighten 1/2-inch cap screws to 85 ft.-lbs. and 3/8-inch cap screws to 35 ft.-lbs. Install the assembled unit as in paragraph 10 and bleed as in paragraph 7.

STEERING CYLINDER
12. REMOVE AND REINSTALL To remove either steering cylinder, first disconnect the pressure hose at cylinder. Remove the cap screws (C—Fig. 15) which retain pins (1) to frame bracket and steering bellcrank; then remove pins using a suitable puller.

NOTE: ID of pin has screw threads for attaching puller.

When reinstalling, make sure step washers (2) are properly installed as shown. Draw the pins into position using longer bolts in place of cap

Fig. 14—Install positioning clamps for valve adjustment.

screws (C). Tighten cap screws in frame bracket and steering bellcrank and secure by bending locking plate (L).

13. OVERHAUL. Refer to Fig. 15 for an exploded view of steering cylinder and attaching parts. To disassemble the removed steering cylinder, unscrew extension (10) using a suitable spanner. Remove snap ring (6) and washer (7) from piston rod (14) and withdraw extension (10) from piston. Bushings (9), packing (11) and seal (13) can be renewed at this time. Renew piston rod (14) if scored or otherwise damaged.

Assemble by reversing the disassembly procedure and install as outlined in paragraph 12.

ENGINE AND COMPONENTS

All tractors are equipped with either a naturally aspirated, or a turbocharged, six cylinder diesel engine having a bore of 4¾ inches, a stroke of 5 inches and a piston displacement of 531 cubic inches.

REMOVE AND REINSTALL
All Models
14. To remove the engine and clutch assembly as a unit, first drain cooling system and, if engine is to be disassembled, drain oil pan. Remove air stack, cab center panel if so equipped, and cowl; then remove hood, grille screens, engine side panels and muffler.

Make a clutch split by disconnecting air conditioning compressor, if so equipped, and placing unit with hoses inside cab. Disconnect cab mounting

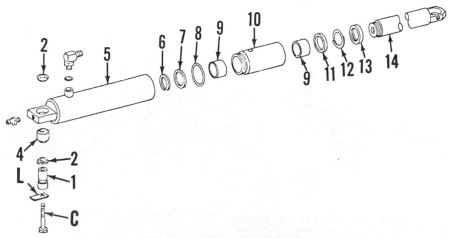

Fig. 15—Exploded view of steering cylinder and associated parts.

C. Cap screw
L. Lock plate
1. Pin
2. Step washer
4. Bushing
5. Cylinder
6. Snap ring
7. Washer
8. "O" ring
9. Bushing
10. Extension
11. Packing
12. Backup ring
13. Oil seal
14. Piston

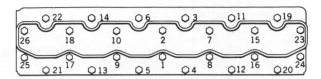

bracket from right frame rail. Disconnect heater hoses, fuel shut-off cable, right hand steering pipe and clamp, temperature gage sensing bulb, tachometer cable, throttle linkage, electric wiring and hydraulic lines on right side of engine. Remove cold starting unit, step and bracket, turbocharger inlet elbow and air cleaner, battery cables, wiring and hydraulic lines on left side. Suitably support both halves of tractor separately, remove cap screws securing engine to clutch housing and side rails to housings. Place wood blocks between front axle and frame rails to prevent tipping. Remove drawbar and support, and install JDG-2M, or other suitable support under transmission case and roll front half of tractor away.

NOTE: When engine is removed, front unit may be heavy in front therefore unstable. Remove front end weights, if used, and securely support front frame before attempting to remove engine.

Shut off fuel valve and remove hydraulic pump drive coupler. Place a wood block under pump for support. Disconnect fuel inlet pipe from fuel pump and hydraulic pump support from engine. Remove upper and lower radiator hoses. Disconnect both steering cylinders from rear brackets. Remove right hand steering pipe and rear portion of left hand steering pipe. Disconnect fuel gage sender wire at connector and remove hydraulic pump discharge pipe. Disconnect fuel leak-off pipe. Swing engine from a hoist and remove cap screws securing engine to side of frames, make sure front section is well supported, then slide engine rearward out of frame unit.

Install by reversing the removal procedure. Tightening torques are as follows:

Hydraulic pump drive .... 30 ft.-lbs.
Hydraulic pump support ..... 85 ft.-lbs.
Cylinder block to
  clutch housing .... 300 ft.-lbs.

Bleed steering as outlined in paragraph 7.

CYLINDER HEAD

All Models

15. To remove the cylinder head, drain cooling system and remove air stack, hood, side panels, grille screens and muffler. Remove water manifold, by-pass pipe and upper water hose. On naturally aspirated models, remove intake and exhaust manifolds. Disconnect turbocharger oil pipes and remove inter-cooler for access to cap screws to intake manifold. Unbolt and remove

exhaust manifold and turbocharger as a unit.

Remove alternator and fan blades then unbolt and remove water pump. Remove injector lines and injectors. Remove ventilator tube, rocker arm cover, rocker arm assembly and push rods, then unbolt and remove cylinder head.

NOTE: Make sure cylinder liners are held down with cap screws and washers if engine is to be turned.

Install cylinder head gasket dry and dip cylinder head cap screw threads in oil prior to installation. Make sure hardened flat washers are installed on all cap screws and tighten to a torque of 100 ft.-lbs. using the sequence shown in Fig. 16. Retighten to a torque of 170-180 ft.-lbs. using same sequence. Tighten rocker arm clamp bolts to a torque of 65 ft.-lbs. Retorque cylinder head and readjust tappets after tractor has been warmed to operating temperature. Refer to paragraph 17 for tappet gap adjustment procedure.

NOTE: NEVER run engine with turbocharger oil lines disconnected.

VALVES AND SEATS

All Models

16. Hardened steel valve seat inserts are used for both intake and exhaust valves. Valve face angle on exhaust valves is 44½ degrees and seat angle is 45 degrees. Intake valve face angle is 29½ degrees and seat angle is 30 degrees. Recommended seat width is 0.115 inch. Runout of seat should not exceed 0.004. Seats can be narrowed using 15 and 70 degree stones. If

necessary to renew seats, use JDE-77 Seat Puller to remove them, being careful not to damage head. Use JDE-72 and 73 to install new seats and chill both seats and driver in dry ice before installation.

Intake and exhaust valve stem diameter is 0.4335-0.4345 with a recommended clearance of 0.002-0.004 in guide bores. Valves should be renewed if clearance in a new guide exceeds 0.006. Valve tappet gap should be adjusted using the procedure outlined in paragraph 17.

17. **TAPPET GAP ADJUSTMENT.** The two-position method of valve tappet gap adjustment is recommended. Refer to Fig. 17 and proceed as follows:

Turn engine crankshaft by hand until "TDC" mark on pulley is aligned with reference mark on timing gear housing, then check the valves to determine whether front or rear cylinder is at top of compression stroke. (Exhaust valve on adjacent cylinder will be partly open). Use the appropriate diagram (Fig. 17) and adjust the indicated valves; then turn crankshaft one complete turn until "TDC" timing mark is again aligned. Adjust remainder of valves using the other diagram. Recommended valve tappet gap is 0.018 for intake valves and 0.028 for exhaust valves (hot or cold).

VALVE ROTATORS

All Models

18. Positive type valve rotators are used on intake and exhaust valves on turbocharged models and only on exhaust valves on naturally aspirated models. Normal service consists of renewing the complete unit. Rotators can

Fig. 16—Use the indicated sequence and tighten cylinder head cap screws to a torque of 100 then 175 ft.-lbs. in two steps.

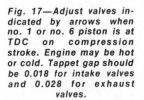

Fig. 17—Adjust valves indicated by arrows when no. 1 or no. 6 piston is at TDC on compression stroke. Engine may be hot or cold. Tappet gap should be 0.018 for intake valves and 0.028 for exhaust valves.

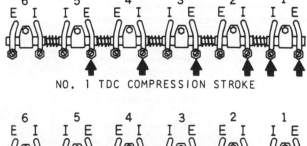

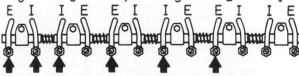

be considered satisfactory if the valve turns a slight amount each time it opens.

VALVE GUIDES AND SPRINGS

All Models

19. Valve guides are renewable in the cylinder head. Valves with oversize stems are not available for service replacement. The manufacturer recommends knurling as a means of resizing the guide to accept the standard stem if wear does not exceed 0.008 inch. Renew any guide that is worn beyond that figure. Use JDE-75 driver to install new guide 1.875 inches from head gasket surface of head. New guides can be knurled to compensate for valve stem wear up to 0.006 inch.

Standard valve guide bore diameter is 0.4365-0.4375 inch and normal stem clearance is 0.002-0.004.

Turbocharged models are equipped with inner and outer valve springs, while other models use single springs. Intake and exhaust valve springs are interchangeable and may be installed either end up. Renew any spring which is distorted, rusted or discolored, or does not meet the test specifications which follow:

Free Length (approx.)
Outer . . . . 2.46 in.
Inner . . . . 2.32 in.
Lbs. test @ length (inches)
Closed
Outer . . . . 70-78 @ 2.01
Inner . . . . 42-46 @ 1.77
Open
Outer . . . . 162-188 @1.48
Inner . . . . 84-98 @1.24

ROCKER ARMS

All Models

20. The rocker arm shaft attaches to bosses which are cast into cylinder head and is held in place by clamps. Turbocharged engines use two double clamps and two single clamps to retain rocker arm shaft. Other models use six single clamps. Shaft rotation is prevented by a spring pin in cylinder head which enters a hole in shaft for positive positioning of lubrication passages.

Rocker arms are right hand and left hand assemblies. Recommended clearance of rocker arms is 0.0005-0.0035 on the ¾-inch shaft. Bushings are not available; if clearance is excessive, renew rocker arms and/or shaft. If only the valve end of rocker arm is worn, the end can be reground and arm reused. Make sure lubrication holes are clear in arms and shaft.

When reassembling, make sure spring pin aligns with locating hole in shaft, tighten clamp screws to a torque

of 65 ft.-lbs. and adjust tappet gap as outlined in paragraph 17.

CAM FOLLOWERS (TAPPETS)

All Models

21. The cam followers can be removed from above without removing camshaft. The cam followers operate in unbushed bores in engine block and are available in standard size only.

If camshaft and/or followers are giving indications of wear, use a dial indicator on valve spring retainer to measure valve lift, which should be 0.517 inch for intake and 0.453 inch for exhaust. New followers should be installed whenever camshaft is renewed.

TIMING GEAR COVER AND CRANKSHAFT FRONT OIL SEAL

All Models

22. To remove the timing gear cover, first drain cooling system and remove hood, grille screens and engine side panels. Remove radiator and fan shroud from left side after disconnecting oil cooler from radiator.

Remove pressure and return lines from hydraulic pump, disconnect pump drive shaft and coupler and remove pump and support. Loosen fan belts, remove crankshaft damper pulley using a suitable puller, remove cover retaining cap screws and lift off the cover. The lip type front oil seal is supplied in a kit which also includes a steel wear sleeve which is pressed on crankshaft in front of gear as shown in Fig. 18. Score old sleeve lightly with a blunt chisel and pry sleeve from shaft. Coat inner surface of new sleeve with a nonhardening sealant and install with a suitable screw-type installer such as JDE-3, or a tube which will just slip over snout of crankshaft. Install oil seal in cover, with sealing lip 0.010 inch the rear and seal 0.010 inch from inner edge of bore.

When installing timing gear cover, tighten retaining cap screws to a torque of 30 ft.-lbs. and damper pulley retaining cap screw to 170 ft.-lbs. Com-

plete the assembly by reversing the disassembly procedure.

TIMING GEARS

All Models

23. The timing gear train consists of crankshaft gear and camshaft gear as shown in Fig. 19. Gears are available in standard size only. If backlash is excessive, renew the parts concerned.

The engine is properly timed when "V" mark on camshaft and crankshaft gears are aligned as shown. The timing of the injection pump gear should be checked at this time. Pump is properly timed when the "V" mark on pump gear aligns with mark on camshaft gear.

The combination camshaft/injection pump drive gear can be renewed after removing camshaft as outlined in paragraph 24. To renew crankshaft gear, remove wear sleeve as outlined in paragraph 22, and use a suitable puller to remove gear. When installing crankshaft gear, heat gear to approximately 350 degrees F. using a hot plate or oven and install with a press or JDH-7 driver, with timing mark to front.

CAMSHAFT AND BEARINGS

All Models

24. To remove the camshaft, first remove timing gear cover as outlined in paragraph 22, and oil pump as in paragraph 33. Remove rocker arm cover, rocker arms assembly and push rods. Raise and secure cam followers just far enough to clear camshaft, using magnetic holders or other suitable means. If cylinder head is off, and followers are removed, be sure they are marked for reinstallation into the same bores. Before camshaft is removed, use a dial indicator to check camshaft end play, which should be 0.004-0.010, but not more than 0.015 inch. Excessive end play indicates a worn thrust plate. Thickness of plate should be 0.1185-0.1215, with a minimum of 0.1135 inch. Working through openings provided in camshaft gear, remove the four cap screws securing camshaft thrust plate

Fig. 18—Score oil seal wear sleeve as shown using blunt chisel, then pry from shaft when renewal is indicated.

Fig. 19—Timing gears with cover removed, showing timing marks on turbocharged model. Other models are similar.

to front face of engine block, then withdraw camshaft and gear assembly forward out of engine.

The 2.3745-2.3755 camshaft journals should have a clearance of 0.002-0.005 in bushings. The pre-sized copper lead camshaft bushings are interchangeable, but only the two front bushings can be renewed without separating engine from clutch housing.

To install the two rear bushings after camshaft is out, detach cylinder block from clutch housing, remove clutch, flywheel and camshaft bore plug, then pull bushings into block bores using a piloted puller. Make sure oil supply holes in block are aligned with holes in bushings.

Camshaft/injector pump drive gear can be removed with a press when camshaft is out, after removing the retaining cap screw and washer. Align key slot in gear with Woodruff key in shaft, make sure thrust plate and spacer are installed and press gear on shaft until it bottoms. Tighten gear retaining cap screw to a torque of 85 ft.-lbs. and thrust plate retaining cap screws to a torque of 20 ft.-lbs.

ROD AND PISTON UNITS

All Models

25. Connecting rod and piston units are removed from above after removing cylinder head, oil pan and rod bearing caps. When reinstalling, correlation numbers, small and large slots and tangs on rod and cap must be in register. Rods and head of piston are stamped "FRONT" for proper installation. Tighten connecting rod cap screws to a torque of 155-165 ft.-lbs.

NOTE: Do not rotate crankshaft with head removed nor attempt to remove rod and piston units without first bolting liners down using washers and short cap screws.

PISTONS, RINGS AND SLEEVES

All Models

26. The aluminum alloy, cam ground pistons are fitted with two compression rings and one oil control ring. The top two compression rings are of keystone

design and their ring grooves contain cast iron inserts.

The wet cylinder sleeves have square rubber packing rings which seal at a shoulder on sleeve at bottom edge of water jacket in cylinder block. Backup rings fit in block grooves below the shoulder to aid in sealing. Top of sleeve extends above top gasket face of block a slight amount as shown in Fig. 20 and is sealed by cylinder head and head gasket.

When installing sleeves, make sure that seal ring grooves are clean. Lubricate sealing rings with liquid soap or other suitable lubricant and make sure rings are not twisted. Work sleeves gently into place by hand as far as possible, then seat the sleeve using a wooden block and hammer.

The manufacturer recommends cleaning pistons using Immersion Solvent "D-PART" and Hydra-Jet Rinse Gun or Glass Bead Blasting Machine. A wear gage (JDE-55) is provided for checking keystone ring groove wear. Measure sleeves in a front-to-rear, and side-to-side direction, at the top and bottom. Renew sleeve if taper is excessive. Other specifications are as follows:

Sleeve inside diameter . . . 4.7493-4.7507
Piston skirt clearance 0.0058-0.0082
(bottom of skirt)

PISTON PINS

All Models

27. The full floating type piston pins are retained in piston bosses by snap rings. Pins are available in 0.003 oversize (marked yellow) and 0.005 oversize (marked red) as well as standard. Always renew snap rings when reinstalling pins.

The recommended fit of piston pins is a hand push fit in piston bores and a slip fit in connecting rod bushings. Standard diameter and clearances are as follows:

Piston pin diameter 1.8739-1.8745
Clearance (rod) 0.0007-0.0023
Clearance (piston) 0.0003-0.0011

CONNECTING RODS AND BEARINGS

All Models

28. Connecting rod bearings are steel-backed aluminum inserts. Bearings are available in standard size as well as undersizes of 0.002, 0.010, 0.020 and 0.030.

Mating surfaces of rod and cap have milled tongues and grooves which positively locate cap and prevent it from being reversed during installation. Check drilled hole to pin bushing for

cleanliness. Connecting rods are marked "FRONT" for proper installation. Check the connecting rods, bearings and crankpin journals for excessive taper and against the values which follow:

Crankpin diameter 3.4980-3.4990
Regrind if out-of-round 0.004
Crankpin diametral
    clearance. 0.0024-0.0054
Rod bolt tightening
    torque . . . . .155-165 ft.-lbs.

CRANKSHAFT AND BEARINGS

All Models

29. The crankshaft is supported in seven main bearings. Crankshaft end play of 0.004-0.010 is controlled by the flanged fifth main bearing. Upper and lower bearing shells of flanged bearing ONLY are not interchangeable, the upper half containing an oil hole.

All main bearing caps can be removed from below after removing oil pan and oil pump. When renewing bearings, make sure that locating lug on bearing shell is aligned with milled slot in cap and block bore. After caps are loosely installed, bump crankshaft forward and rearward to align thrust flanges; then tighten main bearing cap screws to a torque of 205-215 ft.-lbs. Main bearings are available in undersizes of 0.002, 0.010, 0.020 and 0.030. The fifth (thrust) bearing is available in all undersizes with standard flange width; and in 0.010 undersize and 0.007 oversize flange width.

To remove the crankshaft, first remove engine as outlined in paragraph 14 and proceed as follows: Remove flywheel and crankshaft rear oil seal retainer. Remove crankshaft pulley, timing gear cover, oil pan and oil pump. Remove rod and main bearing caps and lift out crankshaft. The hardened crankshaft rear wear sleeve is a press fit on flywheel flange and is renewable. To remove score lightly with a blunt chisel. When installing, place JDE-34 GUIDE on crankshaft, drive wear sleeve on shaft with JDE-68 driver until rear edge is flush with guide. Make sure the chamfer on the O.D. of wear sleeve is to the rear of engine. Refer to paragraph 30 for procedures to renew rear oil seal. Check crankshaft and bearings against the values which follow:

Crankpin diameter 3.498-3.499
Main journal diameter 3.748-3.749
Regrind if out-of-round 0.004
Main bearing
    diametral clearance 0.0019-0.0049
Cap screw torque . . . . .210 ft.-lbs.

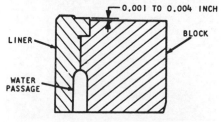

Fig. 20—Cylinder sleeves should stand out 0.001-0.004 inch when properly installed. Refer to paragraph 26.

CRANKSHAFT AND REAR OIL SEAL

All Models

30. The crankshaft rear oil seal is contained in a retainer plate which is attached to rear face of cylinder block by cap screws. Seal is available only in a kit which also includes the steel wear sleeve. To renew the seal, first detach (split) engine from clutch housing as outlined in paragraph 14 and remove clutch and flywheel.

Unbolt and remove oil seal retainer plate and score wear sleeve lightly. Pry from crankshaft using a screwdriver or pry bar. Drive or press the seal from retainer, being careful not to damage retainer plate. Install the seal sleeve with JDE-68 driver as outlined in paragraph 29. Attach a dial indicator with a magnetic base to crankshaft flange face as shown in Fig. 21. After seal has been installed in oil seal housing with lip toward engine and seal housing has been installed with cap screws snug but not tight, turn crankshaft and tap seal housing to center it. Runout should not exceed 0.006 inch. Tighten cap screws to a torque of 30 ft. lbs. To renew crankshaft front seal refer to paragraph 22.

FLYWHEEL

All Models

31. Flywheel is doweled to crankshaft flange and retained by six cap screws.

To install flywheel ring gear, heat gear evenly to 300°F. and position gear so that chamfered end of gear teeth face toward front of engine. When installing flywheel align dowel hole in flywheel over dowel in flange, coat threads of retaining cap screws with sealant and tighten evenly to a torque of 130 ft.-lbs.

OIL PAN

All Models

32. Engine oil pan can be removed without interference from other components on turbocharged models. Naturally aspirated models have the oil filter

mounted on left side of oil pan and filter must be removed before oil pan can be lowered. Use a floor jack or other lifting means to remove and install the cast pan. When installing, tighten the 3/8-inch cap screws to a torque of 35 ft.-lbs. and 1/2-inch cap screws to a torque of 85 ft.-lbs.

OIL PUMP

All Models

33. REMOVE AND REINSTALL. To remove the engine oil pump, first remove oil pan as outlined in paragraph 32. Remove cap screws holding pump to engine, carefully pry on pump to loosen from engine and remove pump. To reinstall reverse removal procedure. Be sure dowel pin hole is aligned on naturally aspirated models before tightening cap screws. Tighten to a torque of 30 ft.-lbs. (all models).

34. OVERHAUL. To overhaul the removed oil pump, first remove pump cover, remove idler gear and examine pump gears and cover for wear or scoring. Gears are available as a matched set only. Check for wear or excessive looseness between shaft and housing at upper (drive gear) end.

Driven gear is retained to drive shaft by a Woodruff key as shown in inset—Fig. 24 and is a press fit on shaft; removal procedure is as follows:

Using a press and a suitable mandrel, press shaft downward out of gear ONLY 1/4-inch. (This will prevent the Woodruff key from damaging the housing.) Release the press, raise gear and place 1/4-inch key stock spacers in

pocket of housing on each side of shaft. Press shaft downward another 1/4-inch and insert 1/2-inch spacers as shown. Repeat the procedure until gear is removed, extract the Woodruff key and withdraw drive shaft and gear assembly.

Assemble the pump by reversing the disassembly procedure and install as outlined in paragraph 33.

PRESSURE RELIEF VALVE

All Models

35. The oil pressure relief valve is located in the tachometer drive cover

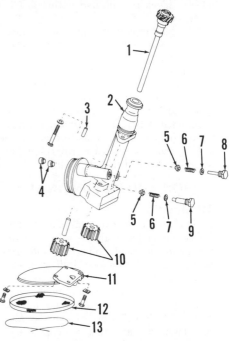

Fig. 23—Exploded view of engine oil pump used on naturally aspirated models.

1. Shaft	8. Relief plug
2. Pump body	9. Bypass valve cap
3. Dowel pin	10. Pump gears
4. Bushings	11. Intake cover
5. Valve	12. Intake screen
6. Spring	13. Safety wire
7. Washer	

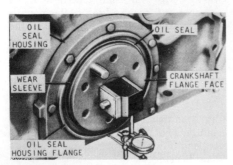

Fig. 21—Check rear oil seal housing runout with dial indicator as shown.

Fig. 22—Exploded view of engine oil pump used on turbocharged models showing component parts.

1. Drive shaft	4. Pump gears
2. Woodruff key	5. Cover
3. Pump body	6. Intake screen

Fig. 24—Oil pump parts will be damaged by drive shaft key (inset) unless oil pump is disassembled as outlined in paragraph 34.

on naturally aspirated models (Fig. 25). On turbocharged models, the valve is located in filter housing as is the filter by-pass valve and oil cooler by-pass valve. Refer to Fig. 26 for exploded view.

Only the oil used to lubricate the engine passes through the filter; excess oil passes through the regulating valve (8) and back to the sump. Regulated pressure at operating temperature and 1900 engine rpm should be 45-55 psi. Pressure can be adjusted by adding or removing shims (9). Adding or removing one shim will change regulated pressure by about 5 psi. Oil filter by-pass valve (3) operates on a pressure differential of approximately 30 psi. Filter by-pass valve opens for cold starting, or if filter becomes plugged, to assure continuing lubrication. Oil cooler by-pass valve (11) opens for cold starting, or in case of cooler restriction.

OIL COOLER

All Models

36. Turbocharged models are equipped with an engine oil cooler of the type shown exploded in Fig. 27. On naturally aspirated models the oil cooler by-pass valve is housed in cooler body (Fig. 28). The oil is cooled by the engine coolant liquid which circulates through tubes in cooler body.

Disassembly for cleaning or other service is normally not required except in cases of contamination of cooling or lubrication systems.

AIR INTAKE SYSTEM

All Models

37. The turbocharged model pre-cleaner contains a centrifugal discharge chute which is continuously evacuated by a suction tube leading to the upright muffler extension. The naturally aspirated model may use the same basic system, except for the suction tube,

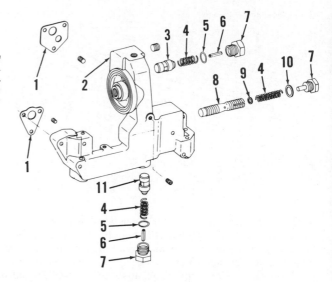

Fig. 26—Exploded view of filter housing and associated parts used on turbocharged models, showing relief and regulating valves.

1. Gasket
2. Housing
3. Filter bypass valve
4. Spring
5. "O" ring
6. Spring pin
7. Plug
8. Regulating valve
9. Shim
10. Washer
11. Cooler bypass valve

which is not used. The partially cleaned air then passes through a dual element dry type filter into the turbocharger or intake manifold; then on into the engine. A vacuum switch, located between the air cleaner and turbocharger or intake manifold, connects to a warning indicator lamp which lights to warn the operator if the filters are restricted.

Because of the balanced system and the large volume of air demanded by a turbocharged diesel engine, it is of utmost importance that only the approved parts which are in good condition be used.

To test for air intake vacuum, install a "T" fitting in intake tube in warning switch mounting port. Mount both the switch and a water manometer or a vacuum gage calibrated in inches of water, such as JDST-11, to the "T".

Normal vacuum at port for switch with clean filters and 2100 rpm (full load) is 13-14 inches of water. Warning switch closes at a vacuum of 24-26 inches of water.

Intake manifold pressure can be checked at the ether cold starting aid plug in intercooler manifold. Install a low pressure air gage. Pressure, checked at 2100 rpm, full load, should be 28-34 inches **mercury** (14-17 psi). A lower than specified manifold pressure may indicate intake manifold air leaks, restricted air intake or air cleaner, exhaust leaks or a malfunctioning turbocharger.

Aspirator vacuum can be checked by installing a plug with a manometer fitting in the tube which is part of the muffler extension on turbocharged models. Remove hose between pre-cleaner and aspirator, install manometer plug and hose and run engine at 2100 rpm. Reading should be 19-20 inches of water.

Restriction in the pre-cleaner screen can be checked by installing the manometer tube to the pre-cleaner hose fitting. At 2100 engine rpm the reading should be no more than 1.9-2.1 inches of water; if reading is higher, clean screen and re-test.

The intercooler (Fig. 29) must be free of dirt, and water passages must be clean to insure proper cooling. Be sure all gaskets and seals are tight and in good condition.

TURBOCHARGER

OPERATION

All Models

38. The exhaust driven tubocharger supplies air to the intake manifold at above normal atmospheric pressure. The additional air entering the com-

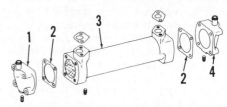

Fig. 27—Exploded view of engine oil cooler used on turbocharged models.

1. Water inlet cap
2. Gasket
3. Cooler body
4. Water outlet cap

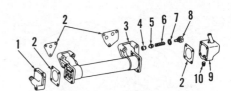

Fig. 28—Exploded view of engine oil cooler and bypass valve used on naturally aspirated models.

1. Water inlet cap
2. Gasket
3. Cooler body
4. Bushing
5. Bypass valve
6. Spring
7. Washer
8. Plug
9. Water outlet cap
10. Drain plug

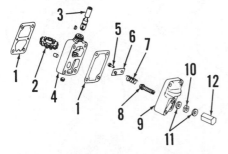

Fig. 25—Exploded view of tachometer drive housing used on naturally aspirated models. The cover on turbocharged models does not contain pressure relief valve.

1. Gasket
2. Gear
3. Shaft
4. Housing
5. Rivet
6. Relief valve
7. Spring
8. Adjusting screw
9. Cover
10. Jam nut
11. Washers
12. Cap

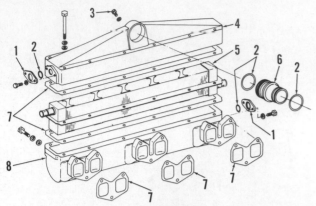

Fig. 29—Exploded view of intercooler used on turbocharged models.

1. Cover plate
2. "O" ring
3. Cap screw
4. Manifold cover
5. Intercooler
6. Coupler
7. Gasket
8. Intake manifold

bustion chamber permits an increase in the amount of fuel burned, and increased power output over an engine of comparable size not so equipped.

The use of the engine exhaust to power the compressor increases engine flexibility, enabling it to perform with the economy of a smaller engine on light loads yet permitting a substantial horsepower increase at full load. Horsepower loss because of altitude or atmospheric pressure changes is also largely reduced.

Because a turbocharger compresses the incoming air, the heat of compression causes the air to expand and become less dense than it would be at a lower temperature. The turbocharged model is equipped with an intercooler which uses coolant that has passed through the tractor radiator to lower the temperature of the air before it enters the engine, and so causes the density of the air to increase. This allows a greater volume of fuel to be used to produce more power. Since a greater power output causes more heat, the turbocharger is driven faster, which produces more manifold pressure, and even greater power.

The turbocharger contains one moving part, a rotating shaft which carries an exhaust turbine wheel on one end and a centrifugal air compressor on the other. The rotating member is precisely balanced and capable of rotative speeds up to 100,000 rpm. Bearings are of the floating sleeve type and the unit is lubricated and cooled by a flow of en-

gine oil under pressure.

Refer to Fig. 30 for a schematic view of turbocharger and to Fig. 31 for an exploded view.

NOTE: Individual parts are serviced on special order, but disassembly is not recommended. Exchange units are usually more economical than an overhaul and are available from the tractor manufacturer.

SERVICE

All Models

39. In a naturally aspirated diesel engine (without turbocharger) an approximately equal amount of air enters the cylinders at all loads, and only the amount of fuel is varied to compensate for power requirements. Turbocharging may supply up to 3 times the normal amount of air under full load.

All diesel engines operate with an excess of air under light loads. In a naturally aspirated engine, most of the air is used at full load, and increasing the amount of fuel results in a higher smoke level with little increase in power output. Turbocharging provides

a variation of air delivery, and a turbocharged engine operates with an excess of air up to and beyond the design capacity of the engine. When more fuel is provided, the turbocharger speed and air delivery pressure increase, resulting in additional horsepower and heat, with little change in smoke level. Smoke cannot, therefore, be used as a guide to safe maximum fuel setting in a turbocharged engine. DO NOT increase horsepower output above that given in CONDENSED SERVICE DATA at the front of this manual.

The turbocharger is serviced only as a new or rebuilt unit, and disassembly is not recommended. Turbine and compressor wheel blades can be inspected through exhaust pipe and air cleaner hose openings after unit is removed.

CAUTION: DO NOT operate the turbocharger without adequate lubrication. Before installation, prime turbocharger with engine oil through drain hole and turn by hand to lubricate bearings. When turbocharger is first installed, turn engine over with starter with engine shut-off knob out until oil pressure indicator light goes out, then start engine. Run engine at slow idle speed for at least two minutes before opening throttle or putting engine under load.

Some other precautions to be observed in operating and servicing a turbocharged engine are as follows:

Do not operate at wide-open throttle immediately after starting. Allow engine to idle until turbocharger slows down before stopping engine. These precautions will insure adequate lubrication to the shaft bearings at all times.

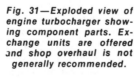

Fig. 31—Exploded view of engine turbocharger showing component parts. Exchange units are offered and shop overhaul is not generally recommended.

1. Turbine housing
2. Turbine wheel
3. Bearing
4. Center housing
5. Thrust collar
6. Impeller
7. Clamp
8. Compressor housing

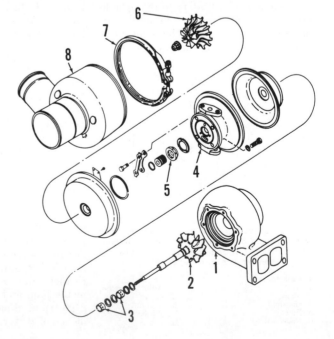

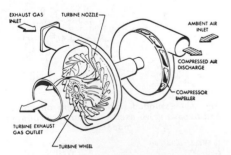

Fig. 30—Schematic view of engine turbocharger.

Because of increased air flow, care of air cleaner and connections is of added importance. Check the system and condition of restriction indicator whenever tractor is serviced. Make sure exhaust pipe opening is closed and air cleaner connected whenever tractor is transported, to keep turbocharger from turning due to air pressure. If exhaust outlet is equipped with weathercap, tape cap closed. If weathercap is missing, tape exhaust pipe opening.

40. REMOVE AND REINSTALL. To remove the turbocharger, first remove exhaust elbow (1—Fig. 32) and adapter (2), oil lines (4 & 5), intake hose connections; then unbolt and remove turbocharger (3) from exhaust manifold.

When installing, attach turbocharger to manifold using a new gasket and tighten stud nuts to a torque of 35 ft.-lbs. Install inlet oil line (5), outlet line (4), adapter (2) and exhaust elbow (1) after first making sure parts are perfectly aligned. Adapter (2) must have a minimum of 1/16 inch end play and be free to rotate. Undue stress on turbocharger at installation may cause bearing failure.

41. INSPECTION. To inspect the removed turbocharger unit, examine turbine wheel and compressor impeller for blade damage, looking through housing end openings. Using a dial indicator with plunger extension, check radial bearing play through large (outlet) oil port while moving both ends of turbine shaft equally. Check shaft end play with dial indicator working from either end. If end or side play exceeds 0.007; or if any of the blades are broken or damaged, renew the turbocharger unit.

Turbocharger oil supply pressure at 2100 engine rpm should be within 10 psi of engine oil pressure but never less than 30 psi. Minimum return oil

flow from turbocharger is ½ gpm at 2100 engine rpm.

DIESEL FUEL SYSTEM
FUEL FILTER, PRIMARY PUMP AND LINES

All Models

42. Turbocharged models are equipped with dual combination fuel filters and sediment bowls as shown in Fig. 33 and naturally aspirated models have a single combination unit. Drain and bleed filters if water or dirt is present, or renew as necessary.

Turbocharged models have an additional sediment bowl as a part of the fuel pump. The bowl should be checked daily, and removed and cleaned whenever water or foreign matter is present. Close the fuel shut-off valve before removing bowl and be sure the bowl gasket is in good condition before reinstalling bowl. The fuel tank is mounted vertically in front of radiator, and shut-off valve is mounted in the lowest part of fuel tank, forward of the front axle.

43. BLEEDING. To bleed the system, refer to Fig. 33 & 34. Open filter

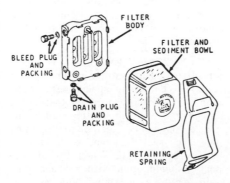

Fig. 33—Exploded view of renewable two-stage filter and sediment bowl element used on naturally aspirated models. Turbocharged models are equipped with dual filters.

bleed plug on naturally aspirated models and actuate hand primer lever on fuel pump until air-free fuel flows out bleed plug opening. Tighten bleed plug. Make sure hand primer lever is in down position before attempting to start engine. For turbocharged models, unscrew hand primer on fuel transfer pump (Fig. 35), open filter bleed plug and pump hand primer until air-free fuel flows out bleed plug opening. If entire system must be bled, loosen injection pump bleed plug, pump hand primer until air-free fuel flows and tighten plug. On all models, loosen the pressure line connections at injector assemblies and, with throttle open, turn engine over with starter until fuel flows from all injector lines. Tighten the connections and start engine. If engine will not start or misses, repeat the above procedure until system is free of trapped air.

FUEL LIFT PUMP

Naturally Aspirated Models

44. The fuel lift pump (Fig. 34) is mounted on right side of engine block and is driven by the camshaft.

The pump has no sediment bowl and is available as an assembly only.

INJECTOR NOZZLES

Naturally Aspirated Models

These tractors use Roosa-Master 9.5

Fig. 35—View of injection pump, transfer pump and bleed screw used on turbocharged models.

Fig. 34—View of injection system components used on naturally aspirated models.

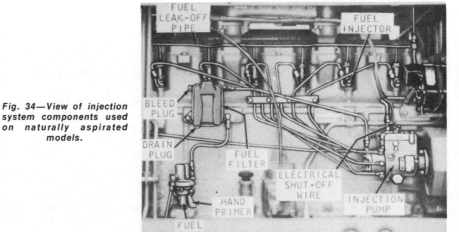

Fig. 32—Turbocharger assembly and associated mounting components.

1. Exhaust elbow
2. Adapter
3. Turbocharger
4. Outlet oil line
5. Inlet oil line
6. Heat shield

millimeter (pencil) nozzles. Refer to Fig. 36 for exploded view.

45. TESTING AND LOCATING A FAULTY NOZZLE. If rough or uneven engine operation or misfiring indicate a faulty injector, the defective unit can usually be located as follows:

With engine running at the speed where malfunction is most noticeable (usually slow idle speed), loosen the compression nut on high pressure line for each injector in turn, and listen for a change in engine performance. As in checking spark plugs, the faulty unit is the one which, when its line is loosened, least affects the running of the engine.

If a faulty nozzle is found and considerable time has elapsed since the injectors have been serviced, it is recommended that all nozzles be removed and checked, or that new or reconditioned units be installed. Refer to the following paragraphs for removal and test procedure.

46. REMOVE AND REINSTALL. Wash injector, lines and surrounding area with clean diesel fuel to remove any accumulation of dirt or foreign material. Remove cowl and hood. Disconnect leak-off pipe at return line fitting above injection pump. Expand lower clamp on each leak-off boot and move clamp upward next to top clamp; then remove leak-off pipe and all boots as a unit.

Disconnect high-pressure line, remove nozzle clamp cap screw, clamp

Fig. 36—Exploded view of Roosa Master Injector used on naturally aspirated models.

1. Adjusting screw assembly
2. Spring
3. Spring seat
4. Injector/valve assy.
5. Upper washer
6. Seal

and spacer; then withdraw injector assembly.

NOTE: If injector cannot be easily withdrawn by hand, the special OTC puller, JDE-38 will be required. DO NOT attempt to pry nozzle from its bore.

Before reinstalling the injector nozzle, clean nozzle bore in cylinder head using OTC Tool JDE-39, then blow out foreign material with compressed air. Turn tool clockwise only when cleaning nozzle bore. Reverse rotation will dull tool.

Renew carbon seal at tip of injector body and seal washer at upper seat whenever injector has been removed.

NOTE: Nozzle tip may be cleaned of loose or flaky carbon using a brass wire brush. DO NOT use a brush, scraper or other abrasive on Teflon coated surface of nozzle body between the seals. The coating may become discolored by use, but discoloration is not harmful.

Insert the dry injector nozzle in its bore using a twisting motion. Tighten pressure line connection finger tight; then install hold-down clamp, spacer and cap screw. Tighten cap screw to a torque of 20 ft.-lbs. Bleed the injector if necessary, as outlined in paragraph 43 then tighten pressure line connection to approximately 35 ft.-lbs. Complete the assembly by reversing disassembly procedure.

47. NOZZLE TESTER. A complete job of testing and adjusting an injector requires the use of special test equipment. Only clean approved testing oil should be used in tester tank. The nozzle should be tested for opening pressure, seat leakage, back leakage and spray pattern. When tested, the nozzle should open with a sharp popping or buzzing sound and cut off quickly at end of injection with a minimum of seat leakage and a controlled amount of back leakage.

Use the tester to check injector as outlined in the following paragraphs:

CAUTION: Fuel leaves the nozzle tip with sufficient force to penetrate the skin. Keep unprotected parts of body clear of nozzle spray when testing.

48. OPENING PRESSURE. Before conducting the test, operate tester lever until fuel flows, then attach the injector using No. 16492 Special Adapter. Close the valve to tester gage and pump the tester lever a few quick strokes to be sure nozzle valve is not plugged, that all spray holes are open and that possibilities are good that injector can be returned to service without overhaul.

Open valve to tester gage and operate tester lever slowly while observing gage reading. Pressure should rise steadily then fall off rapidly when opening pressure is reached. Opening pressure should be 3000 psi, if it is not, adjust opening pressure and valve lift as follows:

Loosen lock nut while holding pressure adjusting screw (1 outer—Fig. 36) from turning, then back out lift adjusting screw (inner) at least two turns to insure against bottoming. Turn adjusting screw (outer) until specified opening pressure is obtained. While holding adjusting screw from turning and before tightening lock nut, turn lift adjusting screw (inner) until it bottoms; then back out ¾ turn. Tighten lock nut and recheck opening pressure.

NOTE: When adjusting a new injector or an overhauled injector with a new pressure spring, set the pressure at 3200 psi to allow for initial pressure loss as the spring takes a set.

49. SPRAY PATTERN. The finely atomized nozzle spray should be evenly distributed around the nozzle. Check for clogged or partially clogged orifices or for a wet spray which would indicate a sticking or improperly seating nozzle valve. If the spray pattern is not satisfactory, disassemble and overhaul the injector as outlined in paragraph 52.

50. SEAT LEAKAGE. Pump the tester handle slowly to maintain a gage pressure of 2400 psi while examining nozzle tip for fuel accumulation. If nozzle is in good condition, there should be no noticeable accumulation for a period of at least 10 seconds. If a drop or undue wetness appears on nozzle tip, renew the injector or overhaul as outlined in paragraph 52.

51. BACK LEAKAGE. Loosen compression nut and reposition nozzle so that spray tip is slightly higher than adjusting screw end of nozzle, then maintain a gage pressure of 1500 psi. After the first drop falls from adjusting screw, leakage should be at the rate of 3-10 drops in 30 seconds. If leakage is excessive, renew the injector.

52. OVERHAUL. First clean outside of injector thoroughly. Place nozzle in a holding fixture and clamp the fixture in a vise. NEVER tighten vise jaws on nozzle body without the fixture. Refer to Fig. 36. Loosen locknut on adjusting screw assembly (1) and back out pressure adjusting screw (outer) containing lift adjusting screw. Slip nozzle body from fixture, invert the body and allow spring and seat to fall from nozzle body into your hand. Catch nozzle valve by its stem as it slides from body. If nozzle valve will not slide from body, use the special retractor (16481) or discard the injector assembly.

Nozzle valve and body are a matched set and should never be intermixed. Keep parts for each injector separate and immerse in clean diesel fuel in a compartmented pan as injector is disassembled.

Clean all parts thoroughly in clean diesel fuel using a brass wire brush and lint-free wiping towels. Hard carbon or varnish can be loosened with a suitable non-corrosive solvent.

NOTE: Never use a steel wire brush or emery cloth on spray tip.

Clean the spray tip orifices using the appropriate size cleaning needle. Four hole nozzles have 0.014-inch diameter orifices.

Clean the valve seat using a Valve Tip Scraper and light pressure. Use a Sac Hole Drill to remove carbon from inside of tip.

Piston area of valve can be lightly polished by hand if necessary, using Roosa Master No. 16489 lapping compound. Use the valve retractor to turn valve. Move valve in and out slightly while turning but do not apply down pressure while valve tip is in contact with seat.

Valve and seat are ground to a slight interference angle. Seating areas may be cleaned up if necessary using a small amount of 16489 lapping compound, very light pressure and no more than 3 to 5 turns of valve on seat. Thoroughly flush all compound from valve body after polishing.

When assembling, back out lift adjusting screw (inner), and reverse the disassembly procedure using Fig. 36 as a guide. Adjust opening pressure and valve lift as outlined in paragraph 48 after valve is assembled.

Turbocharged Models

Turbocharged tractors use Robert Bosch 21 millimeter nozzles, which have 3 spray holes. Refer to Fig. 37 for exploded view.

53. REMOVE AND REINSTALL. Refer to paragraph 45 for testing procedures to determine if injector nozzle shows indication of a malfunction before removing for service.

If nozzle to be removed is near the alternator, disconnect battery ground strap to prevent a short circuit through tools. Wash injector and surrounding area with clean diesel fuel, disconnect leak-off line and fuel pressure line and use special tool JDE-69 to remove gland nut (9—Fig. 37). Use care not to bind tool against head casting. The

gland nut will raise the nozzle out of cylinder head as it is removed.

When reinstalling, make sure injector and hole in cylinder head are clean and dry, apply Never-Seez to gland nut threads, renew nozzle gasket and tighten gland nut to 35 to 45 ft.-lbs. Renew leak-off line gaskets if necessary, hold fitting and tighten 12 millimeter head screw to 20 ft.-lbs. Bleed injectors as outlined in paragraph 43 and tighten pressure lines to 35 ft.-lbs.

54. NOZZLE TESTER. A complete job of testing and adjusting an injector requires the use of special test equipment. Only clean approved testing oil should be used in tester tank and tester connectors should be in good condition to avoid high pressure leaks, which are almost invisible and can penetrate the skin. The nozzle should be tested for opening pressure, seat leakage and spray pattern. When tested, the nozzle should open with a soft chatter, and then only when the lever is moved very rapidly. A bent or binding nozzle valve can prevent chatter. Spray will be broad and well atomized if injector is working properly.

Use the tester to check injector as

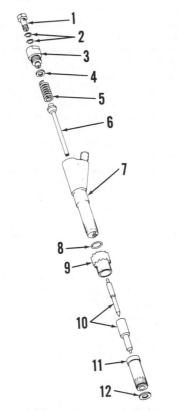

Fig. 37—Exploded view of injector nozzle used on turbocharged models.

1. Screw	7. Body
2. Gasket (2)	8. "O" ring
3. Plug	9. Gland nut
4. Adjusting shim	10. Nozzle
5. Spring	11. Retaining nut
6. Spindle	12. Washer

outlined in the following paragraphs:

CAUTION: Fuel leaves the nozzle tip with sufficient force to penetrate the skin. Keep unprotected parts of body clear of nozzle spray when testing.

55. OPENING PRESSURE. Before conducting the test, operate tester lever until fuel flows, then attach the injector using the proper adapter. Close the valve to tester gage and pump the tester lever a few quick strokes to be sure nozzle valve is not plugged, that all spray holes are open and that possibilities are good that injector can be returned to service without overhaul.

Open valve to tester gage and operate tester lever slowly while observing gage reading. Opening pressure should be 3100 psi; if it is not, recheck by releasing pressure and retest. If pressure is still not correct, remove plug (3—Fig. 37) and change shim (4) until opening pressure is correct. Use only John Deere shims as they are specially hardened. Shims are available in 0.002 inch steps from 0.043 to 0.059 inch. Each 0.002 inch step varies the pressure by about 75 psi. Opening pressure should not vary more than 50 psi between nozzles. If pressure is not correct after changing shims, disassemble injector and recondition.

NOTE: When adjusting a new injector or an overhauled injector with a new pressure spring, set the pressure at 3200 to 3350 psi to allow for initial pressure loss as the spring takes a set.

56. SPRAY PATTERN. The finely atomized nozzle spray should be evenly distributed around the nozzle. Check for clogged or partially clogged orifices or for a wet spray which would indicate a sticking or improperly seating nozzle valve. If the spray pattern is not broad and even, and very rapid stroking of the tester handle does not cause injector to chatter softly, disassemble and overhaul the injector as outlined in paragraph 58.

57. SEAT LEAKAGE. Pump the tester handle slowly to maintain a gage pressure of 2800 psi while examining nozzle tip for fuel accumulation. If nozzle is in good condition, there should be no noticeable accumulation for a period of at least 10 seconds. If a drop or undue wetness appears on nozzle tip, renew the injector or overhaul as outlined in paragraph 58.

58. OVERHAUL. First clean outside of injector thoroughly and place in a soft jawed vise. Relieve pressure on the nozzle spring by loosening the leak-off screw plug (1—Fig. 37). Remove nozzle retaining nut (11) with a box end wrench. DO NOT use a pipe-wrench. Remove nozzle valve assembly (10) and reinstall retaining nut on body (7) to protect the lapped end surface. If nozzle valve can not be removed easily, soak assembly in carburetor cleaner, acetone or other commercial solvent intended to free stuck valves. Use care to keep parts clean and free from grit by submerging in a pan of clean diesel fuel, and handle only with hands that are wet with fuel. Avoid mixing of parts with another injector, and do not allow any lapped surface to come in contact with a hard object.

Valves should be cleaned of all carbon and washed in diesel fuel. Hard carbon may be cleaned off with a brass wire brush. NEVER use a steel wire brush or emery cloth on valve or tip. Use a cleaning wire 0.003 to 0.004 inch smaller than nozzle orifices to clean nozzle tips. The orifices are 0.017 inch in diameter. Using a stone to provide flat surfaces on two sides of cleaning wire will aid in reaming carbon from orifices. Finish cleaning orifices by using a wire 0.001 smaller than hole diameter. A pin vise should be used to hold cleaning wires and wire should extend only about 1/16-inch from vise to prevent breakage. Clean seat in nozzle (10) with sac hole drill furnished with cleaning kit. When held vertically, a valve that is wet with fuel should slide down to the seat in nozzle under its own weight.

Inspect all lapped and seating surfaces for excessive wear or damage. Check spindle, spring, shims and seats. Renew any parts in question and reinstall shims only if they are smooth and flat. Edge type filter in fuel inlet passage of body (7) can be cleaned by blowing air through passage from nozzle end of body. This will provide a reverse flushing action in filter.

Assemble in reverse order of disassembly. Submerge valve and nozzle in fuel while assembling, and make sure all other parts are wet with fuel. Do not dry parts with air or towels before assembly. With injector body clamped in a soft jawed vise, tighten screw plug (3) to 36-44 ft.-lbs. and retaining nut (11) to 44-58 ft.-lbs. Retest injector as outlined in paragraphs 55, 56, and 57, and use new gasket when reinstalling in engine.

INJECTION PUMP

Naturally Aspirated

These engines are equipped with a Roosa-Master Model DGB fuel injection pump which is flange mounted and camshaft driven. Proper injection pump

timing depends on correct installation as outlined in paragraph 59.

Injection pump service requires the use of specialized skill, training and equipment. This section therefore will cover only the information required for removal, installation and field adjustment of the pump.

59. REMOVE AND REINSTALL. To remove the DGB fuel injection pump, first shut off fuel supply and thoroughly clean dirt from pump, lines and connections.

NOTE: Do not steam clean or pour water on a pump while it is warm or running, as this could cause pump to seize. Cap all fittings as they are disconnected to prevent dirt entry.

Remove timing hole cover located on lower right side of clutch housing. Turn engine crankshaft until "TDC" timing mark on flywheel is aligned with mark at rear of timing window and No. 1 piston is on compression stroke.

NOTE: Pump can be removed and reinstalled without regard to crankshaft timing position, however, positioning crankshaft at TDC is recommended so timing can be properly checked and/or adjusted when pump is reinstalled. If timing is not to be checked, scribe timing marks on injection pump mounting flange and engine front plate which can be aligned when pump is reinstalled.

Disconnect fuel inlet line, fuel return line and throttle link rod from injection pump. Disconnect pressure lines from injectors and pump and remove the lines. Remove pump mounting stud nuts and pull pump straight to rear off pump shaft.

To install the pump without changing the timing, refer to Fig. 38. Punch mark on tang of exposed end of pump drive shaft must align with similar mark on pump rotor. Align marks and, using a seal compressor, carefully install the injection pump. Make sure scribe lines on pump flange and engine front plate are aligned, then tighten pump mounting studs securely.

To install and time the injection pump, recheck to be sure that No. 1

piston is at TDC on compression stroke. Remove timing hole cover from side of injection pump and turn pump rotor, if necessary, until timing scribe line on governor weight retainer aligns with scribe line on pump cam. Install the pump using a seal compressor. Seals must not be curled or damaged, since fuel leakage into engine crankcase might result if seals do not hold. Recheck alignment of scribe lines as stud nuts are tightened.

Reinstall timing hole cover and complete pump installation by reversing the removal procedure. Bleed pump and lines as outlined in paragraph 43 and adjust pump linkage, if necessary as outlined in paragraph 65.

60. TIMING. To check injection pump timing without removing injection pump, first turn engine until No. 1 piston is at TDC on compression stroke.

Shut off fuel and remove timing hole cover from injection pump. The timing lines on governor weight retainer and cam should be in perfect alignment. If they are not, loosen pump mounting stud nuts and rotate pump housing until alignment is obtained. Hold pump in this position and tighten mounting stud nuts securely. Turn engine two full turns in direction of rotation and recheck.

61. ADVANCE TIMING. The injection pump is provided with automatic speed advance which is factory set and will not normally need to be checked or reset. Minor adjustments can, however, be made without removal or disassembly of the pump. To check the advance mechanism, proceed as follows:

Shut off fuel, remove pump timing hole cover and install timing window Fig. 39. Note where the line on pump cam ring is positioned in relation to the lines on timing window. Turn on fuel

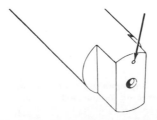

Fig. 38—End view of pump drive shaft on DBG pump. Arrow points to timing mark on tang.

Fig. 39—Injection pump timing window installed showing timing marks aligned. Refer to paragraph 61.

and bleed fuel system, then start and run engine.

If maximum advance of 5 degrees (plus or minus $\frac{1}{2}$ degree) is not correct when tested at 2200 engine rpm, renew or overhaul the pump. If maximum advance was correct, reduce engine speed to 1300 rpm. Timing line should move in window to 2 degrees advance (1 mark) with no load. If dynamometer is available, engine should be run at 1900 rpm (full load) and timing mark should move two lines, which would be 4 degrees advance. If intermediate advance is not correct, remove seal cap from advance timer screw located on engine side of pump. Loosen locknut and turn adjusting screw with an allen wrench, while engine is being run at 1300 rpm (no load) and 1900 rpm (full load), until intermediate advance is correct. Tighten locknut and reinstall seal cap. Remove timing window and reinstall timing hole cover.

Turbocharged Models

The Robert Bosch injection pump used on turbocharged tractors is model number PES6P, and is a multiple plunger in-line pump, with a governor and an aneroid control. It is equipped with an externally mounted fuel transfer pump which includes a hand primer pump, and is lubricated by engine oil pressure. Injection pump service requires the use of specialized skill, training and equipment. This section therefore will cover only the information required for removal, installation and field adjustment of the pump.

62. REMOVE AND REINSTALL. To remove the fuel injection pump, first shut off fuel supply and thoroughly clean dirt from pump, lines and connections.

NOTE: Do not steam clean or pour water on a pump while it is warm or running, as this could cause pump to seize. Cap all fittings as they are disconnected to prevent dirt entry. Remove access plate from front of timing gear cover and remove timing hole plug which is just to the rear of engine oil filler cap (Fig. 40).

(Fig. 40). Rotate engine in normal rotation direction until the mark on pump drive hub lines up with pointer mark (Fig. 41), and "INJ" mark on crankshaft pulley is at notch on timing cover. This is 27 degrees BTDC, not TDC (Fig. 40). Remove the three cap screws holding pump drive gear to pump. Disconnect the fuel and oil lines to pump and the pipe to aneroid. Disconnect the injector lines, return hose, speed control rod, aneroid de-activator linkage and fuel shut-off cable. Remove four mounting nuts which hold pump to engine and withdraw pump.

When installing injection pump, No. 1 piston must be coming up to TDC on compression stroke, and crankshaft pulley "INJ" mark must be aligned with notch on timing cover. Timing marks on pump drive hub and pointer must be aligned (Fig. 41) for proper timing. Compression stroke can be found with rocker cover removed, by turning engine until No. 1 intake valve is closing. Renew "O" ring on front bearing plate on pump, lubricate liberally and slide pump onto mounting studs. Make sure that the three drive gear slots are nearly centered with pump drive hub holes and drive hub mark is aligned with pointer mark before tightening the three cap screws. Tighten mounting stud nuts and drive gear cap screws to 35 ft.-lbs. torque. Reverse disassembly procedure and add $\frac{1}{2}$ to $\frac{3}{4}$ U.S. pint of engine oil through filler plug in pump. Check timing as outlined in paragraph 63 and bleed system as outlined in paragraph 43.

63. TIMING. To check injection pump static timing, proceed as outlined in paragraph 62 but without removing pump, lines or linkage.

If adjustment is required, loosen the three cap screws on pump drive gear until the drive hub can be moved in the slotted holes in the drive gear. When timing marks are aligned (Fig. 40 & 41), retighten three cap screws to 35 ft.-lbs. torque, rotate engine two complete revolutions and recheck timing marks and condition of drive gear teeth. No further timing checks are necessary with this type of pump.

64. ANEROID. Turbocharged tractors are equipped with a diaphragm type control unit, which fits on top of

governor assembly on the Robert Bosch fuel injection pump (Fig. 35). This diaphragm is operated by positive pressure from the intake manifold, which results from the turbocharger producing boost pressure on a hard pull. Until positive pressure is built up by the turbocharger, the vertical shaft which extends down into governor housing provides a stop to limit the travel of the fuel control rack. This allows the engine to acelerate without producing black smoke unnecessarily. When the turbocharger builds sufficient pressure to depress the diaphragm spring in the aneroid, the aneroid shaft moves down and allows the fuel control rack to move farther open, which delivers more fuel at the time the engine can burn it.

Aneroids must be adjusted to the pump on a test stand, so if it becomes necessary to renew the diaphragm or the entire assembly, pump should be recalibrated on a test stand.

All Models

65. LINKAGE AND SPEED ADJUSTMENT. The following adjustments should only be made with engine warm and a master tachometer with JDE-28 speed indicator adapter installed.

Hand speed control lever (Fig. 42) should not move with less than 15 lbs. of force at end of lever. Adjust friction device nut to obtain desired resistance.

Depress foot pedal to floor and adjust the speed control rod until arm on pump contacts fast idle stop screw. Speed control rod can be adjusted for length with jam nuts or yoke at forward end of rod.

Move hand lever up to slow idle until it stops. Adjust hand speed control rod turnbuckle to provide 800 rpm and tighten jam nut.

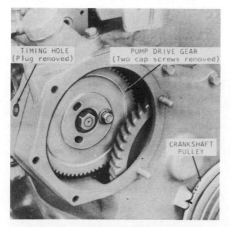

Fig. 40—View of Bosch injection pump drive gears, timing hole and timing marks on turbocharged tractors.

Fig. 41—Timing marks on pointer and pump drive hub must be aligned when "INJ" mark on crankshaft pulley [Fig. 40] is at notch on timing cover.

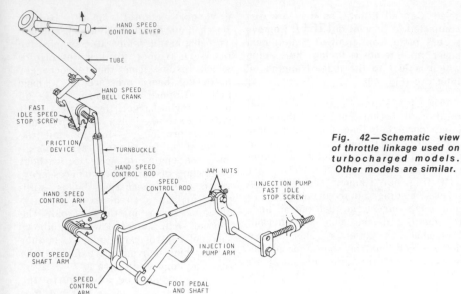

RADIATOR

All Models

68. To remove radiator, drain cooling system and remove air stack, muffler, side panels, grilles, screens, cowl and hood. Remove braces attaching fuel tank to radiator, oil cooler to radiator and remove the screws retaining the fan shroud. Remove air cleaner hose if necessary, and disconnect radiator hoses. Remove radiator retaining cap screws and slide radiator out left side of tractor. Install by reversing the removal procedure. Radiator cap should test from 6.25 to 7.50 psi.

FAN AND WATER PUMP

All Models

69. **REMOVE AND REINSTALL.** To remove fan and/or water pump, drain cooling system and remove air stack, muffler, cowl, hood, side panels and grilles. Remove the screws attaching fan shroud to radiator and cap screws attaching fan to pump hub; then, slide fan and shroud together out left side of tractor. Loosen fan belts, disconnect wiring and remove alternator. Remove by-pass hose and lower radiator hose. Unbolt and remove the water pump. Install by reversing the removal procedure.

70. **OVERHAUL.** To disassemble the removed water pump unit, first remove pulley using a suitable puller which attaches to two fan screw holes; then remove cover (3—Fig. 44). Press shaft and bearing assembly (7) out of impeller toward the FRONT ONLY and remove ceramic sealing insert and rubber cup (5) from impeller bore.

All parts are available individually except the ceramic sealing insert and cup for impeller are sold as a part of seal kit and may come in one or two pieces.

Insert is installed with polished side out; back side is marked with a groove

Fig. 42—Schematic view of throttle linkage used on turbocharged models. Other models are similar.

Rotate hand lever down until arm is against fast idle stop screw on pump, then adjust fast idle speed stop screw on hand speed bell crank until it makes contact. No load fast idle should be 2300 rpm on turbocharged models and 2400 rpm on other models.

Turbocharged Models Only

66. **ANEROID LINKAGE ADJUSTMENT.** Pull cable (1—Fig. 43) through bell crank (4) moves lever (8), stops the engine, and spring (5) forces aneroid de-activator inward to the start position. Spring stays over center rearward until pump arm (Fig. 42) pushes it forward over center as engine is speeded up. The aneroid de-activator is then allowed to move outward, giving aneroid control of maximum fuel delivery.

To adjust linkage, disconnect pull cable (1—Fig. 43) from bellcrank (4). Disconnect rod (7) from stop lever (8). With hand speed lever at slow idle, move cam (4) and spring rearward until spring locks over center. Rotate bellcrank backward until it contacts spring, make sure stop lever (8) is against stop and adjust rod (7) until it will enter hole in stop lever. Hold stop lever all the way down, make sure cable (1) is all the way in and tighten nut (3).

Naturally Aspirated Models

67. **FUEL PUMP SHUT-OFF.** The Roosa Master fuel injection pumps are equipped with a fuel solenoid which must be energized to provide fuel flow. The shut-off valve is spring loaded in the closed position and is electrically opened when key switch is turned to "ON" or "START" positions.

If tractor will not start, turn switch to "ON" position and check continuity of solenoid lead. If solenoid must be re-

moved, remove pump and have it repaired by a competent injection pump serviceman.

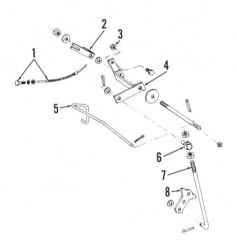

Fig. 43—Exploded view of aneroid linkage and spring. [Turbocharged models].

1. Pull cable	5. Spring
2. Cam	6. Swivel
3. Nut	7. Rod
4. Bellcrank	8. Stop lever

Fig. 44—Exploded view of water pump used on turbocharged models. Other models are similar.

1. Gasket
2. Spacer
3. Cover
4. Impeller
5. Seal assembly
6. Body
7. Shaft and bearing
8. Fan pulley

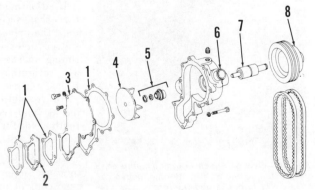

or three dots for identification. Place both the cup and insert at an angle in impeller bore and snap into place by hand. Coat bearing with light oil and press into housing bore until bearing is flush with front edge of housing. Coat outside of seal with Permatex and press seal into housing until it bottoms.

Apply a light coat of Permatex to shaft bore of impeller. Make sure seal lip and ceramic insert face are perfectly clean, apply a coat of light oil to insert (do not use grease), then press impeller onto shaft until vanes of impeller are approximately 0.025 inch from housing seat (Fig. 45). Turn shaft by hand to make sure there is no drag on housing.

NOTE: Support front end of shaft on bed of press when installing impeller and impeller end of shaft when installing fan pulley.

Install fan pulley and rear cover, then reinstall water pump and associated parts by reversing the removal procedure.

Fig. 45—Press impeller to within 0.025 inch of inside of housing as shown.

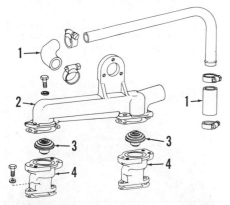

Fig. 46—Exploded view of water manifold and thermostat housings used on naturally aspirated models.

1. Bypass hose   3. Thermostats
2. Water manifold   4. Housings

THERMOSTAT AND WATER MANIFOLD

All Models

71. The thermostats are contained in a thermostat housing in water manifold on naturally aspirated models, or in thermostat cover on turbocharged models. Refer to Fig. 46 or Fig. 47 for application. Turbocharged models use three thermostats. To remove thermostats, detach upper radiator hose at front and remove housing cap screws, then remove hose and housing as a unit. On naturally aspirated models, remove rear bypass hose before removing water outlet manifold. Thermostats can be tested in heated water to determine opening temperature, which should be 177° to 182°F.

ELECTRICAL SYSTEM

ALTERNATOR AND REGULATOR

Motorola Models

72. A Motorola alternator and transistorized regulator are used on turbo-

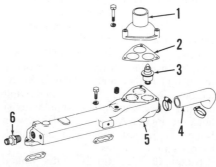

Fig. 47—Exploded view of water manifold and thermostat cover used on turbocharged models.

1. Cover   4. Bypass hose
2. Gasket   5. Water manifold
3. Thermostats(3)   6. Nipple

Fig. 48—Exploded view of Motorola Alternator of the type used.

1. Pulley
2. Fan
3. Spacer
4. Front end frame
5. Bearing
6. Snap ring
7. Rotor
8. Bearing
9. Stator
10. Retainer
11. Rear end frame
12. Brush holder
13. Voltage regulator
14. Insulating washers
15. Insulating sleeve
16. Isolation diode
N. Negative heat sink
P. Positive heat sink

charged and naturally aspirated models with 55 ampere rating. Naturally aspirated models with air conditioning use a 72 ampere unit. Turbocharged models with air conditioning use a Delco Remy 72 ampere unit with a sealed, solid state internal regulator. Refer to paragraph 74 for Delco Remy alternator.

Refer to Fig. 48 for an exploded view of Motorola alternator assembly. Isolation diode (16) is mounted on one terminal screw for positive heat sink (P) and one screw for negative heat sink (N), therefore insulators (14 & 15) must be used. The primary purpose of the isolation diode is to permit use of charging indicator lamp.

The voltage regulator, isolation diode and brush holder can be renewed without removal or disassembly of alternator; all other alternator service requires disassembly.

Failure of the isolation diode is usually indicated by charging indicator lamp, which glows with key switch off and engine stopped if shorted; or with engine running if diode is open.

Failure of a rectifying diode can be indicated by a humming noise when engine is running if diode is shorted; or by a steady flicker of charge indicator light at slow idle speed when diode is open. Either fault will reduce generator output.

To check the charging system, refer to Fig. 49 and proceed as follows:

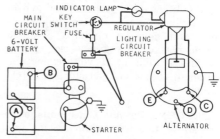

Fig. 49—Schematic diagram of typical charging circuit showing main components and test connections. Refer to paragraph 72 for procedure.

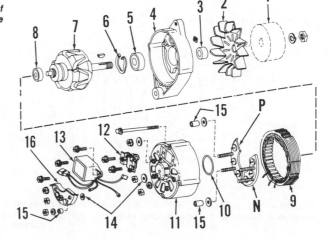

(1). With key switch and all accessories off and engine not running, connect a low reading voltmeter to terminals C-E. Reading should be 0.1 volt or less. A higher reading would indicate a short in isolation diode, key switch or wiring.

(2). Turn key switch on and again check voltmeter. A reading above 3.0 volts or below 1.5 volts may indicate a defective alternator or regulator; or defective wiring. Turn alternator shaft slightly if reading is high, and recheck.

(3). Start and run engine at approximately 1400 rpm and, with all accessories off, again check voltmeter reading, which should be 15 volts. A lower reading could indicate a discharged battery or a defective alternator.

(4). Move voltmeter lead from regulator terminal (E) to output terminal (D). Reading should drop one volt from reading in test (3), reflecting the resistance built into isolation diode (16—Fig. 48). If battery voltage (12 volts) is obtained, isolation diode is open and must be renewed.

(5). If a reading lower than the specified 15 volts was obtained when checked as outlined in test (3), stop engine and disconnect (or remove) the voltage regulator. Connect a jumper wire between output terminal (D—Fig. 49) and the exposed blade (field) terminal on alternator brush holder. Connect a suitable voltmeter to terminals (C-E) on alternator. Start engine and slowly increase engine speed while watching voltmeter. If a reading of 15 volts can now be obtained at 1300 engine rpm or less, renew the regulator. If a reading of 15 volts cannot be obtained, renew or overhaul the alternator.

CAUTION: Do not allow voltage to rise above 16.5 volts when making this test.

73. OVERHAUL. The regulator, isolation diode and brush holder can be removed without removing alternator from tractor. Regulator and brush holder should be removed before attempting to separate frame units.

Exposed length of brushes in removed brush holder should be ¼-inch or more. Brushes are available only in an assembly with holder (12—Fig. 48).

To disassemble the removed alternator unit, remove through bolts and separate slip ring end frame (11) from drive end frame (4). Rotor (7) will remain with drive end frame and stator (9) with slip ring end frame. Be careful not to damage stator windings when prying units apart.

Examine slip ring surfaces of rotor for scoring or wear and field windings for overheating or other damage. Check bearing surfaces or rotor shaft for visible wear or scoring. Check rotor for grounded, shorted or open circuits using an ohmmeter as follows:

Refer to Fig. 51 and touch the ohmmeter probes to points (1-2 and 1-3); a reading near zero will indicate a ground. Touch ohmmeter probes to the two slip rings (2-3); reading should be 5 ohms. A higher reading will indicate an open field circuit, a lower reading will indicate a short. If windings are satisfactory, mount rotor in a lathe and check runout at slip rings using a dial indicator. Runout should not exceed 0.002. Slip ring surfaces can be trued if runout is excessive or if surfaces are scored. Finish with 400 grit or finer polishing cloth until scratches and machine marks are removed.

Make a continuity test of brushes and brush holder (Fig. 52). Poor or intermittent connections can be found by wiggling each unit and terminal while making test. Continuity should exist between Points A and C, and between B and D. If bad connection is found, renew unit as an assembly.

Stator (9—Fig. 48) is delta wound and leads are externally wrapped and soldered. To check for shorts, grounds or opens, first mark carefully then unsolder the connections. Continuity should exist between the two ends of each particular coil but not between adjacent coils or stator frame. Because of the low resistance, shorted windings within a coil cannot be checked. Refer to Fig. 53. After checking stator, connect leads A to B, C to AA and BB to CC. Three positive diodes are located in slip ring heat sink and three negative diodes in grounded heat sink. Diodes should test at or near infinity in one direction when checked with an ohmmeter, and at or near zero when meter leads are reversed. Renew any diode with approximately equal meter readings in both directions. Diodes must be removed and installed using an arbor press or vise and a suitable tool which contacts only outer edge of diode. Do not attempt to drive a faulty diode out of heat sink as shock may cause damage to other good diodes. If all diodes are being renewed, make certain the positive diodes (marked with red printing) are installed in heat sink marked (P—Fig. 48) and negative diodes (marked with black printing) are installed in heat sink (N). Use a pair of needle nose pliers as a heat sink when soldering diode leads; use only rosin core solder and an iron instead of a torch. Excess heat can damage a good diode while it is being installed.

NOTE: A battery powered test light can be used instead of an ohmmeter for all electrical tests except shorts in rotor winding. However, when checking diodes, test light must not be of more than 12 volts.

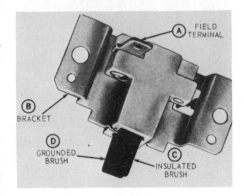

Fig. 52—For continuity tests, check points A and C and B and D. Wiggle each unit to locate poor connections while testing.

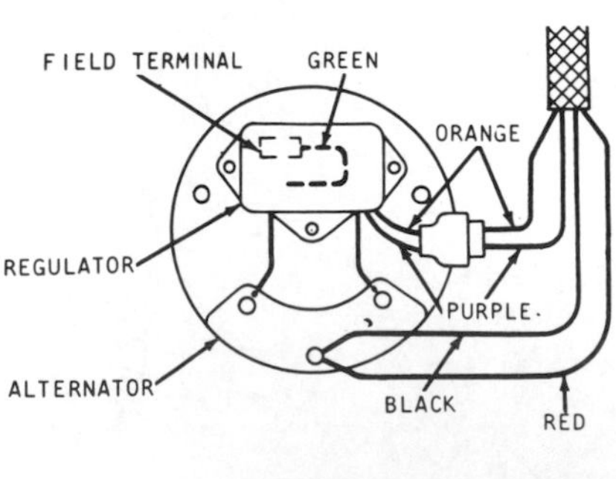

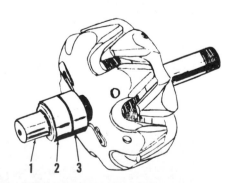

Fig. 50—Rear schematic view of Motorola alternator showing correct wiring connections.

Fig. 51—Removed rotor assembly showing test points to be used when checking for grounds, short and opens.

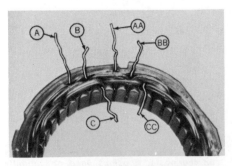

Fig. 53—When renewing a stator or after testing, connect leads A and B, C and AA and BB and CC as shown.

When reinstalling alternator, refer to Fig. 50 for proper wiring diagram. Do not attempt to polarize alternator.

Delco-Remy Models

74. Delco-Remy "Delcotron" alternators are used on turbocharged models with Sound-Gard body and air conditioning. It is a 72 ampere unit, with a solid state regulator mounted inside the rear end frame, and has no provision for adjustment.

CAUTION: Because certain components of the alternator can be damaged by procedures that will not affect a D.C. generator, the following precautions MUST be observed.

a. When installing batteries or connecting a booster battery, the negative post of battery must be grounded.

b. Never short across any terminal of the alternator or regulator unless specifically recommended.

c. Do not attempt to polarize the alternator.

d. Disconnect all battery ground straps before removing or installing any electrical unit.

e. Do not operate alternator on an open circuit and be sure all leads are properly connected before starting engine.

75. TESTING AND OVERHAUL. The only test which can be made without removal and disassembly of alternator is the regulator. If there is a problem with the battery not being charged, and the battery and cable connector have been checked and are good, check the regulator as follows: Attach an accurate voltmeter to output (battery) terminal of alternator, and ground the negative lead of meter to alternator case. Make sure alternator has a good, clean ground to engine and start engine and run at approximately 2000 rpm. The voltage reading should be as listed in following table. Temperatures shown are regulator case temperatures within alternator. If batteries are not in a good state of charge, it may be necessary to charge them before an accurate reading can be obtained.

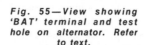

Fig. 55—View showing 'BAT' terminal and test hole on alternator. Refer to text.

85°F .... 14.5—15.0 volts
105°F .... 14.3—14.8 volts
125°F .... 14.1—14.7 volts
145°F .... 13.9—14.5 volts

Recheck battery and cables. If voltage does not read close to rated output, ground the field winding by inserting a screwdriver into test hole (Fig. 55). If voltage is then close to rated output, renew the regulator.

CAUTION: When inserting screwdriver in test hole the tab is within ¾-inch of casting surface. Do not force screwdriver deeper than one inch into end frame.

If voltage is still not close to rated output the alternator will have to be disassembled. Check the field windings, diode trio, rectifier bridge and stator as follows:

To disassemble the alternator, first scribe matching marks (M—Fig. 56) on the two frame halves (4 and 16), then remove the four through-bolts. Pry frame apart with a screwdriver between stator frame (12) and drive end frame (4). Stator assembly (12) must remain with slip ring end frame (16)

when unit is separated.

NOTE: When frames are separated, brushes will contact rotor shaft at bearing area. Brushes MUST be cleaned of lubricant if they are to be reused.

Clamp the iron rotor (13) in a protected vise, only tight enough to permit loosening of pulley nut (1). Rotor end frame can be separated after pulley and fan are removed. Check bearing surface of rotor shaft for visible wear or scoring. Examine slip ring surface for scoring or wear, and rotor winding for overheating or other damage. Check rotor for grounded, shorted or open circuits using an ohmmeter as follows:

Refer to Fig. 51 and touch the ohmmeter probes to points (1-2) and (1-3); a reading near zero will indicate a short circuit to ground. Touch ohmmeter probes to the slip rings (2-3); reading should be 5.3-5.9 ohms. A higher reading will indicate an open circuit and a lower reading will indicate an internal short. If windings are satisfactory, mount rotor in a lathe and check run-

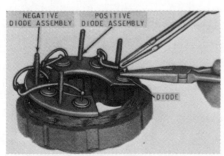

Fig. 54—Use needle nose pliers as a heat sink, and an iron only, when soldering diode connections.

Fig. 56—Exploded view of "DELCOTRON" alternator used with internal mounted solid state regulator. Note matching marks (M) on end frames.

1. Pulley nut
2. Washer
3. Spacer
4. Drive end frame
5. Grease slinger
6. Ball bearing
7. Spacer
8. Bearing retainer
9. Bridge rectifier
10. Diode trio
11. Capacitor
12. Stator
13. Rotor
14. Brush holder
15. Solid state regulator
16. Slip ring end frame
17. Bearing and seal

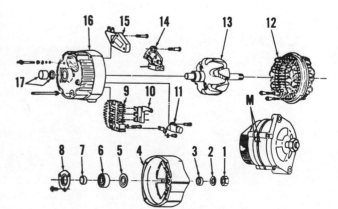

out at slip rings using a dial indicator. Runout should not exceed 0.002. Slip ring surfaces can be trued if runout is excessive or if surfaces are scored. Finish with 400 grit or finer polishing cloth until scratches or machine marks are removed.

Before removing stator brushes or diode trio refer to Fig. 57 and check for grounds between points A to C and B to C with an ohmmeter, using the lowest range scale. Then reverse the lead connections. If both A to C readings or both B to C readings are the same, the brushes may be grounded because of defective insulating washer and sleeve at the two screws. If the screw assembly is not damaged or grounded, the regulator is defective.

To test the diode trio, first remove the stator. Then remove the diode trio, noting the insulator positions. With an ohmmeter, check between points A and D (Fig. 58) and then reverse the ohmmeter lead connections. If diode trio is good it will give one high and one low reading. If both readings are the same, the diode trio is defective. Repeat this test at points B and D and at C and D.

The rectifier bridge (Fig. 59) has a grounded heat sink (A) and an insulated heat sink (E) that is connected to the output terminal. Connect ohmmeter to the grounded heat sink (A) and to the flat metal strip (B). Then reverse the ohmmeter lead connections. If both readings are the same, the rectifier bridge is defective. Repeat this test between points A and C, A and D, B and E, C and E, and D and E. Capacitor (11—Fig. 56) connects to the rectifier bridge and grounds to end frame, and protects the diodes from voltage surges.

Test the stator windings (Fig. 60) for grounded or open circuits as follows: Connect ohmmeter leads successively between any two leads. A high reading would indicate an open circuit.

NOTE: The three stator leads are 'DELTA' connected in the windings. Connect ohmmeter leads between each stator lead terminal and stator frame. A very low reading in only one lead would

indicate a grounded circuit. A shorted circuit within the stator windings cannot be readily determined by test because of the low resistance of the windings, but a shorted lead will probably be discolored and will have a strong odor.

Brushes and springs are available only as an assembly which includes brush holder (14—Fig. 56). If brushes are reused, make sure all grease is removed from surface of brushes before unit is reassembled. When reassembling, first install regulator and then brush holder, springs and brushes. Push brushes up against spring pressure and insert a short piece of straight wire through the hole and through end frame to outside. Be sure that the two screws at points A and B (Fig. 57) have insulating washers and sleeves.

NOTE: A ground at these points will cause no output, or controlled output. Withdraw the wire under brushes only after alternator is assembled.

Remove and inspect ball bearing (6—Fig. 56). If bearing is in satisfactory condition, fill bearing ¼-full with Delco-Remy lubricant No. 1948791 and reinstall. (Renew any bearing that is not in excellent condition.) Inspect needle bearing (17) in slip ring end frame. This bearing should be renewed if its lubricant supply is exhausted; no attempt should be made to relubricate and reuse the bearing. Press old bearing out towards inside and press new bearing in from outside until bearing is flush with outside of end

frame. Saturate felt seal with SAE 20 oil and install seal.

Reassemble alternator by reversing the disassembly procedure. Tighten pulley nut to a torque of 50 ft.-lbs.

Refer to Fig. 61 for proper wiring diagram. Do not attempt to polarize alternator.

STARTING MOTOR
Models

76. Either a Delco-Remy or a John Deere starting motor is used. Both types are of the solenoid shifted, positive engagement, over-running clutch type.

The solenoid switch on starting motor not only closes the circuit between battery and starting motor but also shifts starter drive pinion into mesh with flywheel ring gear. The switch contains a high current-draw pull-in winding which is shorted out when contact points are closed; and a low current-draw hold-in coil which remains active until key switch is released. An open circuit in the pull-in winding will prevent the solenoid from shifting the drive and actuating the motor. An open circuit in hold-in winding (or a low battery) will cause solenoid to chatter.

Specifications of the starting motor are as follows:

Delco-Remy Model
Volts 12.0
Brush spring tension, oz. 80
Minimum brush length, inch. 7/16

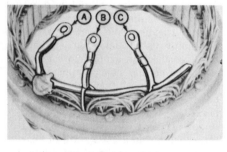

Fig. 60—Use ohmmeter to check stator windings for open or grounded circuits.

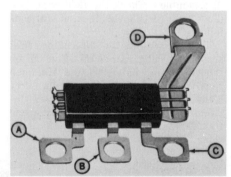

Fig. 58—Diode trio test points. Refer to text.

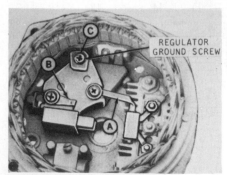

Fig. 57—Test points for brush holder. Refer to text.

Fig. 59—Bridge rectifier test points. Refer to text.

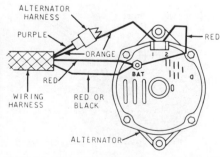

Fig. 61—Rear schematic view of Delco-Remy alternator showing correct wiring connections.

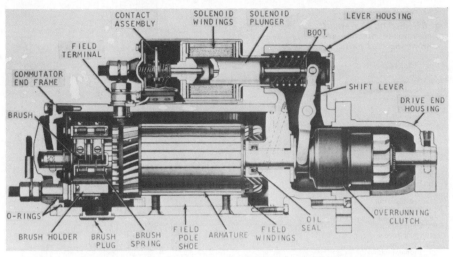

Fig. 62—Cross sectional view of Delco-Remy starting motor and solenoid assembly.

CIRCUIT DESCRIPTION

All Models

77. The electrical system is equipped with a 50 ampere Main Circuit Breaker. Lighting and accessory circuits are shown in Fig. 64. The two lighting circuits use 20 ampere circuit breakers, as do the accessory circuit, cab blower fan motors, wiper, dome lamp and relay control circuits. The ignition circuit is equipped with a 10 ampere breaker while the condenser fan motors and compressor clutch use a 30 ampere breaker.

ENGINE (TRANSMISSION) CLUTCH

The engine clutch is a double disc unit which includes a separator plate between the discs. The power take-off is driven by splines in the clutch cover hub. Refer to the appropriate following paragraphs for adjustment procedure and overhaul data.

No-load test (includes solenoid)
Volts 9.0
Amperes 140-190
RPM 4000-7000
John Deere Model
Volts 12.0
Brush spring tension, oz. 40
Minimum brush length, inch 5/8
No-load test (includes solenoid)
Volts 9.0
Amperes 70-110
RPM 2500-4500

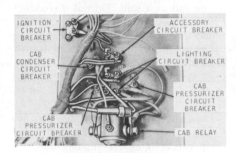

Fig. 64—View of circuit breakers showing individual circuits.

LINKAGE ADJUSTMENT

78. **TRANSMISSION CLUTCH.** Total pedal travel may be adjusted at stop screw (Fig. 65) by shortening or lengthening screw to limit clutch pedal travel to 6-7/8 inches when measured at foot pad. Free play at pedal should be 1½ inches with engine running at 1900 rpm.

To adjust free play at clutch release bearing, refer to Fig. 66. Remove inspection cover on bottom of clutch

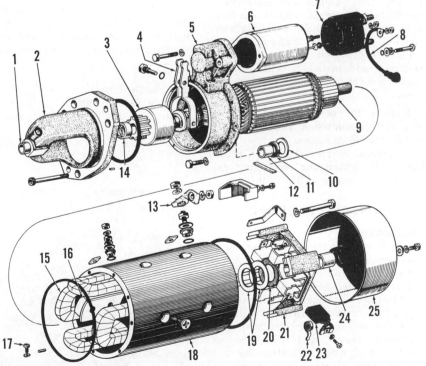

Fig. 63—Exploded view of John Deere starting motor and solenoid assembly.

1. Nose end bushing	8. Shunt lead	15. Field coil	21. Commutator end frame
2. Drive housing	9. Armature	16. Pole shoe	22. Brush spring
3. Drive and clutch	10. Thrust washer	17. Grounding screw	23. Brush
4. Shift yoke	11. Seal	18. Motor housing	24. Commutator end bushing
5. Center housing	12. Center Bushing	19. Shim washers	25. End cover
6. Solenoid	13. Field Connector	20. Thrust washer	
7. Solenoid cover	14. Pinion stop		

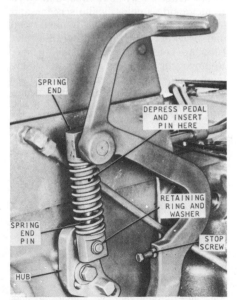

Fig. 65—Clutch pedal and spring, showing hole for pin to aid in spring removal.

housing and measure clearance between clutch fingers and release bearing. If clearance is less than 0.060 inch, use a screwdriver to loosen lock screw about ¼ inch and use adjusting wrench (Part No. R36621) to rotate release bearing sleeve (16—Fig. 67) counter-clockwise to a clearance of 0.102-0.112 inch, which is the thickness of wrench handle. After adjustment is made, align the nearest slot on bearing sleeve with the lock screw and tighten lock screw until it bottoms. Loosen lock screw until the flat side of screw aligns with rivet head (14), which is spring loaded in a blind hole in bearing carrier (15). This acts as a detent to prevent lock screw from loosening.

79. POWER TAKE-OFF CLUTCH.
To adjust the power take-off clutch linkage, be sure transmission oil is warm. Refer to Fig. 68. Disconnect push-pull cable from outer valve lever and loosen cable housing clamp. Remove upper plug of the two test holes in lower left corner of transmission case and install a 0-100 psi pressure gage. Run engine at 1900 rpm and engage pto clutch by turning outer valve lever. If pressure is not 90-100 psi, adjust rear stop screw on lever to obtain correct pressure, then tighten jam nut. Reinstall push-pull cable and tighten clamp. For pto clutch overhaul, refer to paragraph 112.

TRACTOR SPLIT

All Models

80. To detach (split) engine from clutch housing for access to engine clutch and flywheel, proceed as follows:

Drain cooling system and remove air stack, muffler, hood, cab center panel (if so equipped), left hand operator's shield (tractors without cabs) and cowl. Remove side shields, grille screens and steering support cover. Remove step and bracket from left side frame, and cab bracket from right side frame. Remove air cleaner and inlet elbow from turbocharger. If equipped with air conditioner, loosen belt, remove compressor and lay compressor and hoses

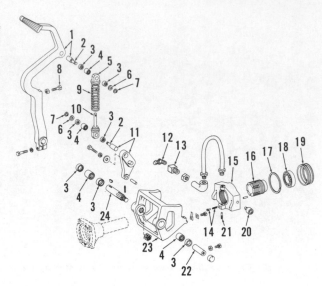

Fig. 67—Exploded view of clutch release mechanism and associated parts.

1. Pedal w/pin
2. Pin
3. Oil seal
4. Needle bearing
5. Spring end
6. Washer
7. Retaining ring
8. Stop screw
9. Spring
10. Spring end
11. Adjusting hub
12. Grease fitting
13. Connector
14. Rivet & spring
15. Bearing carrier
16. Bearing sleeve
17. Sealing ring
18. Ball bearing
19. Bearing collar
20. Cam follower
21. Lock screw
22. Pivot shaft
23. Release cam
24. Pedal shaft

in the cab as a unit. Do not disconnect hoses unless absolutely necessary. Disconnect batteries and capscrew holding battery compartment to top of frame. Disconnect starter solenoid bracket from engine so that steering pipe can be disconnected. Remove heater hoses at front of engine.

Disconnect throttle rod, wiring, hydraulic lines, tachometer cable and temperature indicator sending unit. Be sure to cap all disconnected hydraulic fittings to prevent dirt entry. If tractor is equipped with front end weights, remove the weights and place wood blocks between front support and front axle to prevent tipping to either side. Remove drawbar and drawbar support. Support engine and front axle assembly securely, and install JDG-2M support stand or floor jack under transmission case.

Remove the connecting cap screws and roll transmission assembly rearward away from engine.

To attach, reverse the above procedure and tighten the connecting cap screws to a torque of 300 ft.-lbs.

R&R AND OVERHAUL

All Models

81. To remove the clutch after engine has been detached from clutch housing, proceed as follows: Remove three alternate clutch cover retaining screws and install full thread jack screws (Fig. 68) which must be provided with nuts and flat washers. Screws must be 3 inches or longer, and should be bottomed in flywheel. Tighten the nuts finger tight against clutch cover and move the three operating levers toward outer rim of clutch and push out the pins connecting levers to links. Remove the three remaining capscrews. Back off jack screw nuts evenly, allowing clutch unit to separate. Refer

to Fig. 69. Remove clutch cover (5), spring retainers (6), springs (7 & 8), pressure plate (11), rear clutch disc (12), separator plate (13) and front clutch disc from flywheel.

Examine friction surfaces of flywheel, separator plate and pressure plate. Surfaces must not be scored, grooved or out-of-true more than 0.006. Drive pins and pressure plate must not be excessively worn or damaged. If clutch face of flywheel is remachined, do not remove more than 0.035 of the surface to true the flywheel.

Clutch springs consist of an inner and outer spring. Examine springs for rust, pitting or distortion, and check springs against the specifications which follow:

Test length (inner)
  Before Ser. No. 30952 . . . . 1¾ in.
  After Ser. No. 30951 . . . . 2-3/8 in.
Test length (outer)
  All Models . . . . 2-3/8 in.

Fig. 66—Bottom view of clutch housing, showing adjustment points for clutch free play.

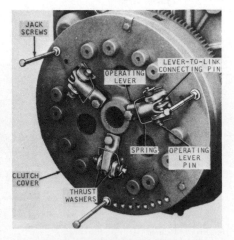

Fig. 68—Use three full-thread jack screws as shown to remove clutch unit from flywheel. Refer to text.

Pressure at test length (lbs.)
  Outer spring (all) 129-157
  Inner spring
    Early 79.2-96.8
    Late 63.7-77.9

To assemble the removed unit, position rear clutch disc on aligning tool (Fig. 70) with long hub to rear. The aligning tool consists of JDE-52 shaft, JDE-52-4 adapter and JDE-76 pilot bearing adapter. Position separator plate (13—Fig. 69) in front of rear disc, then place front disc on aligning tool with long hub forward. Install the assembled driving unit in flywheel, fitting separator plate over drive pins and alignment tool in pilot bearing bore.

NOTE: Front clutch disc is not cushioned and new thickness is 0.380-0.394 inch. Rear disc is cushioned and new thickness is 0.402-0.432 inch. If the later type button discs are being used, links (9) and drive pins (14) MUST also be changed to prevent clutch damage.

Place the pressure plate on a bench, friction surface down, install the 12 outer and 12 inner clutch springs in spring cups; then position drive plate and connect release lever links. Install the assembled drive plate and pressure plate over end of aligning tool and on drive pins. Install jack screws in three alternate holes and tighten nuts evenly. Tighten the clutch retaining cap screws to a torque of 35 ft.-lbs.

If clutch shaft (28) is to be inspected, remove pto shaft quill (19) from clutch housing. Drive spring pin (25) from shaft (24) while supporting both shafts.

Inspect clutch shaft oil seal (26) which is installed in pto and transmission oil pump drive shaft and gear (24). Oil seals should be renewed any time unit is disassembled.

NOTE: Seal should not be installed deeper than 0.91 inches (1-29/32") from end of shaft or it may be damaged (see Fig. 73). Be careful not to damage seal when clutch shaft is installed.

82. CLUTCH ADJUSTMENT. Whenever clutch has been disassembled, release levers must be adjusted as follows: Use adjusting gage JDE-32. Measuring from rear face of pto drive hub in clutch cover as shown in Fig. 72, adjust the screw in each release lever to clear gage by 0.005. Tighten adjusting screw jam nuts securely. If gage is not available, make a gage as shown in

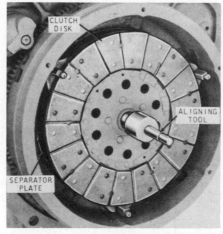

Fig. 70—Use clutch aligning tool to install both clutch discs and cover. See text.

Fig. 71, or adjust levers equally within 0.005 to extend 0.810 beyond rear surface of pto drive hub in clutch cover.

NOTE: Be sure new discs are used to make this adjustment, or shim worn disc to same thickness as new disc. The pedal free play adjustment will be extended to its limit before disc wear allows damage to other parts if the above adjustment procedures are followed.

Check and adjust operating linkage as outlined in paragraph 78, after tractor is assembled. The hub (Fig. 65) below pedal return spring should be adjusted so that as pedal is released, spring will return pedal solidly into engagement position.

83. CONTROL LINKAGE To overhaul the clutch control linkage, first detach (split) engine from clutch housing as outlined in paragraph 80.

Before any disassembly is attempted, depress clutch pedal until the holes in spring end (5—Fig. 67) and spring end (10) are aligned, and insert an 1/8-inch steel pin or nail through the parts to retain the spring. Remove retaining ring and washer securing spring end (10) to lower pin (2); loosen clamp bolt in clutch pedal (1) and remove pedal and spring as an assembly. Remove grease fitting and nut from release bearing lube line and free the line from housing. Rotate actuating cam (23) downward and withdraw release bearing and carrier as a unit. Remove expansion plug at outer end of cam pivot shaft (22) from right side of

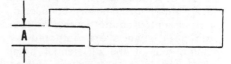

Fig. 71—A gage for clutch release lever adjustment can be made from flat stock. Dimension 'A' is 0.810 inch. See Fig. 72.

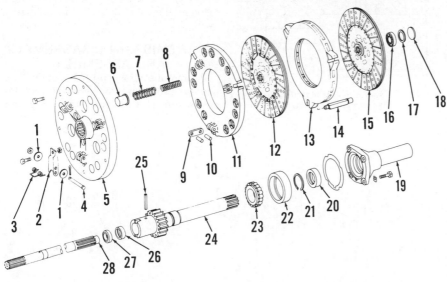

Fig. 69—Exploded view of clutch and associated parts.

1. Thrust washer
2. Operating lever
3. Spring
4. Pin
5. Clutch cover
6. Spring retainer
7. Outer spring
8. Inner spring
9. Link
10. Dowel pins
11. Clutch plate
12. Rear disc
13. Separator plate
14. Drive pin
15. Front disc
16. Pilot bearing
17. Thrust washer
18. Thrust plate
19. PTO shaft quill
20. Oil seal
21. Snap ring
22. Outer bearing race
23. Roller bearing
24. PTO & trans. pump shaft
25. Spring pin
26. Oil seal
27. Spacer
28. Clutch shaft

Fig. 72—Use special tool as shown to adjust clutch release levers to within 0.005 inch of each other.

housing and loosen set screw holding pivot shaft. Pull cam pivot shaft (22), using a 3/8-inch bolt as a puller. Outer end of shaft is threaded to receive puller bolt.

Remove cotter pin and nut from inner end of actuating shaft (24) and remove shaft and actuating cam (23).

Needle bearings (4) should be packed with Lubriplate or similar high-temperature grease when linkage is overhauled or bearings are renewed. Lips of seals should face toward bearings, except on pivot shaft (22) where lip should face away from bearing. Cam follower rollers (20) are retained in release bearing carrier (15) by roll pins which must be removed if disassembly is indicated. Detent plunger (14) fits against a flat on set screw (21) to prevent screw from loosening. Plunger and spring are free when set screw is removed.

Disassemble and/or assemble return spring (9) and spring ends in a suitable press. Spring free length should be approximately 6-1/32-inches, and spring should test 231-281 lbs. when compressed to a length of 4 inches.

Assemble by reversing the disassembly procedure and adjust linkage as outlined in paragraph 78.

SYNCRO-RANGE TRANSMISSION

The "Syncro Range" transmission is a mechanically engaged transmission consisting of three transmission shafts and a single, mechanically connected, remote mounted control lever. The four basic gear speeds are selected by coupling one of the differential bevel pinion shaft idler gears to the splined main drive bevel pinion shaft, and can only be accomplished by disengaging the engine clutch and bringing the tractor to a stop. The high and low speed ranges within the four basic speeds are selected by shifting only the synchronized coupler on the transmission drive shaft and, because of the design of the coupler, can be accomplished by disengaging the engine clutch and moving the control lever, without bringing the tractor to a halt. All reverse speeds are engaged with a sliding toothed collar, which necessitates stopping tractor before any reverse gear is selected.

NOTE: The rotating speeds of the transmission drive shaft and its idler gears are automatically equalized by the synchronizing clutches. All other phases of shifting are under the direct control of the operator. The fact that clashing of gears is eliminated by the synchronizing clutches does not relieve him of

the responsibility of using care and judgment in re-engaging the clutch after the gears have been shifted.

The idler gears and bearings on the main shaft and bevel pinion shaft are pressure lubricated by a separate transmission oil pump. Refer to paragraph 90 for service procedures on transmission pump.

INSPECTION

All Models

84. To inspect the transmission gears, shafts and shifters, first drain the transmission and hydraulic fluid and remove the operator's platform, then remove the transmission top cover. Examine the shaft gears for worn or broken teeth and the shifter linkage and cam slots for wear.

CONTROL QUADRANT

This section covers disassembly and overhaul of the shifter controls mounted in tractor steering support. Removal, inspection, overhaul and adjustment of shift mechanism inside the transmission housing is included with transmission gears and shafts.

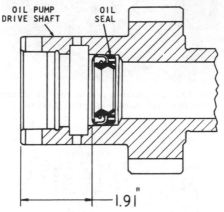

Fig. 73—Oil seal must be installed no deeper than dimension shown.

All Models

85. R&R AND OVERHAUL. To overhaul the control quadrant, remove the cowl and raise the dash enough to clear quadrant. Right mounting handle may be removed for convenience. Disconnect the shifter rods at upper ends. Remove outer end of quadrant shaft (10) and, while holding shaft with a wrench, remove the nut and washer securing shaft to bracket (1). Quadrant assembly may then be completely disassembled.

If lever (6) or pivot (8) are damaged and need to be renewed, proceed as follows: Clamp the lower curved portion of lever (6) in a soft-jawed vise and slip a 5/32-inch cotter pin inside the two roll pins (7). Grasp roll pins with a good vise-grip plier and extract with a twisting motion. When reassembling, leave at least 3/16-inch of roll pin protruding from lever (6). To drive pin farther will damage bushings (4).

After quadrant is reassembled, check the action of the shift lever to make sure transmission is shifting properly. Adjust shift rods as necessary to provide smooth shifting action.

86. NEUTRAL-START SWITCH. The neutral-start switch is mounted in the left top side of clutch housing, behind the control support cover. When shift lever is in neutral or park position, the switch is closed by the lug which protrudes from top of speed range shifter cam arm (5—Fig. 75).

To adjust the switch, add one washer under switch at a time until switch will not close when in neutral, then remove one washer and reinstall. This will prevent starting in gear, and will prevent lug on arm from damaging switch.

TRANSMISSION DISASSEMBLY AND ASSEMBLY

Paragraphs 87A and 88 outline the general procedure for removal and installation of the main transmission

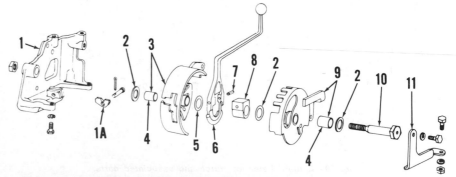

Fig. 74—Exploded view of the dash mounted shifter controls. The notch in latch (1) fits around the lower rocker of lever (6) and moves to positively lock the opposite quadrant (3 or 9) when the other is shifted.

1. Bracket
2. Washer
3. Quadrant
4. Bushing
5. Washer
6. Lever
7. Roll pin
8. Pivot
9. Quadrant
10. Shaft
11. Outer Support

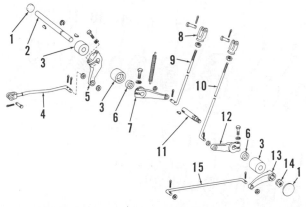

1. Plug
2. Shaft
3. Bushing
4. Speed range cam rod
5. Speed range cam arm
6. Oil seal
7. Speed range shift arm
8. Yoke
9. Shift rod
10. Shift rod
11. Speed change shaft
12. Speed change shift arm
13. Speed change cam arm
14. Lock nut
15. Speed change cam rod

components. Disassembly, inspection and overhaul of the removed assemblies is covered in overhaul section beginning with paragraph 89 which also outlines those adjustment procedures which are not an exclusive part of assembly.

87. TRACTOR SPLIT. To renew either the transmission or pto clutch shaft, first detach (split) tractor between clutch housing and transmission case. Refer to the following:

First relieve all pressure in main hydraulic system and brakes, drain hydraulic system and remove batteries, tool box door and operator shields.

NOTE: To relieve pressure in brake system accumulator, open right hand brake bleed screw and depress right hand brake pedal with engine not running.

Remove right, front platform extension, disconnect hydraulic pipe from fitting at bottom of accumulator.

Remove operators' platform and disconnect light wires. Disconnect both hydraulic brake lines and remote cylinder hydraulic lines on tractors so equipped. Disconnect main hydraulic pump inlet and remote selective control valve return lines. Remove pto clutch control valve.

Place wooden blocks between front axle and front support to prevent tipping sideways. Loosen differential lock valve and pry outward. Push toolbox downward and remove tool box and accumulator. Disconnect differential lock rod and all hydraulic lines and pipes on right side. Remove rockshaft cover and quick coupler.

Remove transmission top cover and the three top flange cap screws. Disconnect front shifter rods from both cam arms. Support the two tractor halves separately, remove the securing capscrews and separate the units.

The clutch shaft and transmission oil pump drive shaft (28 & 24—Fig. 69) can be withdrawn as a unit.

When reconnecting, tighten ¾-inch capscrews to a torque of 300 ft.-lbs. and

7/8-inch capscrews to a torque of 445 ft.-lbs.

87A. DISASSEMBLY. To disassemble the shafts, gears and controls located in transmission housing, detach

(split) transmission from clutch housing as outlined in paragraph 87. Transmission must be disassembled in the approximate sequence outlined, but may stop at the point where all damaged parts have been removed.

Remove rockshaft housing or transmission rear cover. If the differential drive shaft is to be removed, remove differential as outlined in paragraph 97. Remove transmission oil pipes and countershaft front bearing housing and adjusting shims (Fig. 76). Move first-third speed change shifter collar (7—Fig. 80) rearward to engage first-third gear (5). Move synchronizer range shifter to high range position. Remove rear countershaft bearing cup snap ring (1—Fig. 86). Use JDT-19-1 driver (Fig. 78) to drive rear bearing cup rearward

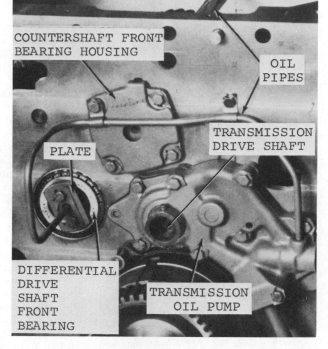

Fig. 76—View of front of transmission housing with clutch housing removed.

COUNTERSHAFT FRONT BEARING HOUSING

OIL PIPES

PLATE

TRANSMISSION DRIVE SHAFT

DIFFERENTIAL DRIVE SHAFT FRONT BEARING

TRANSMISSION OIL PUMP

Fig. 77—Schematic view of Syncro-Range transmission gears, shafts and associated parts. Item 9 is the only unit that may be shifted without stopping tractor.

1. First-third gear
2. Sixth-eighth gear
3. Second-fifth gear
4. Fourth-seventh gear
5. Countershaft
6. Reverse range pinion
7. Reverse shift collar
8. Low range pinion
9. Low-high synchronizer
10. High range pinion
11. Trans. drive shaft
12. Fourth-seventh diff. gear
13. Front shift collar
14. Second-fifth diff. gear
15. Sixth-eighth diff. gear
16. Rear shift collar
17. First-third diff. gear
18. Diff. drive shaft
19. Bevel pinion gear

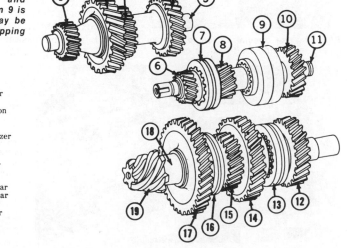

until it contacts ring gear (if differential is still in place). Refer to Fig. 79. (Step one); move countershaft rearward and, using JDT-19-3 bearing removal tool, drive front bearing and cup forward as far as is practical. (Step two); move countershaft rearward enough to install JDT-19-2 tool beside JDT-19-3 tool. With both tools installed, continue to drive cup forward. Lift out countershaft gear.

Remove transmission oil filter and oil filter relief valve housing from left side of transmission housing. Remove the snap ring, shifter pawl retainer, spring and pawl from each side of transmission. Remove the retaining snap rings, thrust washers and roll pins; then remove both shifter cam shafts from transmission housing. Drive out the roll pins retaining the two upper shifter rails; remove the three rails and lift out shifter forks and cams. Remove transmission oil pump from front of transmission housing and rear bearing quill from transmission drive shaft. Tape or clip synchronizer together to prevent separation, then lift out the shaft.

Remove the differential assembly as outlined in paragraph 97. Remove bevel pinion shaft front bearing retainer plate (24—Fig. 80). Remove bearing cone (23) and cup (22) from front of shaft by driving shaft rearward. Remove shim pack (21) and thrust washer (20), keeping shim pack together for

Fig. 78—Use special driver to remove rear countershaft bearing cup.

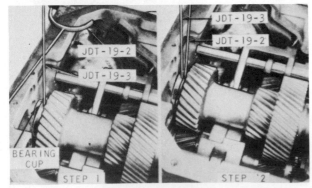

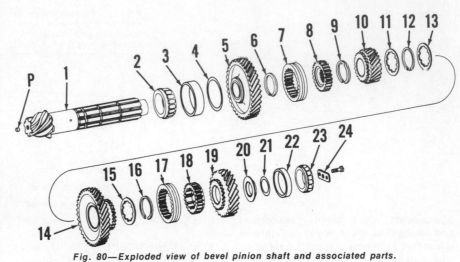

Fig. 80—Exploded view of bevel pinion shaft and associated parts.

1. Bevel pinion shaft	8. Shifter gear	14. Second-fifth gear	20. Thrust washer
2. Bearing cone	9. Snap ring	15. Thrust washer	21. Shim
3. Bearing cup	10. Sixth-eighth gear	16. Snap ring	22. Bearing cup
4. Shims	11. Thrust washer	17. Front shift collar	23. Bearing cone
5. First-third gear	12. Snap ring	18. Shifter gear	24. Retainer plate
6. Snap ring	13. Thrust washer	19. Fourth-seventh gear	P. Dowel pin
7. Rear shift collar			

reinstallation when unit is assembled. Slide gear (19), shift collar (17) and shifter gear (18) forward, unseat and expand snap ring (16) and move it forward on shaft. Expose and unseat snap rings (9 and 6) and move them forward out of their grooves; then drift the shaft rearward, removing the gears as they are free.

Overhaul the transmission main components as outlined in paragraphs 89 through 96; assemble as outlined in paragraph 88.

88. ASSEMBLY. If bevel pinion shaft or transmission housing were renewed; or if shaft bearings are not properly preloaded, adjust shaft and bearings as outlined in paragraph 96.

The special John Deere Snap Ring Expanding Cone (JDT-11) and Snap Ring Retainers (JDT-3) (See Fig. 81) should be used when installing the bevel pinion shaft. Slide the snap rings (6, 9 & 16—Fig. 80) down the cone and insert the retainer (Fig. 81) to hold rings open, and leave retainers in place while shaft is being installed.

Insert the shaft (1—Fig. 80) and bearing cone (2) through bearing cup (3) in transmission housing and install the larger gear (5) on shaft. Place the

thicker snap ring (6) on shaft and continue to insert the shaft with gears and washers in proper order until the front gear (19) is installed. Remove the snap ring retainers (Fig. 81) and seat the snap rings in their grooves, starting with rear snap ring. Installation of bearings can be facilitated by heating them to a temperature of 300°F. Tighten front bearing plate retaining cap screws to a torque of 35 ft.-lbs.

Install speed change shifters (Fig. 82). Install the transmission drive shaft front bearing cup and spacer, if removed, then install transmission drive shaft assembly and transmission oil pump. tighten oil pump retaining capscrews to a torque of 35 ft.-lbs. Install rear bearing quill and shims and check drive shaft end play using a dial indicator. Adjust end play to 0.004-0.006, if necessary, by adding or removing shims behind rear bearing quill. Tighten rear bearing retaining cap screws to a torque of 35 ft.-lbs. when bearings are properly adjusted.

Install speed range shifters in transmission by reversing the disassembly procedure, using Fig. 83 as a guide. Place speed range shifter cam in any detent position except reverse, and adjust reverse shifter stop screw (8) to provide 0.005-0.015 clearance between screw and case. Tighten the lock nut.

Install countershaft and adjust countershaft end play to 0.001-0.004 by varying the thickness of shim pack behind front bearing retainer. Tighten capscrews to a torque of 35 ft.-lbs.

Complete the assembly by reversing disassembly procedure.

OVERHAUL

To overhaul the transmission, first disassemble the unit as outlined in

Fig. 79—Two special drivers are needed for removal of front bearing and cup. Step 1 uses only one driver. Step 2 uses both.

paragraph 87A. Disassemble, overhaul and inspect the components as outlined in the appropriate following paragraphs; then reassemble as in paragraph 88.

89. SHIFTER CAMS AND FORKS. Refer to Figs. 82 and 83 for an exploded view of shifter mechanism and to Fig. 84 for an exploded view of park pawl and associated parts. Examine shift grooves and detents in shifter cams for wear or other damage. Parking lock spring (7—Fig. 82) should require a pull of 17-21 lbs. to deflect free end of spring 2-inches.

Speed range detent spring should have a free length of approximately 1-1/8-inches and test 12-14 lbs. when compressed to a height of 13/16-inch. Speed change detent spring should have a free length of 1-15/32-inches and test 27-33 lbs. when compressed to a height of ¾-inch. Approximate free length of park pawl spring is 2-1/8-inches and spring should test 9-13 lbs. when compressed to a height of 1-11/16-inches.

90. TRANSMISSION PUMP. The transmission pump is driven by the pto drive shaft at engine speed, and operates whenever engine is running. The pump cover is retained to housing only by dowels after pump is removed.

To disassemble the removed pump, tap the cover (1—Fig. 85) from its doweled position on housing (8) and remove the pump gears and drive sleeve (10). The drive gear (6) is keyed to drive sleeve but is a slip fit on

sleeve. Drive sleeve bushings (2 & 9) in cover and housing; and idler gear bushing (4) are renewable. Check the pump parts for wear, scoring or other damage, being sure to check pto clutch valve bore in transmission pump housing and the removed pto clutch valve. Specifications are as follows:

Pump gear diameter OD.....2.853-2.854
Pump gear thickness 0.625-0.626
Gear bore in housing 2.8565-2.8585
Pump gear radial
  clearance.............0.0025-0.0055
Pump gear end
  clearance.............0.0012-0.0036
Drive sleeve OD 1.937-1.938
Drive sleeve
  bushing ID 1.9395-1.9415
Idler gear shaft OD 0.7497-0.7503

91. COUNTERSHAFT. Refer to Fig. 86. The countershaft is a one-piece unit except for the bearings and high-speed gear (5). The high-speed gear is keyed to shaft and retained by snap ring (4); and may be removed with a press after removing the snap ring. Countershaft bearings should have 0.001-0.004 end play when shaft is properly installed.

92. TRANSMISSION DRIVE SHAFT. To disassemble the removed transmission drive shaft proceed as follows:

Remove snap ring (2—Fig. 87), remove bearing cone (4) with a press

or bearing puller, then remove reverse range pinion (6).

Remove snap ring (8) and use a press or bearing puller to remove drive collar (9) by pulling on collar ONLY (not low range pinion). Remove snap ring (10), then withdraw low range pinion (11) from shaft.

Remove low range synchronizer drum and plates and the low and high range blocker rearward from shaft being careful not to lose the detent balls and springs.

Remove snap ring (26), bearing cone (24) and high range pinion (23) from front of shaft, then withdraw the remaining synchronizer clutch parts.

The high and low range drive collar can be pressed from shaft after removing snap ring (20), if renewal is indicated.

93. SYNCHRONIZER CLUTCHES. The purpose of the synchronizer clutches is to equalize the speeds of the transmission drive shaft and the selected range pinion for easy shifting without stopping the tractor. The synchronizer clutches operate as follows:

The range drive collar (21—Fig. 87) is keyed to the shaft. Synchronizer clutch drums (12 and 19) are splined to the range pinions. The blocker ring (18) is centered in drive collar slots by the detent assemblies (17). Synchronizer clutch discs (13 and 14) are connected alternately (by drive tangs) to the

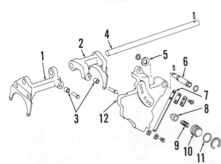

Fig. 81—Use JDT-1 cone and JDT-3 snap ring retainers to install rings on bevel pinion shaft.

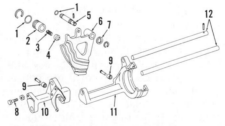

Fig. 83—Exploded view of speed range shifter cam and associated parts.

1. "O" ring
2. Retainer
3. Spring
4. Pawl
5. Pivot shaft
6. Shifter cam
7. Spacer
8. Stop screw
9. Roller
10. Reverse shift yoke
11. High-low shifter
12. Shaft

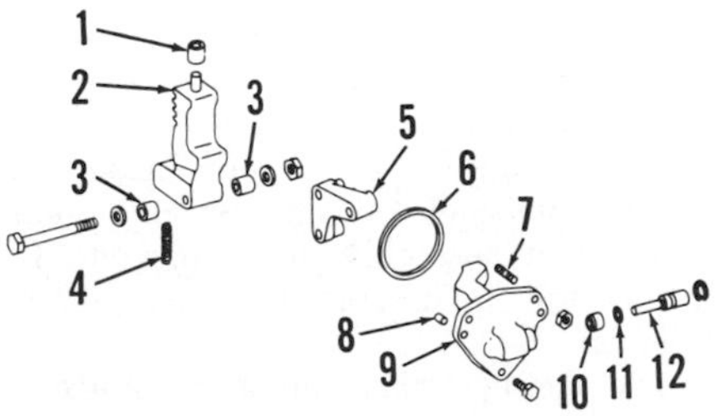

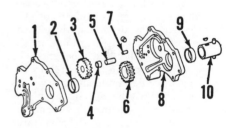

Fig. 85—Exploded view of transmission pump.

1. Cover
2. Bushing
3. Idler gear
4. Bushing
5. Shaft
6. Drive gear
7. Dowel
8. Body
9. Bushing
10. Drive shaft

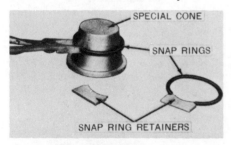

Fig. 82—Exploded view of speed change shifter cam and associated parts.

1. Rear shift yoke
2. Front shift yoke
3. Roller
4. Shaft
5. Spacer
6. Pivot shaft
7. Spring
8. Pawl
9. Spring
10. Retainer
11. "O" ring
12. Shifter cam

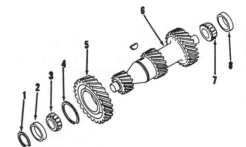

Fig. 84—Exploded view of parking lock pawl.

1. Roller
2. Park pawl
3. Bushing
4. Spring
5. Operating arm
6. Packing
7. Stud
8. Dowel pin
9. Cover
10. Bushing
11. "O" ring
12. Pivot shaft

Fig. 86—Exploded view of transmission countershaft.

1. Snap ring
2. Bearing cup
3. Bearing cone
4. Snap ring
5. Gear
6. Countershaft
7. Bearing cone
8. Bearing cup

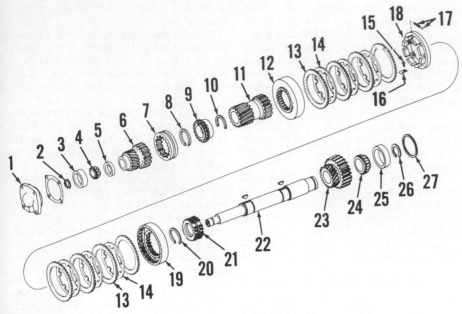

Fig. 87—Exploded view of transmission drive shaft showing component parts.

1. Bearing quill
2. Selective snap ring
3. Bearing cup
4. Bearing cone
5. Spacer
6. Reverse pinion
7. Reverse shift collar
8. Snap ring
9. Reverse drive collar
10. Snap ring
11. Low range pinion
12. Synchronizer drum
13. Synchronizer discs (4)
14. Synchronizer plates (3)
15. Drive stud (2)
16. Spring (2)
17. Spring/ball detent (4)
18. Synchronizer blocker
19. Synchronizer drum
20. Snap ring
21. Snychronizer collar
22. Drive shaft
23. High range pinion
24. Bearing cone
25. Bearing cup
26. Selective snap ring
27. Spacer

clutch drum and blocker ring. When the engine clutch is disengaged and the control lever moved to change gear speeds, the first movement of shifter linkage moves clutch discs (13 and 14) into contact. The difference in rotational speeds of shaft and pinion causes the blocker to try to rotate on the drive collar. The drive lugs inside the blocker ring ride up the ramps in the drive collar causing the blocker to be **temporarily locked in the center of the** drive collar. The clutch discs are thus compressed and the speeds of pinion and shaft are equalized. When the speeds become equal, the thrust force on blocker is relieved and shifting pressure causes the drive lugs to move back down the ramps to a center position in drive collar slot. The synchronizer drum is now permitted to move toward drive collar until the splines are engaged and the range pinion securely coupled to shaft.

94. INSPECTION AND ASSEMBLY. Inspect the transmission drive shaft for scoring or wear in areas of range pinion rotation and make sure oil passages are open and clean. Carefully inspect the blocker rings for damage to the drive lugs and inspect friction faces of blocker rings and synchronizer drums. Check synchronizer discs for wear, using a micrometer. The thickness of a new disc is 0.123, and a new plate should measure 0.096 inch. Renew any disc which measures less than 0.103 and any plate which is scored or damaged.

Reassemble the transmission drive shaft by reversing the disassembly procedure. A special installing cone (JDT-10) is required to install detent assemblies in blocker rings and to install blocker assemblies on drive collar. See Fig. 88.

95. BEVEL PINION SHAFT. Except for bearing cups in housing, and rear bearing cone on shaft, the bevel pinion shaft is disassembled during removal. Refer to Fig. 80 for exploded view and to paragraph 87 for removal procedure.

The bevel pinion shaft is available only as a matched set with the bevel ring gear. Refer to paragraph 99 for information on renewal of ring gear. Refer to paragraph 96 for mesh position adjustment procedure if bevel gears and/or housing are renewed.

96. PINION SHAFT ADJUSTMENT. The cone point (mesh position) of the main drive bevel gear and pinion is adjustable by means of shims (4—Fig. 80) which are available in thicknesses of 0.003, 0.005 and 0.010. The cone point will only need to be checked if the transmission housing or ring gear and pinion assembly are renewed. To make the adjustment, proceed as follows:

The correct cone point of housing and pinion are factory determined and assembly numbers are etched on rear face of pinion. To determine the shim pack thickness, subtract the number etched on pinion from the standard guide number 9.5295.

The result is the correct shim pack thickness. To add or remove shims, or to check shim pack thickness, use a punch and drive out rear bearing cup (3).

The bevel pinion bearings are adjusted to a pre-load 0.004-0.006 by means of shims (21). If adjustment is required, it should be made before installing the gears as follows:

First make sure that cone point is correctly adjusted as outlined above. Install shaft (1), cone (2), washer (20), the removed shim pack (21) plus one 0.010 shim, and bearing cone (23), plate (24) and the retaining cap screws. Measure the shaft end play using a dial indicator, then when disassembling, remove shims equal to the observed end play plus 0.005. Assemble the shaft and gears as outlined in paragraph 88.

DIFFERENTIAL AND MAIN DRIVE BEVEL RING GEAR

REMOVE AND REINSTALL

All Models

97. To remove the differential assembly, first drain transmission and hydraulic fluid.

On models with 3-point hitch, remove

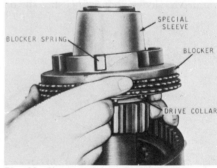

Fig. 88—Use special tool JDT-10 to assemble synchronizer detents into blocker. Transfer loaded blocker to drive collar.

Fig. 89—View of differential equipped with differential lock with cover removed.

rockshaft housing as outlined in paragraph 126.

On models without 3-point hitch, remove seat and differential case top cover.

On all models, block up tractor and remove both final drive units as outlined in paragraph 104. Remove brake backing plates and brake discs and withdraw both differential output shafts. Remove lubrication tube and oil pressure inlet pipe (Fig.. 89). On tractors with rockshaft, drive the cam follower pivot pin out of transmission case so that cam follower casting and arm can be removed.

Place a chain around differential housing as close to bevel gear as possible, attach a hoist and lift the differential enough to relieve the weight on carrier bearings. Remove both bearing quills using care not to lose, damage or mix the shims located under bearing quill flanges. Differential assembly may now be removed.

Overhaul the removed differential as outlined in paragraph 98.

When installing, place an additional 0.010 shim on left bearing quill, tighten the retaining cap screws and measure differential end play using a dial indicator. Preload the carrier bearings by removing shims equal in thickness to the measured end play plus 0.002-0.005. Shims are available in thicknesses of 0.003, 0.005 and 0.010.

After the correct carrier bearing preload is obtained, attach a dial indicator, zero dial indicator button on one bevel ring gear tooth and check the backlash between bevel ring gear and pinion. Proper backlash is 0.011-0.015. Moving one 0.005 shim from one bearing quill to the other will change backlash by about 0.010.

DIFFERENTIAL OVERHAUL

All Models

98. To overhaul the removed differential assembly, index the differential case halves, bend down the locking tabs, remove the retaining cap screws and separate the two halves. Differential gears, pinions, thrust washers and spider can now be removed.

Examine the parts for wear, scoring or other damage. If spider or any of the differential pinions are unserviceable, all five parts should be renewed. New differential pinion thrust washers are 0.030-0.032 in thickness, renew if excessively worn or scored. Renew differential case or axle gears if worn, scored or otherwise damaged.

When reassembling, make sure index marks on case halves are aligned, tighten retaining cap screws to a

torque of 85 ft.-lbs. and bend up lock plates.

Refer to paragraph 99 for installation of main drive bevel gear if renewal is required.

NOTE: Some tractors are equipped with a hydraulically actuated differential lock. Refer to paragraphs 100 through 104 for information on differential lock.

BEVEL RING GEAR

All Models

99. The main drive bevel ring gear and pinion are available as a matched set only. To renew the ring gear and pinion, first split tractor at clutch housing as outlined in paragraph 80 and remove the differential assembly as outlined in paragraph 97.

Remove the cap screws retaining

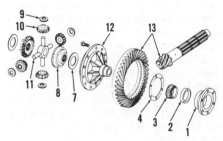

Fig. 90—Exploded view of differential without differential lock. Items 1 through 6 on the left side are not shown, but are identical to items 1 through 6 in Fig. 91.

7. Thrust washer (2)	11. Spider
8. Bevel gear (2)	12. R. differential housing
9. Thrust washer (4)	13. Ring & pinion gears
10. Bevel pinion (4)	

main drive bevel gear to differential housing and remove gear using a drift and heavy hammer. The main drive bevel gear is a press fit on housing. Heat ring gear evenly to a temperature of approximately 300°F. and position the gear. Install the retaining cap screws and lock plates. Tighten cap screws to a torque of 170 ft.-lbs. Renew pinion shaft as outlined in paragraph 95 and reinstall assembly as outlined in paragraph 97.

DIFFERENTIAL LOCK

Tractors may be optionally equipped with a hydraulically actuated differential lock which may be engaged to insure full power delivery to both rear wheels when traction is a problem. The differential lock consists essentially of a foot operated control and regulating valve and a multiple disc clutch located in differential housing, which locks the right axle gear to differential case.

OPERATION AND ADJUSTMENT

100. Refer to Fig. 91. When differential lock pedal is depressed, pressurized fluid from the main hydraulic system is directed to clutch piston (19), locking bevel lock gear (12) to differential case, and both differential output shafts and main drive bevel gear turn as a unit.

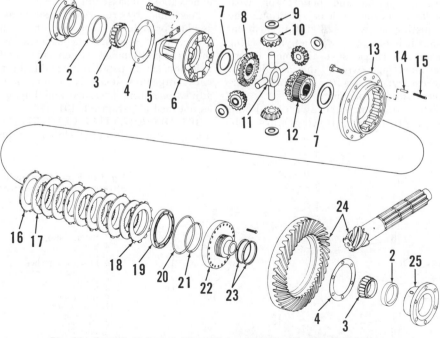

Fig. 91—Exploded view of differential equipped with differential lock.

1. Bearing quill	8. Left bevel gear	14. Spring pin (3)	20. "O" ring
2. Bearing cup	9. Thrust washer (4)	15. Spring	21. Packing
3. Bearing cone	10. Bevel pinion (4)	16. Separator plate (5)	22. Housing cover
4. Shim	11. Spider	17. Drive discs (5)	23. Sealing ring (2)
5. Lock plate (4)	12. Bevel lock gear	18. Backing plate	24. Ring & pinion gears
6. L. differential housing	13. R. differential housing	19. Piston	25. R. bearing quill
7. Thrust washer			

Equal power is this transmitted to both rear wheels despite variations in traction.

To release the differential lock, slightly depress either brake pedal which releases system pressure by acting through linkage (11—Fig. 92).

101. ADJUSTMENT. Differential lock valve and linkage should be adjusted for operating pressure and pedal free play, and release linkage adjusted for length. Proceed as follows:

Operating pressure will only need to be checked if valve has been overhauled, system does not operate properly, or if incorrect pressure is suspected. To check the pressure, refer to Fig. 92. Install a 0-1000 psi pressure gage in valve port after removing test hole plug (12). If tractor has no test plug, or if it is hard to reach, attach a 'T' fitting at pressure outlet (18), reconnect pipe and install gage to 'T' fitting. With engine operating at rated speed, depress differential lock actuating pedal. Gage pressure should be 420-500 psi; if it is not, disassemble valve as outlined in paragraph 102 and add or remove adjusting shims (5) as required.

After valve has been disassembled, or if linkage adjustment is required, disconnect release link swivel (9) from pedal and depress pedal until heavy spring pressure is encountered, but not far enough for valve linkage to snap over-center. Release the pedal and turn adjusting screw (19) in or out until a clearance of 0.085 to 0.105 inch exists between pedal and pedal stop with pedal in normal release position. The adjusting screw should stay in position, since it is equipped with a nylon locking insert. Recheck several times before reconnecting swivel, and readjust if necessary, to assure minimum clearance without touching.

Reconnect release swivel (9) and back off both adjusting nuts (10)

several turns. With engine running, fully depress pedal (8) until control valve is in locked position. Lightly pull release link (11) to rear until front ends of slots in release link contact release pins in brake pedals and all slack is removed; then turn both nuts (10) into contact with swivel without moving linkage or pedal. Tighten both nuts securely. When properly adjusted, the lightest movement of either brake pedal will release the differential lock, but valve will not release unintentionally because of linkage vibration.

OVERHAUL

102. CONTROL VALVE. To remove the differential lock control valve, disconnect release linkage and the three oil pipes at the valve. Remove the two retaining nuts and lift off the valve, noting the location of the two spacer bushings on retaining studs.

Disconnect pedal to valve linkage, remove pedal pivot pin and lift off pedal; then withdraw return valve (3—Fig. 92), spring (4), shim pack (5), flow control valve (6), bottom plug (14) check valve (16) and spring (15) from housing bore.

All parts are renewable individually. Examine parts for wear or scoring and springs for distortion, and renew any parts which are questionable. Keep shim pack (5) intact for use as a starting point when readjusting operating pressure. Renew "O" rings whenever valve is disassembled.

Assemble by reversing the disassembly procedure, using Fig. 92 as a guide. Make sure spacer bushings are properly positioned between valve and axle housing, reinstall valve and tighten retaining nuts securely. Check and adjust operating pressure and linkage as outlined in paragraph 101.

103. DIFFERENTIAL CLUTCH. The multiple disc differential clutch can be

overhauled after removing the unit as outlined in paragraph 97. Remove the cover retaining cap screws and lift off cover and the three piston return springs as shown in Fig. 93. Clutch discs, plates and splined bevel gear can now be lifted from housing. Remove piston (19—Fig. 91) with air pressure or by grasping two opposing strengthening ribs with pliers. Renew sealing "O" rings (20 and 21) and blow out oil passage in differential housing whenever piston is removed. Overhaul differential assembly as outlined in paragraph 98, if required, while differential clutch is disassembled.

Examine sealing surface in bore of right bearing quill and renew quill if sealing area is damaged. Renew the cast iron sealing rings (23) on differential housing if broken, scored or badly worn. Check clutch plates and discs and renew if scored, warped, discolored by heat, or worn to a thickness of 0.100 or less.

The five clutch separator plates (16) to be installed next to piston have nine external lugs; the backing plate (18) has twelve lugs. When assembling clutch, first install splined bevel gear and one plate with nine lugs, then alternately install the internally splined clutch discs and remainder of clutch plates, either side up. Align the external lugs on clutch plates, leaving three blank spaces for clutch return springs. Assemble springs over guide dowels, in right hand housing half (Fig. 93). Insert three steel pins into the drilled cross holes as shown, while holding springs compressed. This will hold springs depressed far enough to allow backing plate to be installed so that the twelve lugs can fit into the slots in housing while the differential housing cover is being installed. Reinstall cover, tighten cover retaining cap screws to a torque of 85 ft.-lbs. and bend up lock plates. Remove steel pins.

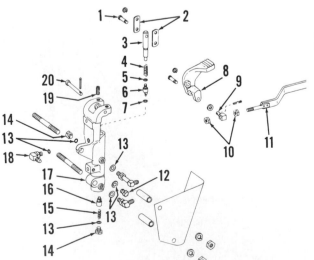

Fig. 92—Exploded view of differential lock control valve.

1. Pin (2)
2. Link (2)
3. Return valve
4. Spring
5. Shim
6. Flow control valve
7. "O" ring
8. Pedal
9. Swivel
10. Lock nuts
11. Release link
12. Test hole plug
13. "O" ring
14. Plug (3)
15. Spring
16. Check valve
17. Valve housing
18. Pressure outlet
19. Set screw
20. Pin

Fig. 93—Install piston return springs over dowel pins, compress springs and insert steel cross pins to hold springs compressed.

REAR AXLE
AND
FINAL DRIVE

All models have a planetary reduction final drive gear which is located at inner ends of rear axle housings.

All Models

104. REMOVE AND REINSTALL. To remove the final drive unit, first drain the transmission and hydraulic fluid, suitably support rear of tractor and remove rear wheel or wheels.

Remove fenders and light wiring on models so equipped. On tractors with breakaway couplings for hydraulic remote cylinders, remove operator's platform and interfering hydraulic lines. If right-hand housing is to be removed, remove oil pipe to differential lock if so equipped.

Support final drive assembly with a hoist, remove the attaching cap screws and swing the unit from transmission housing. When reinstalling, use care to prevent the sun pinion from coming out far enough to allow brake disc to drop behind sun pinion teeth. Tighten the retaining capscrews to a torque of 170 ft.-lbs.

105. OVERHAUL. To disassemble the removed final drive unit, remove lock plate (22—Fig. 94) and capscrew (21) at center of planet pinion carrier and withdraw the planet carrier assembly.

Planet pinion shafts (16) are retained in carrier by snap ring (14). To remove, contract the snap ring and, working around the carrier, tap each shaft out while snap ring is contracted. Withdraw the parts, being careful not to lose the 25 loose bearing rollers in

each planet pinion. Examine shaft for wear and renew if indicated. New dimensions are as follows:

Planet pinion ID 2.2467-2.2481
Carrier pinion shaft
　bore ID 1.7467-1.7487

After planet carrier has been removed, axle shaft (2) can be removed from inner bearing and housing by pressing on inner end of axle shaft. Remove bearing cone (6) and spacer (5) if they are damaged or worn. When assembling, heat spacer (5) and bearing (6) in oil to approximately 300 degrees F. and install on axle shaft making sure they are fully seated. Heat inner bearing cone (12) in oil to a temperature of 300 degrees F. Have planet carrier assembled and ready to install. Insert bearing on shaft and install and partially tighten the retaining capscrew (21). Leave a barely noticeable end play in axle bearings. Temporarily install lock plate (22) and, using a torque wrench calibrated in inch-pounds, check and record the rolling torque of the axle and planetary assembly. Remove lock plate (22) and tighten capscrew (21) until a rolling torque 20 to 70 inch-pounds greater than the previously recorded figure is obtained. Reinstall lock plate (22). Reassemble on tractor as outlined in paragraph 104.

BRAKES

OPERATION AND ADJUSTMENT

All Models

106. The hydraulically actuated single disc brakes are located on the differential output shafts (planetary sun pinions) and are accessible after removing the final drive units as outlined in

paragraph 104 and the output shaft and backing plate.

Power is supplied by the system hydraulic pump through foot operated control valves when engine is running or manually by means of master cylinders when hydraulic system pump is inoperative.

NOTE: Tractor is equipped with a nitrogen-filled accumulator to assist in providing pressure fluid when main hydraulic system is inactive.

The only adjustment provided is an adjustable yoke on upper end of each operating valve. This adjustment is used to equalize the height of the brake pedals. To adjust, loosen the locknut and turn operating rod in or out until both pedals are equal in height and are fully retracted.

For service of the hydraulic pump, refer to paragraph 124. To service control valve, actuating cylinders and brake discs, refer to the appropriate following paragraphs.

NOTE: Leakage or wear of the brake control valves can usually be determined by performing the leakage test outlined in paragraph 122. Other tests of the main hydraulic system which might affect brake operation are covered in paragraphs 118 through 121.

BLEEDING

All Models

107. When brake system has been disconnected or disassembled, bleed the system as follows:

Start the engine, loosen lock nut on bleed screw (Fig. 95) and back out the bleed screw two full turns. Bleed return passage is internal. Retighten lock nut to prevent external leakage, depress brake pedals and hold in depressed position for approximately two minutes to flush air from system. Tighten bleed screw first, then release the pedal. Bleed opposite brake following the same procedure.

To test the system, stop the engine, wait 15 minutes and depress each brake pedal once. Solid pedal action should be obtained on next application.

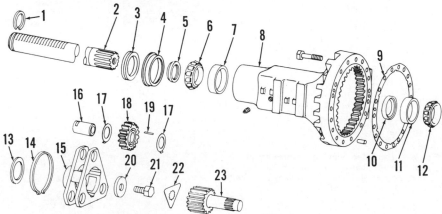

Fig. 94—Exploded view of planetary type final drive.

1. Snap ring	7. Bearing cup	13. Thrust washer	19. Roller bearing (75)
2. Axle	8. Housing & gear	14. Snap ring	20. Washer
3. Oil seal	9. Gasket	15. Planet carrier	21. Special capscrew
4. Seal cup	10. Oil seal	16. Pinion shaft (3)	22. Lock plate
5. Spacer	11. Bearing cup	17. Thrust washer (6)	23. Sun pinion
6. Bearing cone	12. Bearing cone	18. Planet pinion (3)	

Fig. 95—To bleed brakes, turn out bleed screw two turns and depress pedal. Bleeding action is internal.

If pedal action is spongy or pedal travel exceeds three inches, repeat the bleeding operation. Run engine at 1900 rpm to charge accumulator after testing.

CONTROL VALVE

All Models

108. To remove the control valve, disconnect the pressure and discharge lines at valve housing. On all models, remove right platform extension and the attaching cap screws and lift off the control valves and pedals as a unit.

To disassemble the removed unit, remove the two operating rod to pedal connecting pins and drive the roll pin (1—Fig. 96) from the pedal shaft (10). Tap out the shaft and remove the two pedals. Remove the cap screws attaching pedal bracket (12) to valve body and lift off the bracket and operating rod assemblies. It will not be necessary to disassemble operating rods unless seal or parts renewal is indicated.

Use Fig. 97 as a guide when disassembling the control valve assembly. Manual brake pistons (3), brake valve plungers (4) and plunger return spring (5) can be lifted out. Use a deep socket to remove inlet valve nipple (6). Clean the parts thoroughly and check against the specifications which follow:

Valve spring test data
  Valve plunger return
    spring ...... 39-47 lb.@ 1-11/16 in.
Pedal bushing ID 0.688-0.692
Manual piston OD.........0.933-0.935
Manual piston bore.......0.9365-0.9375
Brake valve plunger OD ..0.5595-0.5605

When reassembling, use new "O" rings and gaskets. Tighten the inlet valve nipple to a torque of 40 ft.-lbs. When control valve assembly has been installed, check for equal pedal height and bleed system as outlined in paragraph 107.

DISCS AND SHOES

All Models

109. To remove the brake discs or operating cylinders, first remove the final drive unit as outlined in paragraph 104, remove the backing plate, output shaft and brake disc. The three stationary shoes are riveted to the backing plate. The three actuating shoes are pressed on the operating cylinders which can be withdrawn from transmission housing after disc is removed. Facings are available in sets of three and should only be renewed as a set.

Operating pistons are 2.6245-2.6255 inches in diameter and have a diametral clearance of 0.0025-0.0075 in cylinder bores. Refer to Fig. 100 for breakdown of brake parts.

BRAKE ACCUMULATOR

All Models

110. R&R AND OVERHAUL. To remove the brake accumulator, blood fluid and pressure from accumulator by opening bleed screws (Fig. 95) and depressing brake pedals several times. Remove right platform extension, disconnect hydraulic pipe from bottom side of accumulator; then, unbolt and remove accumulator.

CAUTION: Overhaul of brake accumulator should not be attempted unless charging equipment is available. Gas side of accumulator piston is charged to 475 - 525 psi with NITROGEN gas. The charging valve is required in order to bleed cylinder before disassembly can be safely attempted.

Before attempting to disassemble the accumulator, remove plug (1—Fig. 102), depress valve (8) to exhaust gas from cylinder. The Nitrogen gas is harmless and non-inflammable. After the charging gas is completely exhausted, remove retaining rings (3) from both end caps, remove caps (4 and 13), piston (12) and remove gas valve body from cylinder. Remove packing and "O" rings from piston and end cap and use new parts when unit is reassembled.

Fig. 98—Remove manual brake pistons and plungers. Plungers can be pushed from pistons by hand.

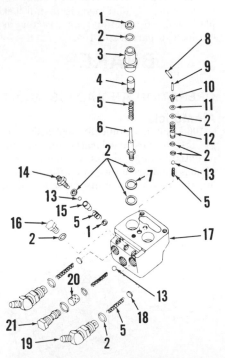

Fig. 97—Exploded view of hydraulic power brake control valve. A master piston (3) applies brakes manually if power is not available.

1. Washer	12. Guide
2. "O" ring	13. Steel Ball
3. Manual piston	14. Connector
4. Plunger	15. Inlet guide
5. Spring	16. Plug
6. Nipple	17. Valve housing
7. Backup ring	18. Check valve
8. Equalizing valve pin	19. Elbow
9. Equalizing valve shaft	20. Filter screw
10. Sleeve	21. Plug
11. Washer	

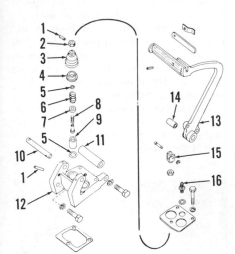

Fig. 96—Exploded view of hydraulic power brake operating pedals and bracket. Refer to Fig. 97 for view of control valve parts.

1. Spring pin	9. Guide
2. Sleeve	10. Pedal shaft
3. Boot	11. Spacer
4. Retainer	12. Pedal bracket
5. "O" ring	13. Pedal
6. Spring	14. Bushing
7. Operating rod stop	15. Yoke
8. Operating rod	16. Connector

Fig. 99—Equalizing valve pins (arrows) must be clean and free in bores.

Check the cylinder, cylinder cap, charging valve and piston for scoring, pitting, dents or wear. Diameters of new parts are as follows:

Cylinder ID 2.998-3.001
Piston packing 3.012-3.022
Piston OD 2.993-2.996

Assemble by reversing the disassembly procedure, using Fig. 102 as a guide; and installing new "O" rings, backing rings and packing. Recharge the cylinder using approved charging equipment and Nitrogen ONLY, to a pressure of 500 psi. Remove charging valve and reinstall plug (1). Install the accumulator by reversing the removal procedure. Start engine, run at 1900 rpm and test accumulator as outlined under "BLEEDING" in paragraph 107.

POWER TAKE-OFF

111. The power take-off shaft is driven at 1000 rpm ASAE speed at approximately 2100 engine rpm by an independently controlled, hydraulically actuated, multiple disc clutch mounted on front end of transmission housing. The gear on this clutch receives power from the pto drive shaft gear. A manually operated hydraulic control valve directs oil to the clutch piston, which will apply clutch any time engine is

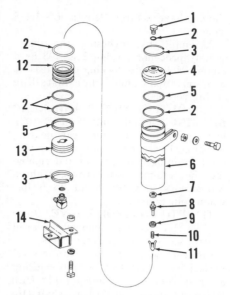

Fib. 102—Exploded view of brake accumulator.

1. Plug
2. "O" ring
3. Retaining ring
4. End cap
5. Backup ring
6. Cylinder
7. Washer
8. Valve
9. Packing
10. Spring
11. Guide
12. Piston
13. Cap
14. Shield

running. The pto drive shaft and gear is also the transmission pump drive shaft and is driven by the engine clutch cover hub. Refer to paragraph 112 for overhaul of the multiple disc clutch and to paragraph 114 for control valve overhaul.

R&R AND OVERHAUL

All Models

112. **PTO DRIVE GEARS AND CLUTCH.** To remove the pto drive gears and multiple disc clutch, first detach (split) transmission from clutch housing as outlined in paragraph 87. Remove pto drive shaft and gear and clutch shaft as a unit.

To remove the pto clutch assembly, remove access plate from bottom of clutch housing, remove the three cap screws retaining front bearing cover (36—Fig. 103) and remove the cover. Unseat and remove snap ring (33) from outer race of bearing (31) and remove clutch assembly and front bearing as a unit.

Remove snap ring (32) from front of

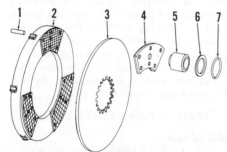

Fig. 100—Exploded view of wet type hydraulic disc brake operating parts located on differential output shaft (final drive sun gears).

1. Dowel
2. Backing plate
3. Brake disc
4. Pressure plate
5. Piston
6. Backup ring
7. "O" ring

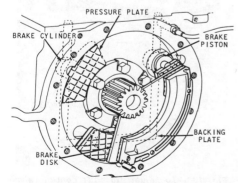

Fig. 101—Cutaway view of assembled brake unit.

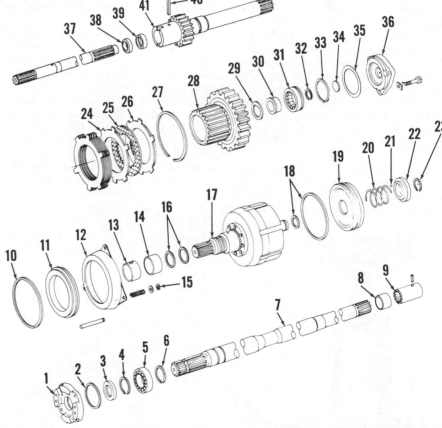

Fig. 103—Exploded view of power take-off clutch and gear train.

1. Seal housing
2. Packing
3. Oil seal
4. Snap ring
5. Ball bearing
6. Snap ring
7. PTO shaft
8. Bushing
9. Coupler
10. Sealing ring
11. PTO brake piston
12. PTO brake shoe
13. Sleeve
14. Bushing
15. Retaining rings (3)
16. Sealing rings
17. PTO clutch drum
18. Sealing rings
19. PTO clutch piston
20. Spring
21. Steel ball
22. Retainer
23. Snap ring
24. Clutch plates (11)
25. Clutch discs (11)
26. Backing plate
27. Snap ring
28. Driven gear
29. Thrust washer
30. Inner bearing race
31. Roller bearing
32. Snap ring
33. Snap ring
34. Thrust disc
35. Gasket
36. Bearing cover
37. Clutch shaft
38. Spacer
39. Oil seal
40. Spring pin
41. PTO drive shaft

clutch hub and remove bearing (31). Inner bearing race (30), thrust washer (29) and drive gear (28) can be lifted off after bearing is removed. Unseat the large snap ring (27) and multiple disc clutch plates and discs.

Clutch piston must be disassembled in a press (Fig. 104), using a suitable straddle-mounted fixture and compressing spring until the retaining snap ring (23—Fig. 103) can be unseated.

NOTE: A fixture can be constructed using about 6 inches of 2-inch ID steel pipe. Make sure both ends of pipe are square with center-line. Machine a 1" x 1½" notch in one end of pipe for working space to unseat snap ring.

Using the fixture, depress spring retainer (22) and remove snap ring (23). Release spring pressure carefully after snap ring is removed, to prevent injury or damage, until all pressure is removed from spring (20). Remove the fixture, retainer, steel ball (21) and spring; then remove piston (19) using compressed air or other suitable means. Steel ball (21) fits a recess in clutch hub and retainer to prevent retainer turning on hub.

Inspect all parts for wear, scoring or other damage. Examine sleeve (13) for wear or grooving at sealing ring area. Make sure oil passage is aligned if sleeve must be renewed. Inspect clutch plates and discs for heat discoloration, warping or scoring. Externally lugged clutch plates are 0.085-0.095 in thickness. Internally splined discs are 0.112-0.118; renew discs if worn to a thickness of 0.099 or less.

Spring (20) should have a free length of 2-7/8 inches and test 513-627 lbs. when compressed to a height of 1-21/32 inches. Brake shoe return springs

should have a free length of 1-1/8 inches.

Assemble by reversing the disassembly procedure, using Fig. 103 as a guide. Make sure locking ball (21) is properly positioned when installing spring retainer.

113. **OUTPUT SHAFT AND REAR BEARING.** Output shaft (7—Fig. 103), bearing (5), seal (3) and rear bearing housing (1) can be removed as a unit after draining transmission and removing the retaining cap screws. Tighten cap screws to a torque of 35 ft.-lbs. when reinstalling.

114. **PTO CONTROL VALVE.** Refer to Fig. 105 for an exploded view of control valve and associated parts.

Control valve assembly can be removed as a unit from clutch housing after disconnecting control cable (13). Drive out spring pin (6) from shaft (10) and remove shaft from arm (7). If spool (2) is damaged, worn or scored, detach clutch housing from transmission as outlined in paragraph 87, and examine spool bore in transmission pump (1). If bore is damaged, renew oil pump as outlined in paragraph 90. Oil seal (9) must be installed with lip toward oil pressure.

To remove valve spool (2) or spring (4), drive out roll pin (3). The variable rate spring (4) controls the pto operating pressure; when reinstalling valve, adjust as outlined in paragraph 115.

115. **PTO VALVE ADJUSTMENT.** The pto control valve linkage should be adjusted to provide 90-100 psi clutch pressure at rated speed when clutch in engaged and sufficient brake pressure to stop pto shaft rotation at slow idle speed with clutch disengaged. To

adjust the linkage, proceed as follows:

With tractor completely assembled and at operating temperature, remove upper oil pressure test hole plug (Fig. 106) and install a suitable pressure gage. With engine speed at 1900 rpm and pto clutch engaged, gage pressure should be 90-100 psi. If pressure is higher or lower than recommended, loosen clamp screw and disconnect push-pull cable from lever arm. While holding outer valve lever in engaged position and while still observing gage, loosen jam nut on rear adjusting screw on outer valve lever and adjust screw in or out to obtain correct pressure.

Stop engine and move gage to the lower oil pressure test hole. With engine again running 1900 rpm, adjust the front adjusting screw so that 16 psi is reached when outer valve lever is moved to disengaged position, then tighten jam nut.

Install push-pull cable and tighten cable housing clamp. Adjust lever stop screw at operating lever so that no strain is on cable when in the disengaged position.

HYDRAULIC SYSTEM

The hydraulic lift system working fluid is supplied by the tractor main hydraulic system which also provides fluid for the power steering and power brakes. Working fluid for the hydraulic units is available at all control valves except pto operating valve at a constant pressure of 2250 psi. The pto operating valve and clutch are supplied by the transmission pump at 95 psi.

MAIN HYDRAULIC SYSTEM

All Models

116 **OPERATION.** The main hydraulic pump is mounted underneath

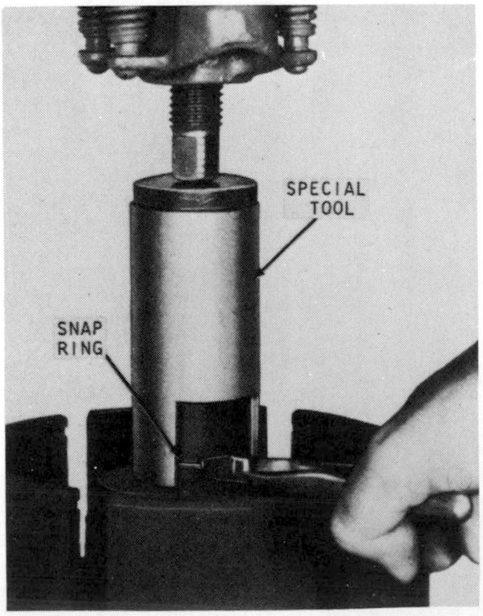

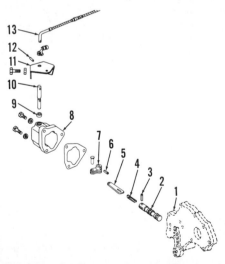

Fig. 105—Exploded view of pto shift control valve and associated parts.

1. Transmission pump
2. Valve spool
3. Roll pin
4. Spring
5. Link
6. Pin
7. Shift arm
8. Housing
9. Oil seal
10. Shift shaft
11. Outer valve lever
12. Pin
13. Shift cable

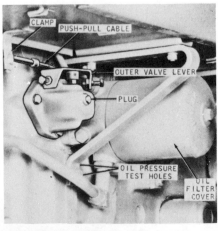

Fig. 106—To adjust pto shift control valve, install a gage in oil pressure test holes. Refer to text.

Fig. 104—Use press and special tool to remove snap ring (23-Fig. 103).

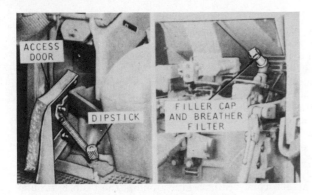

Fig. 107—Hydraulic system dipstick is located under seat behind access door. Filler cap is located on rockshaft control valve housing.

the tractor radiator and coupled to front of engine crankshaft. This variable displacement, radial piston pump provides only the fluid necessary to maintain system pressure. When there are no demands on the system, pistons are held away from the pump camshaft by fluid pressure and no flow is present. When pressure is lowered in the supply system by moving a control valve or by leakage, the stroke control valve in the pump meters fluid from the camshaft reservoir permitting the pistons to operate and supply the flow necessary to maintain system pressure. A maximum of 22 gallons per minute is available at full stroke.

The transmission pump provides pressure lubrication for the transmission gears and shafts and also supplies operating fluid for the power take-off. On all models, excess fluid from transmission pump passes through the full system filter to the inlet side of the main hydraulic system pump. If no fluid is demanded by the main pump, the fluid passes into the oil cooler then back to reservoir in transmission housing. Transmission pump capacity at 1900 engine rpm is 5 to 8 gallons per minute.

The oil cooler is mounted in front of the tractor radiator and contains cooling fins to control fluid temperature.

117. RESERVOIR AND FILTER. The hydraulic system reservoir is the transmission housing and the same fluid provides lubrication for the transmission gears and differential and final drive units. The manufacturer recommends that only John Deere Type 303 Special Purpose Oil or it's equivalent be used in the system. Reservoir capacity is 16 gallons. To check the fluid level, stop the tractor on level ground and check to make sure that fluid level is in "SAFE" range on dipstick (Fig. 107). Oil level should be checked with dipstick cap resting on threads. Remove access door under seat to reach dipstick.

The oil filter element is located on left side of transmission housing as shown in Fig. 106. The filter cartridge

may be renewed without draining the fluid reservoir by removing the filter cover and extracting the element.

A return oil (surge pressure) relief valve and a filter relief (bypass) valve are located in the external housing at rear of filter cover. To remove the filter relief valve, remove plug (Fig. 108) and withdraw the valve assembly. Service the return relief valve by removing plug.

118. SYSTEM TESTS. Efficient operation of the tractor hydraulic units requires that each component of the main supply system functions properly.

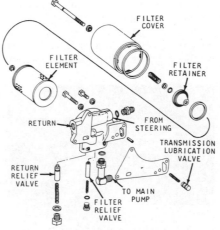

Fig. 108—Exploded view of filter relief valve housing and hydraulic system oil filter.

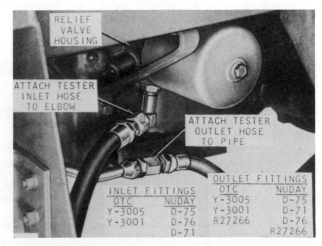

Fig. 109—Remove pipe to main hydraulic pump and attach flow meter as shown.

A logical procedure for testing the system is therefore required. The indicated system tests include transmission pump flow test, system pressure tests and leakage tests as outlined in the following three paragraphs. Unless the indicated repairs of hydraulic units is obvious because of breakage, these tests should be performed before proceeding with repairs on the individual hydraulic units.

119. TRANSMISSION PUMP FLOW. A quick test of transmission pump operation can be performed by removing the fluid filter (Fig. 108) and turning the engine over with the starter. If pump operation is satisfactory, a generous flow of fluid will be pumped into the filter housing.

If a hydraulic tester is available, disconnect pipe to main pump at elbow and install tester hoses as shown in Fig. 109. If necessary the outlet hose may be routed back to transmission filler neck so that all fluid will be returned to reservoir.

NOTE: Run engine for very short periods only and cap the pipe to main pump if the outlet hose is not connected to main pump line. Pump flow should be 5-8 gallons per minute at 1900 engine rpm. Tester control valve should be open for this test.

An alternate method which may be used is as follows: Disconnect the main hydraulic pump supply line at filter relief valve housing. Plug the line to prevent loss of fluid from oil cooler. Loosen lock nut on connector elbow, turn open end of elbow outward and attach a short pipe which contains a 0-250 psi pressure gage and a gate valve. The pressure gage must be located between the gate valve and tractor system. Start the engine and adjust engine speed to 1900 rpm. Place a clean container beneath pipe opening, and check fluid flow for a period of 15 seconds. The amount of flow should be approximately 1½ gallons.

If flow is much less, there is an internal leak, or transmission pump should be overhauled. See paragraph 90.

120. SYSTEM PRESSURE TEST. To check the system pressure, disconnect the steering outlet of the pressure control valve as shown in Fig. 110 and install a 3000 psi pressure gage. Make sure that all hydraulic control levers are in neutral, start tractor engine and operate at 1900 rpm. The pressure gage needle should immediately rise to 2250 psi then remain stationary.

Operate the remote cylinder operating levers and rockshaft lever at the same time. The gage needle should drop momentarily then immediately rise to 2000 psi or above and maintain this pressure while action continues.

If a flow meter is available, connect inlet hose as shown in Fig. 111.

NOTE: Use a 'T' fitting at filter housing to route tester return oil AND transmission pump oil to main pump, so that adequate volume of oil is supplied for this test.

With tester control valve closed, standby pressure at 1900 rpm should be 2200-2300 psi. With valve open, engine at 2100 rpm and 2000 psi pressure, flow should be 22 gallons per minute.

Adjust the standby pressure to the specified 2250 psi as follows: Loosen the locknut on stroke control valve on the bottom of main hydraulic pump, and turn the screw in or out until specified pressure is obtained. See Fig. 112.

If system pressure is below the specified 2000 psi pressure while testing, check and adjust the operating valves in control valve housing as outlined in paragraph 128.

121. PRESSURE CONTROL (PRIORITY(VALVE TEST. Leave gage attached to pressure control valve steering outlet as outlined in paragraph

120. If tractor is equipped for remote cylinder operation, connect a jumper hose from the pressure side of a breakaway coupling to the return side, so as to allow oil to return to reservoir. If a jumper hose is not available, connect a hose to one breakaway coupling and return free end of hose to transmission filler opening. If tractor is not equipped with remote valve, disconnect rockshaft cylinder hydraulic line at either end and attach a jumper hose leading back to reservoir.

Start engine and operate at 800 rpm (Slow Idle), move control lever to pressurize the disconnected hydraulic line and note pressure gage reading. Gage will register priority valve setting which should be 1600-1700 psi. If pressure is not within recommended range, remove and adjust the valve as outlined in paragraph 9.

122. LEAKAGE TEST. To check for leakage at any of the system valves, move all valves to neutral and run the engine for a few minutes at a speed of 1900 rpm. Check all of the hydraulic unit return pipes individually for heating. If the temperature of one return pipe is appreciably higher than the rest of the lines, that valve is probably leaking. Disconnect that return line and measure the flow from the line for a period of one minute. Leakage should not exceed 25cc for any one line, (except selective control valve which can be 40cc) or 150cc for all functions

Fig. 111—Remove main pressure line from pump to priority valve and install inlet hose of tester as shown. See text for return hose installation.

combined; if it does, overhaul the system valves as outlined in the appropriate sections of this manual.

123. LINES AND FITTINGS. Flared, seamless steel tubing is used for all hydraulic system components. Fittings have SAE .Straight Tubing threads with "O" ring seals. Do not attempt to substitute pipe thread fittings for components or test equipment.

124. MAIN HYDRAULIC PUMP. When external leaks, or failure to build or maintain operating pressure indicates a faulty pump, the main hydraulic pump must be removed for service as follows: To remove the pump, first relieve hydraulic pressure by opening a brake bleed screw as shown in Fig. 95 and depressing brake pedal several times. Drain oil cooler and radiator. Remove radiator, oil cooler and frame plate. Disconnect and remove the drive coupler from between crankshaft pulley and pump drive connector (Fig. 113). Disconnect main supply line, oil cooler line and pressure line from pump.

Remove the four pump attaching capscrews, attach a suitable lifting device and lift pump from tractor.

Install the pump by reversing the disassembly procedure and tighten the retaining cap screws as indicated below:

Pump to support screws . . . . .85 ft.-lbs.
Drive coupler cap screws . . . .32 ft.-lbs.
Coupler to pump clamp
   bolts . . . . .30 ft.-lbs.

125. OVERHAUL. Before disassembling the pump, check pump shaft end play using a dial indicator, and record the measurement for convenience in reassembling. End play should be 0.001-0.003 for either pump. End play is adjusted by adding or removing shims (20—Fig. 114) which are available in thicknesses of 0.006 and 0.010. Bearing wear, or wrong number of adjusting shims can cause excessive end play.

To disassemble the pump, remove the four cap screws retaining the stroke control valve housing (16—Fig.

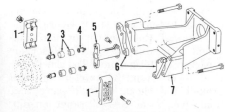

Fig. 113—Exploded view of main pump support and drive connectors used on naturally aspirated models. Turbocharged model is similar.

1. Coupler half
2. Pump drive screw
3. Bushing (4)
4. Pump drive screw
5. Drive shaft
6. Spacer
7. Pump support

Fig. 110—Remove line to steering at pressure control valve and install gage as shown.

Fig. 112—Adjust standby pressure at stroke control valve (arrow) on bottom of main pump.

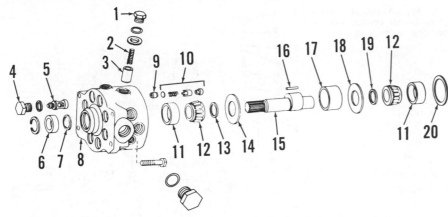

Fig. 114—Exploded view of main hydraulic pump body, shaft and associated parts.

1. Plug	6. Oil seal	11. Bearing cup	16. Roller bearing
2. Spring	7. Packing	12. Bearing cone	17. Race
3. Piston	8. Body	13. Spacer	18. Thrust washer
4. Plug	9. Seat	14. Thrust washer	19. Spacer
5. Inlet valve	10. Discharge valve	15. Pump shaft	20. Shim

115) to front pump, and remove the housing. Withdraw discharge valve plugs, guides, springs and valves (10—Fig. 114). Remove all piston plugs (1), springs (2) and pistons (3). Tap shaft lightly to loosen outer bearing cup. Insert a soft shim material between bearing cone and cup on high cam side of shaft so that race will clear the housing when removed. Carefully withdraw pump shaft (15) together with bearing cones, thrust washers (14), cam race (17) and the loose needle rollers (16).

Although the piston valves and pistons are available as individual parts, it is good shop practice once they have been installed and used, to install them in their original locations. Use a compartmented pan or other means to keep them identified when pump is disassembled. Remove the plugs retaining inlet valve assemblies (5) and check inlet valve lift using a dial indicator. Lift should be 0.060-0.080 inch. If lift exceeds 0.080, spring retainers are probably worn and valves should be renewed. Also check for apparent excessive looseness of valve stem in guide. Do not remove inlet valve assembly unless removal is indicated or discharge valve seat (9) must be renewed. To remove the inlet valve, use a small pin punch and drive valve out, working through discharge valve seat (9). If inlet valve is to be reused, place a flat disc on inlet valve head from the inside, so that all the driving force will not strike valve head in the center. Be sure the disc will drive through the hole without touching. Discharge valve seat can be driven out after inlet valve is removed. To renew discharge valve seat, use a piece of tubing and drive seat in until top of seat is 0.8700-0.8705 below face of bore. Do not bottom in bore.

Press inlet valve assembly into housing until top of guide is 0.533-0.537 below housing surface. Do not bottom valve. Be sure to reinstall pistons, springs, valves and seats to their own respective bores. The piston bores are

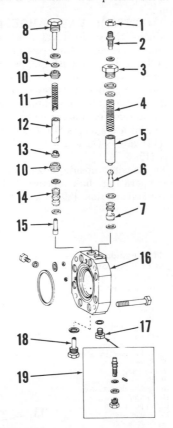

Fig. 115—Exploded view of stroke control valve housing showing component parts. Item 19 might not be used.

1. Nut	
2. Adjusting screw	
3. Bushing	11. Spring
4. Spring	12. Filter
5. Spring guide	13. Guide
6. Stroke control valve	14. Valve sleeve
7. Valve sleeve	15. Outlet valve
8. Plug	16. Housing
9. Shim washer	17. Plug
10. Packing	18. Plug
	19. Lockout valve

lined with a Teflon sheath, so all bores should be carefully inspected. Scored pistons or bores could cause pistons to stick. Check I.D. of bore where packing (7) is installed. This quad-ring packing requires a very close fit between pump shaft and bore in housing. Packing may be forced into this space and be ruptured, if clearance is excessive. This would allow seal (6) to fail since oil pressure from camshaft area would act directly on the seal. I.D. of bore should be 1.385-1.387 inches. Pump piston O.D. should measure 0.8740-0.8744. Piston bore I.D. should be 0.8747-0.8753.

The manufacturer recommends that the eight piston springs (2) test within 1½ lbs. of each other at 34 to 40 lbs. when compressed to a height of 1-5/8 inches. Install seal (6) only deep enough to allow snap ring to enter groove, to avoid blocking the relief hole in body. Identification numbers on the seal go to the outside. Valves located in stroke control valve housing control pump output as follows: The closed hydraulic system has no discharge except through the operating valves or components. Peak pressure is thus maintained for instant use. Pumping action is halted when line pressure reaches a given point by pressurizing the camshaft reservoir of pump housing, thereby holding pistons outward in their bores.

The cutoff point of pump is controlled by pressure of spring (4—Fig. 115) and can be adjusted by turning adjusting screw (2). When pressure reaches the standby setting, valve (6) opens and meters the required amount of fluid at reduced pressure into crankcase section of pump. Crankcase outlet valve (15) is held closed by hydraulic pressure and blocks the outlet passage. When pressure drops as a result of system demands, crankcase outlet valve is opened by the pressure of spring (11) and a temporary hydraulic balance on both ends of valve, which dumps the pressurized crankcase fluid and pumping action resumes. Stroke control valve spring (4) should test 125-155 lbs. pressure when compressed to 3.3 inches, and crankcase outlet spring (11) should test 45-55 lbs. at 2.2 inches. Cutoff pressure is regulated by the setting of adjusting screw (2) and adjustment procedure is given in paragraph 120. Cut-in pressure is determined by the thickness of shim pack (9) and/or pressure of spring (11). A special tool (JDH-19) is available to determine shim pack; refer to Fig. 116 and proceed as follows:

Assemble outlet valve units (8 through 15—Fig. 115), using existing shim pack (9). Install special tool (JDH-19) in place of plug (18), using one

1/8-inch thick washer as shown in Fig. 116. If adjusting shim pack thickness is correct, scribe line on tool plunger should align with edge of tool plug bore as shown; if it does not, remove top plug (8—Fig. 115) and add or remove shim washers (9) as required. Shims (9) are available in 0.030 thichness. If special tool is not available use shim washers of same thickness as those removed, then add shims to raise cut-in pressure, or remove shims to lower pressure.

When installing stroke control housing, add or remove shims (20—Fig. 114) as necessary to obtain specified pump shaft end play of 0.001-0.003 inch. Always use new O-rings, packings and seals. Oil all parts liberally with clean hydraulic system oil. Tighten stroke control valve housing retaining cap screws to 85 ft.-lbs. and tighten piston cap plugs to 100 ft.-lbs. and tighten piston cap plugs to 100 ft.-lbs. torque. Adjust standby pressure as outlined in paragraph 120 after tractor is reassembled.

ROCKSHAFT HOUSING & COMPONENTS

All Models

126. **REMOVE AND REINSTALL.** To remove the rockshaft housing, first remove the rockshaft housing covers, driver's seat and platform. Disconnect the three-point lift links and remove selective control valves on tractors so equipped. Disconnect and remove all interfering wiring and hydraulic lines

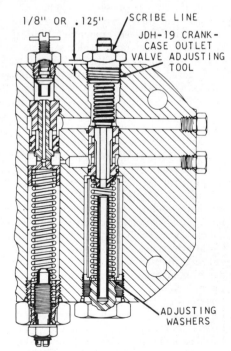

Fig. 116—Use JDH-19 adjusting tool in place of plug with pin (18-Fig. 115) to adjust crankcase outlet valve spring tension which controls cut-in pressure of pump. Refer to paragraph 125.

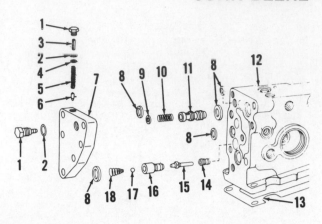

Fig. 117—Exploded view of rockshaft control valve housing, valves and cover.

1. Plug
2. "O" ring
3. Spring
4. Shim
5. Spring
6. Thermal relief valve
7. Cover
8. Packing
9. Washer
10. Spring
11. Flow control valve
12. Housing
13. Gasket
14. Guide
15. Metering shaft
16. Control valve
17. Ball
18. Spring

and disconnect the rockshaft control valve cable. Cap all open connections and pipes, then remove the attaching bolts and lift housing from tractor with a hoist.

When installing, make sure that the cam follower roller in valve housing is positioned to the rear of the follower arm in transmission housing, and lower the unit carefully so as not to damage or bend the draft control mechanism. Tighten the attaching bolts to 85 ft.-lbs. torque.

127. **CONTROL VALVE HOUSING.** To remove the control valve housing, disconnect the operating rod and hydraulic lines. If tractor is equipped with triple breakaway couplers, remove the upper one. Remove capscrews to valve housing and lift off the complete unit.

To disassemble, remove cover (7-Fig. 117), withdraw springs (10 & 18), steel balls (17) valves (11 & 16) and metering shafts (15).

Turn housing upside down and remove operating link spring (22-Fig. 118), operating link (20), load selector

control link (21) and control levers and arms (10, 6, 5, 1, and 4). Remove bearing quill (25) shaft (26) and sector gear (23). Remove control valve arm (19) with pinion (12), then remove levers with pins (16).

Inspect all parts for cracks, wear and scoring. If bores in housing are scored, housing must be renewed. If valve seat ends or valve ends are grooved or nicked, a wooden dowel may be inserted in valve, lapping compound applied to valve, valve inserted in bore and dowel twisted to lap damaged surfaces.

128. **CONTROL VALVE ADJUSTMENT.** When assembling the control valve housing, it is necessary to adjust the operating hinges to a positive but minimum clearance by means of the adjusting screws (15—Fig. 118). To accurately make this adjustment with a minimum of effort requires the use of the special adjusting tools JDH-8, JDH-9, JDH-10 and JDH-11 and a dial indicator as shown in Figs. 120 and 122. To make the adjustment, turn housing

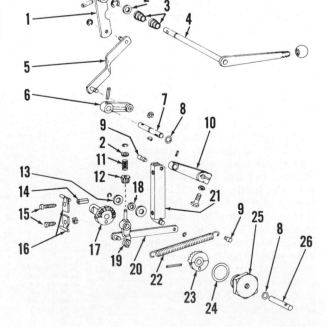

Fig. 118—Exploded view of control valve operating mechanism contained in control valve housing.

1. Upper selector arm
2. Washer
3. Bushing (2)
4. Control lever
5. Control strap
6. Lower selector arm
7. Selector shaft
8. "O" ring
9. Pin
10. Load selector lever
11. Spring
12. Upper pinion
13. Washer (2)
14. Special nut
15. Adjusting screw (2)
16. Lever w/pin (2)
17. L. sector gear
18. Roller
19. Control valve arm
20. Operating link
21. Control link
22. Spring
23. R. sector gear
24. "O" ring
25. Lever quill
26. Control shaft

upside down in a vise and loosen adjusting screws (15—Fig. 118). Install adjusting plate JDH-8 and adjusting cover JDH-10 as shown in Fig. 120. Springs (18—Fig. 117) are removed before adjusting cover JDH-10 is installed. Back out the screws (1 & 2—Fig. 120) in cover to be sure clearance exists and tighten the cover retaining cap screws securely, then tighten the screws (1 & 2) in cover to a torque of approximately 10-12 inch-pounds to be sure valves are tight on their seats. Use screw JDH-9 to secure adjusting tool JDH-11 to valve operating shaft sector gear (17—Fig. 118). Use the tension wire as shown in Fig. 121 to hold adjusting tool away from stop on adjusting plate JDH-8.

With the housing upside down, the lower valve is the return valve and must be adjusted first. With adjusting tool held away from stop (Fig. 121) turn lower adjusting screw clockwise until adjusting tool moves down and just contacts adjusting plate JDH-8. Tighten the jam nut on adjusting screw.

Mount a dial indicator approximately ¼ inch from end of adjusting tool JDH-11 as shown in Fig. 122, with indicator button contacting adjusting tool and remove adjusting plate JDH-8. Turn the inlet (upper) valve adjusting screw until total up and down movement at end of tool JDH-11 is 0.009-0.012 and tighten jam nut. Be sure to use only the tension wire when measuring valve movement. Because of the length of the indicator arm, the valve clearance is barely noticeable at valve hinges and measurement without the aid of the tools is almost impossible.

After valves are adjusted, remove the tools and reinstall valve springs with tapered ends toward the steel balls. Install flow control valve, with small end of spring toward cover. Use new "O" rings and install cover. Install and time control valve arm and pinion (19 & 12-Fig. 118), so that "V" on arm aligns with "V" inside housing, Install valve operating lever, shaft and right sector gear. The linkage is correctly

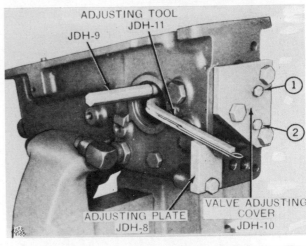

Fig. 120—View of rockshaft control valve housing showing adjusting tools installed.

Fig. 121—Adjusting the return valve in control valve housing.

Fig. 122—Adjusting the pressure valve in control valve housing.

timed when pin in operating lever and "V" mark in housing are aligned. Tighten lever quill (25) securely, install

load selector lever and actuating parts, cam follower link and spring, then install valve housing.

After assembly is completed, move selector lever on rockshaft housing to the lower (D) position and attach control rod. Start engine and check to see that rockshaft moves to full raised and full lowered position when dash control lever is moved the full length of quadrant.

Adjust speed-of-drop of rockshaft and implement by loosening the lock nut on rear of rockshaft cylinder and turn adjusting screw (6—Fig. 123) IN to increase drop speed or OUT to decrease speed.

Fig. 119—Exploded view of rockshaft control lever and associated parts.

1. Valve operating lever
2. Operating rod
3. Yoke
4. Upper arm
5. Bushing (2)
6. Control lever
7. Adjusting screw
8. Stop
9. Special washer (2)
10. Spring washer (2)
11. Special capscrew
12. Washer w/facing (2)
13. Friction plate
14. Spring (3)

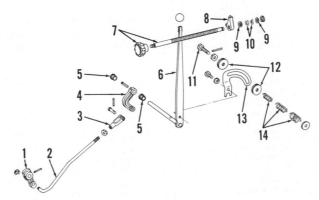

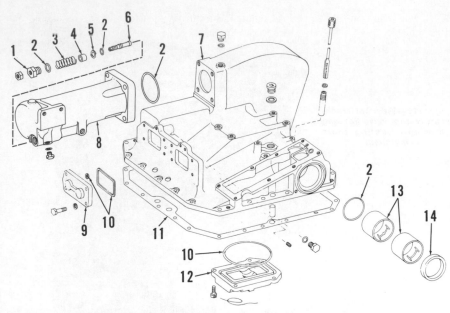

Fig. 123—Exploded view of rockshaft housing, showing removable cylinder and throttle valve parts. Left side bushing & oil seal not shown.

1. Plug
2. "O" ring
3. Spring
4. Throttle valve
5. Backup ring
6. Valve shaft
7. Housing
8. Cylinder
9. Cover
10. Packing
11. Gasket
12. Bottom cover
13. R. bushing (2)
14. R. Oil seal

129. ROCKSHAFT HOUSING OVERHAUL. Remove rockshaft housing as outlined in paragraph 126. Remove rockshaft valve assembly, lift arms and cylinder with piston. place cylinder open end down on wood block or table, CAREFULLY and slowly apply compressed air to oil inlet and remove piston. Remove bottom cover (12—Fig. 123), capscrew in rockshaft cam (8—Fig. 124) and drive rockshaft out right side of housing. Cam and crank arm with piston rod will come off as rockshaft is removed. Remove right and left oil seals and inspect bushings, housings and all parts for wear, cracks and damaged gaskets surfaces. New bushings for left side measure 3.2515-3.2565 inches, whereas the right bushings measure 3.6265-3.6315. Rockshaft left side O.D. should measure 3.247-3.249, and right O.D. should measure 3.662-3.624 inches. Use a suitable driver to renew bushings if clearance is excessive, which might cause oil leaks or erratic rockshaft depth and load control response. Be sure to align bushing oil holes.

Check rockshaft piston and cylinder wall for scoring. Specifications are as follows:

To serial no. 033851
  Piston O.D...........3.370-3.372 in.
  Cylinder I.D. ....3.374-3.376 in.
Serial no. 033852 and later
  Piston O.D...........3.870-3.872 in.
  Cylinder I.D. ....3.8755-3.8785 in.

Lubricate all parts thoroughly during reassembly and install new "O" rings,

packing and gaskets. Make sure the "V" marks on crank arm (7) and shaft are aligned as shaft is installed. Note that backup ring (3—Fig. 124) is installed in piston groove toward open end of cylinder. Tighten cylinder to housing capscrews to 130 ft.-lbs. torque. Be sure to safety wire the capscrews on bottom cover to prevent loss of capscrews. The bottom cover packing seals rockshaft piston return oil in upper housing to direct oil through the drilled rockshaft (10) to the three rockshaft bushings and also to oil filter housing to help charge the main pump during heavy demand. Steel balls (9) are driven into each end of rockshaft to seal ends from leakage. Check return holes in gasket surface of rockshaft housing, which allow lube oil from bushings to return to reservoir. Be sure operating link roller is on the back side of cam follower when housing is installed on transmission case. Tighten housing capscrews to 85 ft.-lbs. torque.

LOAD CONTROL ARM AND SHAFT

130. OPERATION. When the load selector lever is moved to the "L" (Load) position, operating depth is con-

trolled by the draft of the implement and position of control lever.

Refer to Fig. 125 for an exploded view of load control sensing linkage and associated parts. The lower links attach to draft link support (1) which is mounted beneath transmission housing and suspended from the two spring-steel control shafts (7 and 11). Positive or negative draft causes a controlled flexing of the shafts which is transmitted by the front shaft (11) to the pivoted arm (15) and through push rod (17) to the control valve.

131. REMOVE AND REINSTALL. To remove the support (1—Fig. 125), or load control shafts (7 and 11), first drain transmission and remove drawbar and lower links.

Support the frame (1) on a rolling floor jack or other suitable means. Remove rear shaft retainers (8), locking cap screws (9) and front shaft retainers (12); then drive the shafts (7 and 11) either way out of draft link support frame and transmission housing.

Remove shaft washers (6), sealing rings (5) and "O" rings (4) from the two shaft bores on each side of transmission housing and renew both whenever shaft is removed.

Inspect shaft bushings (2, 3, and 10) in transmission housing and draft link support frame. DO NOT renew transmission housing bushings unless necessary, since frequent renewal might enlarge bushing bore in housing and cause a leak. I.D. of small end of bushing is 1.128-1.130 inches. If bushings must be renewed, insert a long rod through one bushing and drive the opposite bushing from housing or support. Inside diameters of shaft bushings in transmission housing are tapered. Late model bushings have a thicker "O" ring and sealing ring for better sealing. The I.D. of rear support bushings is 1.281-1.287, and I.D. of front bushings is 1.179-1.185 inches. Chill, and without "O" ring and sealing ring, install bushings in transmission housing from outside, chamfered end first and sealing ring groove (Fig. 126) to the outside. Install bushings in support frame from outside, chamfered end first. Heat sealing ring in 160°F. water to soften. Collapse sealing ring,

Fig. 124—Exploded view of rockshaft and associated parts.

1. Lift arms
2. "O" ring
3. Backup ring
4. Piston
5. Piston rod
6. Spring pin
7. Crank arm
8. Servo cam
9. Steel ball (2)
10. Rockshaft

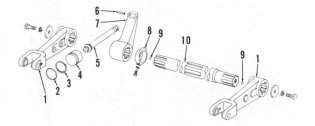

install in groove in bushing with plenty of lubrication after "O"-ring is installed. Press sealing ring round so that load control shaft will go through and install as many selective washers (6—Fig. 125) as necessary to provide minimum clearance between transmission case and support. Washers are available in thicknesses of 0.036, 0.048 and 0.062 inch. Carefully install load control shaft (11) with cutout section down, shaft (7) with adequate lubrication and install retainers (8 & 12).

Sensing linkage can be removed after removing seat, rockshaft housing and operator's platform. Remove locking cap screws (9) and retainers (12), and drive out front shaft (11) only far enough to clear control arm (15). Remove the cap screws securing retainers (13) to transmission housing and lift the load control arm (15) and push rod from housing as a unit. Spring (16) should require a pull of 18-22 lbs. to extend spring to a length of 18-5/8 inches.

Reinstall by reversing the removal procedure, using Fig. 125 as a guide. Use care when installing load control shaft (11) to prevent damage to "O" rings (4) and sealing rings (5) in housing bores.

132. ADJUSTMENT. Start engine and operate at slow speed to prevent too rapid rockshaft operation. Push valve operating lever (1—Fig. 119) for-

ward to completely raise rockshaft. Pull control lever (6) all the way to rear of slot in console. Align yoke (3) with operating upper arm (4), then shorten rod by turning another ½ turn of the yoke so that lever (6) will not touch rear of console slot. Lower rockshaft, then raise it completely and check lever for 1/32 to 1/16 inch clearance in slot.

Set the load selector lever in "L" (Load) position. Remove the hex head plug from top of rockshaft housing at right of seat support and, working through plug hole, turn adjusting screw (20—Fig. 125) until rockshaft will just start to raise when rear edge of control lever is no further forward than the "O" mark on quadrant. Rockshaft should lower when edge of control lever is no further forward than the "1" mark.

Loosen locknut (24) (on top of rockshaft housing rearward from plug hole) and turn stop screw (25) until stop is felt on follower arm (23) and back off ½ turn. Tighten locknut (24) to secure the adjustment.

SELECTIVE (REMOTE) CONTROL VALVES

All Models So Equipped

133. OPERATION. Tractors are optionally equipped with one, two or three selective (remote) control valves for operation of remote cylinders.

Mounting positions of valves are shown in Fig. 127.

As with all other units of the hydraulic system, pressure is always present at the valves but no flow exists until the valve is moved. Refer to Fig. 129 for an exploded view of valve mechanism. Each breakaway coupler is equipped with two return valves (20) and two pressure valves so arranged

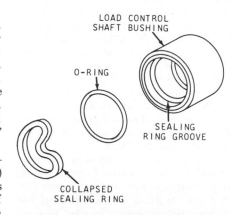

Fig. 126—Collapse sealing ring for ease of installation.

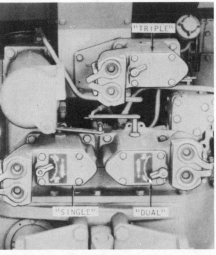

Fig. 127—Rear view of tractor showing location of selective (remote) control valves.

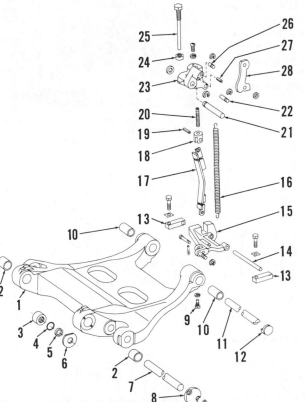

Fig. 125—Exploded view of load control sensing mechanism and linkage.

1. Support
2. Bushing
3. Bushing
4. "O" ring
5. Sealing ring
6. Selective washer
7. Shaft
8. Retainer
9. Capscrew
10. Bushing
11. Load control shaft
12. Retainer
13. Support
14. Pivot shaft
15. Load control arm
16. Spring
17. Push rod
18. Rod end
19. Spring pin
20. Adjusting screw
21. Shaft
22. Pin
23. Follower arm
24. Locknut
25. Stop pin screw
26. Pin
27. Spring pin
28. Cam follower

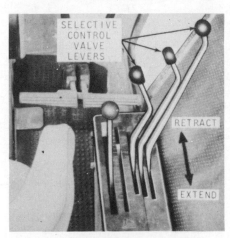

Fig. 128—View of console equipped with triple selective control valves.

that one of each is opened when control lever is moved off center in either direction. Detent piston (16) is actuated by pressure differential across metering valve (26) and released by pressure equalizaiton when flow stops at end of piston stroke. Flow control valve (25) maintains an even flow with varying pressure loads.

134. **OVERHAUL.** Before removing selective control valve, be sure linkage is not broken or binding, which would not allow valve to function properly. Refer to Fig. 129 for an exploded view of the selective control valve. Clamp the unit in a vise and unbolt and remove cap (23) and associated parts carefully as shown in Fig. 130.

CAUTION: Valve guides (21—Fig. 129) are spring loaded and can hang up, then can come loose very rapidly, and with considerable force.

Identify parts as required for later assembly, then remove valves, springs and guides. Outer detent guide (18) may be difficult to remove. If so, remove rocker assembly and follower (13) so that guide may be removed from the inside by pushing on detent pin (15). Refer to following paragraph.

Rotate valve body in vise so that rocker assembly is up as shown in Fig. 131. Rocker arm can be disassembled by driving out spring pin, and removing control arm and shaft. Remove screws holding cam (3 and 7—Fig. 129) and remove rocker (5). Notice how parts are assembled, to aid in reassembly. Inspect all bores, valves and valve seats. Seats (Fig. 133) are nonrenewable, but can be reconditioned by using NJD 150 Valve Seat Repair Kit (Use exactly as directed). Some tractors may be equipped with a spring pin in cover (23) to prevent "O" ring (24)

from being pulled into the passage under sudden oil surges.

Check all valve springs, which should be within the following specifications:

Pressure ....36-44 lbs.@ 1.25 in.
Return ....19-23 lbs.@ 1.25 in.
Flow
control .....40.5-49.5 lbs.@ 2.15 in.

Assemble by reversing the disassembly procedure. If actuating cam was disassembled, refer to Fig. 132 and note that the float cam (7) is shorter than the regular cam (3), and is installed pointed end up on numbered side of housing as shown. Install detent cam (4) in rocker (5) with 2 pins (6) and install rocker assembly and rocker shaft (10).

Adjusting the valve requires use of special adjusting cover (JDH-15C) and a dial indicator. Remove the two adjusting plugs (9) and loosen the two cam locking screws (8). Back out the four adjusting screws (2) at least two turns.

Install pressure and return valves, detent follower, piston, guides and retaining snap ring. Be sure detent follower roller properly rides on detent cam. Also make sure that operating valve rollers are turned to ride properly on ramps of cams and are not turned crosswise. Back out all adjusting screws on special plate and install the plate with the angled screw pointing at detent pin (15—Fig. 129). Carefully FINGER TIGHTEN the four screws contacting operating valves until valves are seated; then while holding operating lever in center position FINGER TIGHTEN the detent locking screw until detent roller is seated in neutral detent on detent cam

(4—Fig. 132). With operating lever in neutral position, refer to Figs. 131 and 134. Turn in the two diagonally opposite Pressure Valve Cam Adjusting Screws until screws, cams and follower rollers are in contact. Install rubber keeper then back out ¼ turn as shown. Turn in the two diagonally opposite Return Valve Cam Adjusting Screws until screws, cams and follower rollers are in contact. Install rubber keeper, then back out 1/8 turn. Move the two cams (3 & 7—Fig. 132) into contact with adjusting screws and tighten screws (8) securely.

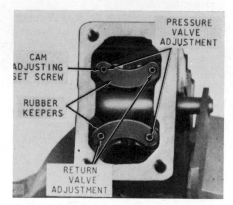

Fig. 130—Clamp housing in a vise to remove rear cap.

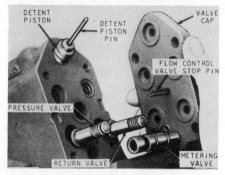

Fig. 131—Front view of housing showing rocker assembly, adjustment screws and associated parts.

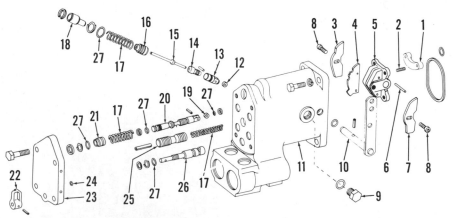

Fig. 129—Exploded view of Selective (Remote) Control valve showing component parts.

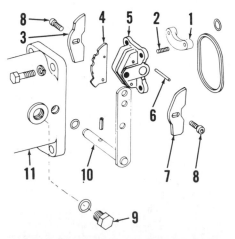

Fig. 132—Exploded view of control rocker. Refer to Fig. 129 for parts identification.

1. Rubber Keeper (2)	
2. Adjusting screw (4)	8. Cam clamp screws
3. Regular cam	9. Plug (2)
4. Detent cam	10. Rocker shaft
5. Rocker	11. Housing
6. Pin (2)	12. Detent roller
7. Float cam	13. Detent follower

14. Inner detent guide	21. Valve guide (4)
15. Detent pin	22. Lever
16. Detent piston	23. Cover
17. Spring	24. "O" ring
18. Outer detent guide	25. Flow control valve
19. Cam roller (4)	26. Metering valve
20. Poppet valve (4)	27. "O" ring

To double-check the adjustment, install a dial indicator 3 inches from center of shaft on operating arm as shown in Fig. 134. Zero dial indicator while locked in neutral detent, then back out the detent locking screw on adjusting cover. Back out the two adjusting cover screws which contact operating valves on lever side and measure rocker movement which should be 0.021 toward return valve or 0.060 toward pressure valve as shown. (Valves contacting cam 3—Fig. 132 opposite lever side are being checked.) Tighten the two adjusting cover screws on lever side and loosen the other two screws, then check adjustment of valves on lever side. Readjust as necessary for correct rocker movement. This procedure will allow return valves to open before pressure valves, when selective control valve is used.

Before installing cover (23—Fig. 129) check length of flow control valve stop pin (Fig. 130). Pin should extend 0.9375 inch from cover. If pin is too short, jerky valve action can result. Tighten cover cap screws to 35 ft.-lbs. torque.

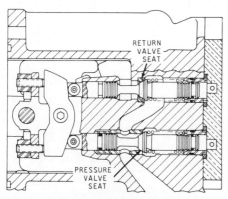

Fig. 133—Valve seats can be reconditioned using special tool NJD 150.

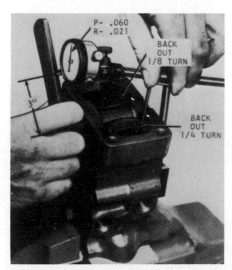

Fig. 134—Use a dial indicator to measure rocker arm movement as shown. Refer to paragraph 134 for procedure.

Fig. 135—Exploded view of breakaway coupler and associated parts. All removable parts are duplicated in adjacent bore.

1. Snap ring
2. Snap ring
3. Backup ring
4. "O" ring
5. Ball
6. Backup ring
7. Receptacle
8. Ball
9. Spring
10. Plug
11. Snap ring
12. Sleeve
13. Operating lever
14. Cam
15. Lock washer
16. Washer
17. Expansion plug

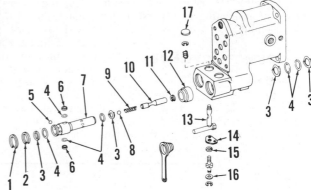

135. BREAKAWAY COUPLER. Drive a punch into the expansion plugs (17—Fig. 135) and pry out of housing. Remove retainer rings and springs. Operating levers can then be removed. Drive receptacle assembly from housing. Check steel balls, springs, and all parts for wear and replace as neces-

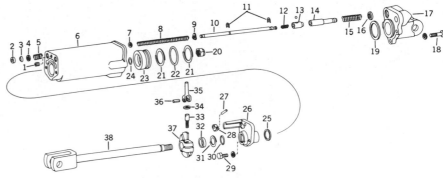

Fig. 136—Exploded view of remote cylinder with hydraulic stop. Items 12 and 13 are not used on some models. Item 25 is an "O" ring and backup ring on some models.

1. Plug	11. Snap ring	20. Lock nut	29. Capscrew
2. Packing adapter	12. Spring	21. Backup ring (2)	30. "O" ring
3. Packing	13. Bleed valve & ball	22. "O" ring	31. Backup ring
4. Packing adapter	14. Stop valve	23. Piston	32. Oil seal
5. Spring	15. Spring	24. "O" ring	33. Stop screw
6. Cylinder	16. Gasket	25. Gasket	34. Washer
7. Washer	17. End cap	26. Piston rod guide	35. Stop lever
8. Spring	18. Capscrews	27. Groove pin	36. Spring pin
9. Stop rod washer	19. Gasket	28. Stop rod arm	37. Piston rod stop
10. Stop rod			38. Piston rod

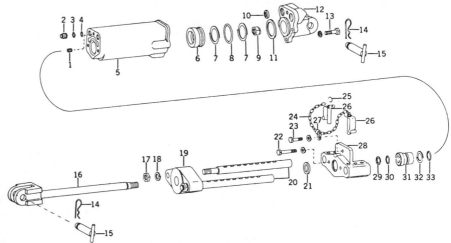

Fig. 137—Exploded view of remote cylinder with mechanical stop. Item 6 may be equipped with an "O" ring.

1. Plug	10. Gasket (2)	18. Lockwasher	26. Stop pin
2. Plug	11. Gasket	19. Piston rod stop	27. Clamp
3. Backup ring	12. Cap	20. Stop rod (2)	28. Stop rod guide
4. "O" ring	13. Cap screw	21. Oil seal	29. Backup ring
5. Cylinder	14. Lock pin	22. Cap screw	30. "O" ring
6. Piston	15. Attaching pin	23. Cap screw	31. Piston rod guide
7. Backup ring (2)	16. Piston rod	24. Chain	32. Backup ring
8. "O" ring	17. Jam nut (2)	25. Link	33. "O" ring
9. Nut			

sary. Renew "O" rings and backup washers. Reassemble in reverse order of disassembly.

REMOTE CYLINDER

All Models

136. Refer to Fig. 136. for exploded view of the double acting, hydraulic stop remote cylinder. To disassemble, remove end cap (17), stop valve spring (15) and valves (14 and 13), using care not to lose ball, if cylinder is equipped with override provision. Fully retract the cylinder and remove nut (20) from piston end of piston rod (38). To remove the stop rod and springs, drive the groove pin (27) from stop rod arm (28).

Install new wiper seal (32) in guide (26) with lip to outside, assemble rod end of cylinder, rod packing and piston rings, and have piston fully inserted in cylinder before installing rod nut. Tighten nut securely. Be sure the piston rod stop (37) is located so that stop lever (35) is opposite the stop rod arm (28). Tighten cap screws securing piston rod guide (26) to a torque of 35 ft.-lbs. and cap screws retaining piston cap (17) to 85 ft. lbs.

Fig. 137 shows parts identification on cylinders with mechanical stop.

NOTES

NOTES

NOTES

NOTES

NOTES

Technical Information

Technical information is available from John Deere. Some of this information is available in electronic as well as printed form. Order from your John Deere dealer or call **1-800-522-7448**. Please have available the model number, serial number, and name of the product.

Available information includes:

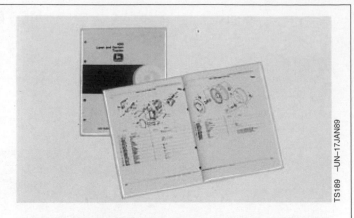

- PARTS CATALOGS list service parts available for your machine with exploded view illustrations to help you identify the correct parts. It is also useful in assembling and disassembling.
- OPERATOR'S MANUALS providing safety, operating, maintenance, and service information. These manuals and safety signs on your machine may also be available in other languages.
- OPERATOR'S VIDEO TAPES showing highlights of safety, operating, maintenance, and service information. These tapes may be available in multiple languages and formats.

- TECHNICAL MANUALS outlining service information for your machine. Included are specifications, illustrated assembly and disassembly procedures, hydraulic oil flow diagrams, and wiring diagrams. Some products have separate manuals for repair and diagnostic information. Some components, such as engines, are available in separate component technical manuals
- FUNDAMENTAL MANUALS detailing basic information regardless of manufacturer:
  - Agricultural Primer series covers technology in farming and ranching, featuring subjects like computers, the Internet, and precision farming.
  - Farm Business Management series examines "real-world" problems and offers practical solutions in the areas of marketing, financing, equipment selection, and compliance.
  - Fundamentals of Services manuals show you how to repair and maintain off-road equipment.
  - Fundamentals of Machine Operation manuals explain machine capacities and adjustments, how to improve machine performance, and how to eliminate unnecessary field operations.